T0206275

Soay Sheep
Dynamics and Selection in an Island Population

Soay Sheep synthesises the results of a unique study of population dynamics, selection and adaptation in a naturally regulated population of mammals. Soay sheep were first introduced to the St Kilda archipelago in the Bronze Age, and to the island of Hirta in 1932, where their numbers are naturally regulated by the availability of resources. Unlike most other large mammals, their numbers show persistent fluctuations, sometimes increasing or declining by more than 60 per cent in a year. *Soay Sheep* explores the causes of these fluctuations and their consequences for selection on genetic and phenotypic variation within the population, drawing on nearly twenty years' study of the life-histories and reproductive careers of many individuals. The study provides important insights into the regulation of other herbivore populations and the effects of environmental change on selection and adaptation. It will be essential reading for vertebrate ecologists, demographers, evolutionary biologists and behavioural ecologists.

TIM CLUTTON-BROCK is Professor of Animal Ecology at the University of Cambridge. His research has explored the evolutionary causes and ecological consequences of animal breeding systems. He has worked with primates, ungulates and carnivores, and has been responsible for establishing several long-term studies where recognition of individuals made it possible to explore the interface between population dynamics and natural selection.

JOSEPHINE PEMBERTON is Reader in Molecular Ecology at the University of Edinburgh. Her research uses modern molecular techniques to help understand the ecology and evolution of natural populations. She specialises in applying DNA profiling to long-term studies of individuals, yielding insights about breeding systems, the consequences of inbreeding and host–parasite coevolution.

Soay Sheep

Dynamics and Selection in an Island Population

Edited by

T. H. Clutton-Brock
University of Cambridge
and
J. M. Pemberton
University of Edinburgh

CAMBRIDGE
UNIVERSITY PRESS

CAMBRIDGE
UNIVERSITY PRESS

University Printing House, Cambridge CB2 8BS, United Kingdom

One Liberty Plaza, 20th Floor, New York, NY 10006, USA

477 Williamstown Road, Port Melbourne, VIC 3207, Australia

4843/24, 2nd Floor, Ansari Road, Daryaganj, Delhi - 110002, India

79 Anson Road, #06-04/06, Singapore 079906

Ruiz de Alarcón 13, 28014 Madrid, Spain

Dock House, The Waterfront, Cape Town 8001, South Africa

Cambridge University Press is part of the University of Cambridge.

It furthers the University's mission by disseminating knowledge in the pursuit of education, learning and research at the highest international levels of excellence.

www.cambridge.org
Information on this title: www.cambridge.org/9780521529907

© Cambridge University Press 2004

This publication is in copyright. Subject to statutory exception and to the provisions of relevant collective licensing agreements, no reproduction of any part may take place without the written permission of Cambridge University Press.

First published 2004

A catalogue record for this publication is available from the British Library

Library of Congress Cataloging in Publication data
Soay sheep : dynamics and selection in an island population / editors, T. H. Clutton-Brock and J. M. Pemberton.
   p. cm.
ISBN 0 521 82300 5 – ISBN 0 521 52990 5 (paperback)
1. Soay sheep – Selection – Scotland – Saint Kilda. 2. Mammal populations – Scotland – Saint Kilda. 3. Soay sheep – Ecology – Scotland –
Saint Kilda. I. Clutton-Brock, T. H. II. Pemberton, J. M. (Josephine M.)
QL737.U53S66 2003
599.649´1788–dc21 2003043936

ISBN 978-0-521-52990-7 Paperback

Cambridge University Press has no responsibility for the persistence or accuracy of URLs for external or third-party internet websites referred to in this publication, and does not guarantee that any content on such websites is, or will remain, accurate or appropriate.

Contents

Appendix 1

Appendix 2

Appendix 3

Colour plate section between pages   20 and 21

Contributors

S. D. Albon
Centre for Ecology and Hydrology, Hill of Brathens, Banchory AB31 4BY, UK

D. R. Bancroft
GPC AG Genome Pharmaceutical Corporation, Lochhamer Strasse 29, D-82152 Munich, Germany

D. R. Bazely
Department of Biology, York University 4700 Keele Street, North York, Ontario M3J 1P3, Canada

H. E. G. Boyd
Department of Zoology, University of Cambridge, Cambridge CB2 3EJ, UK

T. H. Clutton-Brock
Department of Zoology, University of Cambridge, Cambridge CB2 3EJ, UK

D. W. Coltman
Department of Animal and Plant Sciences, University of Sheffield, Alfred Denny Building, Western Bank, Sheffield S10 2TN, UK

T. Coulson
Department of Zoology, University of Cambridge, Cambridge CB2, 3EJ, UK

M. J. Crawley

Department of Biological Sciences, Imperial College, Silwood Park, Ascot SL5 7PY, UK

M. C. Forchhammer

Department of Population Ecology, Zoological Institute, University of Copenhagen, Universitetsparken 15, DK-2100 Copenhagen, Denmark

B. T. Grenfell

Department of Zoology, University of Cambridge, Cambridge CB2 3EJ, UK

F. M. D. Gulland

Marine Mammal Centre, Marin Headlands, 1065 Fort Cronkhite, Sausalito CA 94965, USA

A. W. Illius

Institute of Cell, Animal and Population Biology, University of Edinburgh, West Mains Road, Edinburgh EH9 3JT, UK

L. E. B. Kruuk

Institute of Cell, Animal and Population Biology, University of Edinburgh, West Mains Road, Edinburgh EH9 3JT, UK

J. Lindström

Department of Environmental and Evolutionary Biology, University of Glasgow, Glasgow G12 8QQ, UK

A. D. C. MacColl

Department of Animal and Plant Sciences, University of Sheffield Alfred Denny Building, Western Bank, Sheffield S10 2TN, UK

P. Marrow

BTexact Technologies, Martelsham Heath, Ipswich IP5 3RE, UK

J. M. Milner

Scottish Agricultural College, Kirkton Farm, Crianlarich, FK20 8RU, UK

S. Paterson

School of Biological Sciences, University of Liverpool, Liverpool L69 72B, UK

J. M. Pemberton
> Institute of Cell, Animal and Population Biology, University
> of Edinburgh, West Mains Road, Edinburgh EH9 3JT, UK

J. G. Pilkington
> Institute of Cell, Animal and Population Biology, University
> of Edinburgh, West Mains Road, Edinburgh EH9 3JT, UK

B. T. Preston
> Institute of Biological Sciences, University of Stirling, Stirling
> FK9 4LA, UK

J. A. Smith
> Department of Veterinary Parasitology, University of Glasgow,
> Bearsden Road, Glasgow G61 1QH, UK

I. R. Stevenson
> Institute of Biological Sciences, University of Stirling, Stirling
> FK9 4LA, UK

A. L. Tuke
> Department of Biological Sciences, Imperial College, Silwood
> Park, Ascot SL5 7PY, UK

K. Wilson
> Institute of Biological Sciences, University of Stirling, Stirling
> FK9 4LA, UK

Editors' and Authors' Acknowledgements

Research on the dynamics of fluctuating vertebrate populations has a long and distinguished history and we are well aware that we stand on the shoulders of giants. Much of the original stimulus for our work on St Kilda came from studies of unstable or cyclical populations of grouse, voles, lemmings and lagomorphs and we have important debts to Charles Krebs, Tony Sinclair, Adam Watson, Sam Berry, Bob Moss, Ilka Hanski, Peter and Rosemary Grant and Norman Owen-Smith for pointing the way. In addition, we are very conscious of the influence of the work of David Lack and, later, of Graeme Caughley. Our choice of Soay sheep on St Kilda as a system was inspired by a previous, ground-breaking study of the sheep led by Peter Jewell and John Morton Boyd that culminated in the 1974 book *Island Survivors*, which demonstrated that it was feasible to examine the causes and consequences of population fluctuations at the level of individual animals. Ten years after the publication of *Island Survivors*, Peter Jewell introduced us to St Kilda and helped us to establish and develop the work described here.

A large number of scientists have contributed to the work described in this book. We are deeply grateful to Jerry Kinsley, Richard Clarke, David Green, Owen Price and Tony Robertson who collected data on the island in the early years, establishing many methods which have stood the test of time. We are also especially grateful to Andrew Mac-Coll, who was responsible for collecting the core field data from 1990 to 1993; to Ian Stevenson for his generous and wide ranging contributions to many organisational and technical aspects of the work; and to Jill Pilkington who has now been in charge of all core data collection, as well as all logistics, for more than ten years. Over the years, research students have made a huge contribution to the collection of core data as well as pursuing their own studies; to date, one MSc by Research (Mark Vicari) and eleven PhDs have been completed on the study (Frances Gulland, David Bancroft, Ian Stevenson, Judith Smith, Tamsin Braishier, Steve Paterson, Jos Milner, Ettie Boyd, Amanda Tuke, Stuart Thomas and Brian Preston) and four are ongoing (Barbara Craig, Owen Jones, Dan Nussey and Louisa Tempest). Several colleagues from our network of universities and institutes have also made substantial contributions to the field work over the years, including Bill Amos, Marco Festa-Bianchet, Iain Gordon, Mariko Hiraiwa-Hasegawa, Alison Hester, Fiona Hunter, Andrew Illius, Tristan Marshall, Cathy Rowell, Jon Slate and Giac Tavecchia. Members of the Mammal Conservation Trust were instrumental in developing the summer catching system. Volunteer helpers have been the heroes of the study and we thank them all, even if we have inadvertently omitted them from this list: Jim Alexander, Roslyn Anderson, Jennifer Andrews, Mark Anstee, Helen Armstrong, Gebre Asefa, Elsie Ashworth, Nia Ball, Andrew Balmford, Liesje Birchenough, Colin Bleay, Frank Blowey, Jennifer Bobrow, Vivienne Booth, Matthew Bowell, Neil Brookes, Sue Brown, Tracy Brown, Muirne Buchanan, Colleen Covey (née Burgoyne), Chris Chinn, Dave Christie, Tom Clarke, Victor Clements, Conal Cochrane, Tilly Collins, Gill Cooper, Buster Culverwell, Ailsa Curnow, Ben Dansie, Oliver Dansie, Angus Davidson, Nell Davidson, Colin Delap, Pat Delap, Sarah Delap, Alison Donald, Andrew Dunn, Sam Emmerich, Francis Evans, Marco Festa-Bianchet, Dan Fitton, Kathleen Fraser, Helen Freeston, Liz Garner, Angela Garton, Belen Gimenez, Nicola Goodship, Malcolm Grant, Chris Green, Phil Green, Cy Griffin, Jude Hamilton, Mark Hampton, Laurel Hannah, Ian Hartley, Mary Harman, Matt Heard, Sarah Helyar, Tom Henfrey, Dave Henman, Brian Hodgkins, Nina Hofbauer, Quin Hollick, Lorraine Hoodless, Russell Hooper, David Howell, Elodie Hudson, Nye Hughes, Henry Humphreys, Kerry Hutcheson, Neil Illins, Justin Irvine, Alice Jarvis, Cath Jeffs, Vicky Jeffries, Mark Jennings, Helen Jewell, Jeremy Johns, Kenny Johnston, Robin Jones, Sarah Keer-Keer, Elsie Krebs, Wiebke Lammers, Kirsty Laughlin, Camilla Lawson, Ruth Lawson, Jackie Leah, Roger Leah, Katrina Lee, Suzanne Livingstone, Hywel Lloyd, Mike Lonergan, Sara Lourie, Richard Luxmoore, Lindsay Mackinlay, Karen Macleod, John Maxone, Andrew Mayo, Claire de Mazancourt, Morag McCracken, Mhairi

McFarlane, Colin MacGillivray, Graeme Millar, John D. Milne, Ruaraidh Milne, Alastair Mitchell, Becky Morris, Sean Morris, Kelly Moyes, Scott Newey, Nancy Ockendon, Johan Olofsson, Stephen Oswald, Vicky Parker, Claire Parrott, Ernie Patterson, Richard Payne, Howard Payton, Chris Pendlebury, Javier Pérez-Barbería, Emma Pilgrim, Bruce Pilkington, Gina Prior, John Quayle, Dan Racey, Sophia Ratcliffe, Tim Reeves, Keith Reid, Sarah Ritter, Des Robertson, Doug Ross, Helen Sainsbury, Michelle Simeoni, Harm Smeenge, Alan Smith, Tilly Smith, Diane Srivastava, Rosemary Stetchfield, Leigh Stephen, Gill Telford, Nancy Thackeray, Ben Themen, John Thompson, Will Todd, Lindsay Turnbull, Risto Virtanen, Marcus Wagner, Eddie Wallace, Lesley Watt, Lucy Webster, Linda Wilson, David Wood, Stephen Young and Anna Zakarova.

St Kilda is owned by the National Trust for Scotland and, for most of the project to date, managed by Scottish Natural Heritage (SNH) in partnership with NTS. It is one of the most heavily designated protected areas in the world, being a National Nature Reserve, a Site of Special Scientific Interest, a National Scenic Area, a Special Area of Conservation, a Special Protection Area, a World Heritage Site and (in parts) a Scheduled Ancient Monument. Into this jewel come sheep researchers wanting to stay in the limited accommodation, set traps to catch sheep, tag sheep, sample vegetation with the help of temporary exclosures, put up weather stations and many other potentially disturbing practices. We are extremely grateful to NTS and SNH for permission to research on the island and to members of the SNH South Uist office (especially Rhodri Evans, John Love, Anne MacLellan (née Shepherd) and Mary Harman) and various NTS offices (especially Anne May, Alasdair Oatts, Philip Schreiber and Robin Turner) for their help and support. In addition, we thank the succession of wardens on the island for their support and contributions to our work: David Miller, Peter Moore, Jerry Evans and Jo Babbs, Jim Ramsay, Steve Holloway, Jim Vaughan, Paul Tyler, Stuart Murray and Andy Robinson. Stuart Murray deserves especial mention for extracting sheep counts of other islands from the warden's filing cabinet and establishing more rigorous count methods. We also thank Historic Scotland (especially Sally Foster) for permission to set up our catch-up and site weather stations in Village Bay, and a succession of archaeology wardens for their tolerance of our activities: Lorna Innes (née Johnstone), Marcia Taylor and Susan Bain.

The island would be impossible to get to and hard to live on, were it not for the infrastructure provided by the missile tracking station on the island, in particular the ship and helicopter transports, the provision of running water, electricity and telephones and above all the Puff Inn. At the start of the study in 1985, several army regiments were involved: the Royal Corps of Transport, the Royal Artillery, the Royal Engineers, the Royal Electrical and Mechanical Engineers, the Royal Army Medical Corps and the Royal Signals and a number of Commanding Officers (Hebrides), especially Tony Ball. Over the years and with privatisation, we have been grateful to staff of SERCo and now QinetiQ (formerly the Defence Evaluation and Research Agency), and especially to several of the longstanding members of staff who have outlasted all these changes: Dave Clark, Tony Horne, Greg (Tom) Gregory, and Colin Brown. Thanks too to the base's nursing staff who have patched us up over the years. We are also grateful to Bristows, Bond and PDG, in whose helicopters we have frequently travelled.

When we want to reach the island with larger teams, for catching up the sheep or conducting behavioural studies in the rut, we have chartered boats from Oban. Many of these voyages have been epics in their own right, and it was not until the converted lifeboat MV Poplar Diver came on the scene that we consistently arrived on the day we expected: many thanks to former owner Bob Theakston and his crew Norman Temby and to her present operators at Northern Light charters.

Our research on St. Kilda has been supported by a number of UK funding agencies, primarily the Natural Environment Research Council, but also the Biotechnology and Biological Sciences Research Council, The Wellcome Trust and the Royal Society. Funding from these agencies is won in open competition at three to five year intervals, and the effort involved in maintaining a long term project is substantial. We thank all who have contributed over the years.

A number of colleagues from our and other universities and institutions have been generous with advice, expertise, comments and access to related work. In particular, we are grateful to Andrew Illius, Iain Gordon, Gerald Lincoln, Ted Catchpole, Byron Morgan, Mary Harman and Mariko Hiraiwa-Hasegawa. Marco Festa-Bianchet provided extensive comments on the initial draft of the entire book which were invaluable in revising the text, while Tracey Sanderson at Cambridge University Press provided support and guidance with preparation of the final manuscript and Anna Hodson was an enthusiastic and encouraging copy-editor. We are extremely grateful to them all for their generosity.

1
Individuals and populations

T. H. Clutton-Brock *University of Cambridge*

and

J. M. Pemberton *University of Edinburgh*

1.1 Sheep on St Kilda

Off the north-west coast of Scotland, beyond the protective chain of the Outer Hebrides, lie the islands of St Kilda (Fig. 1.1). The fragmented ring of a tertiary volcano, the four main islands are steep and craggy, their low ground green with lush grass (Fig. 1.2). The steep sea-cliffs of the islands are streaked with the droppings of seabirds for whom the islands are a major breeding base. Over many centuries, the birds have enriched the islands' soil and their eggs and young have attracted humans for more than four thousand years. The earliest hunters left few visible marks on the landscape, but the dry-stone walls built by the farmers that followed still bisect the low ground and the ruins of their cottages are scattered across the lower slopes of the hills.

Through these ruins, Soay sheep wander (Fig. 1.3a, b). Small, horned and mostly brown or black, they are the survivors of the earliest domestic sheep that spread through Europe in the Bronze Age, reaching Britain's remotest islands between three and four thousand years ago. In the course of time, they were replaced by larger, more productive breeds, but a remnant population of original sheep was abandoned on Soay, a 99-ha island where their existence was protected by the difficulty of access (Fig. 1.4).

For many centuries, the human population of St Kilda ebbed and flowed (Harman 1995) until, at the start of the twentieth century, numbers fell to such a low level that continued occupation of the islands was no longer feasible. In 1930, the last islanders left and Soay

1

Soay Sheep: Dynamics and Selection in an Island Population, ed. T. H. Clutton-Brock and J. M. Pemberton.
Published by Cambridge University Press. © T. H. Clutton-Brock and J. M. Pemberton 2003.

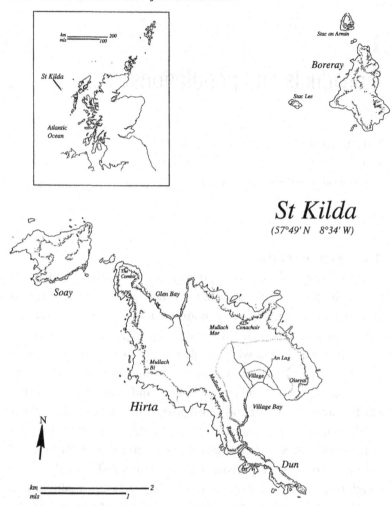

St Kilda
(57°49' N   8°34' W)

FIG. 1.1. St Kilda. The study area is enclosed by the interupted line. Inset shows the position of the islands off the north-west coast of Scotland.

sheep were returned to the largest island, Hirta, in 1932 to maintain the grazings. Their numbers were allowed to expand without human interference until they were limited by the supply of food.

Like populations of many other animals, numbers of Soay sheep on Hirta are limited by the resources available in winter. In years when sheep numbers are high, over half of the sheep can starve in

(a)

(b)

FIG. 1.2. (a) Dun, Hirta and Soay from Boreray, with Stac Lee in the foreground. (Photograph by Ian Stevenson.) (b) Village Bay from the south-west showing the Head Dyke enclosing the village street, the meadows between the street and the sea, and the fields behind the street; Oiseval in the background. (Photograph by Ian Stevenson.)

(a)

(b)

FIG. 1.3. (a) Soay sheep grazing among cleits in the fields behind the street, Village Bay; Calum Mór's house in the foreground. (Photograph by Tim Clutton-Brock.) (b) A typical party of light wild and dark wild ewes with a range of horn types (left to right: scurred, polled, normal-horned, scurred). (Photograph by Sarah Lloyd.)

FIG. 1.4. Soay from Hirta. (Photograph by Ian Stevenson.)

late winter (Grubb 1974c). The occurrence of intermittent population crashes in the sheep makes it possible to investigate the causes and consequences of episodes of acute starvation. As Malthus (1798) and Darwin (1859) originally realised, starvation is one of the principal driving forces both of population regulation and of natural selection. As it is feasible to catch and mark large numbers of individual sheep on St Kilda and to monitor their growth, condition and breeding success throughout their lifespans, we can compare the effects of starvation on different categories of animals, allowing us to examine the interactions between population dynamics and selection. As a result, the Soay sheep population allows us to ask three questions of fundamental importance in understanding the dynamics and evolution of animal populations. First, what are the causes of fluctuations in population density and to what extent do different demographic processes contribute to changes in population size? Second, how do changes in population density affect the intensity and direction of natural and sexual selection? And, third, how does population density affect the costs and benefits of reproduction and the reproductive strategies that would be expected to evolve?

We are not the first biologists to recognise the opportunities offered by island populations. Studies on island populations of invertebrates, reptiles, passerine birds and rodents have all made crucial contributions to our understanding of population dynamics and evolution (Lack 1947, 1968, 1971; Carson 1959, 1982; Mayr 1963; Grant 1965, 1968, 1986; Schoener 1968; Clarke and Murray 1969; Berry *et al.* 1978). Nor are we the first to recognise the opportunities that island populations provide for studying ungulates. Resource-limited populations of ungulates occur on a substantial number of islands in both hemispheres, offering good visibility and easy access. They have been widely used in studies of population dynamics, reproduction and behaviour (Wilson and Orwin 1964; Woodgerd 1964; Clutton-Brock *et al.* 1982a; Orwin and Whitaker 1984; Leader-Williams 1988; van Vuren and Coblentz 1989). Studies of Soay sheep on Hirta were among the first detailed investigations of the population ecology of ungulates. Research on the ecology of sheep on Hirta was started by John Morton Boyd and Peter Jewell in 1955, intensifying between 1960 and 1968 when other scientists, including Peter Grubb, joined the project to investigate different aspects of growth, feeding ecology and reproductive behaviour. These studies led to a monograph on the Soay sheep population of Hirta (Jewell *et al.* 1974) which anticipated detailed studies of other large herbivores by several years, both in the questions that it investigated and in the quality of the answers that it provided. The work showed that sheep numbers on St Kilda were unstable and that the population suffered periodic crashes when large numbers of animals died (Fig. 1.5), though it did not provide an answer to why this occurred. Seventeen years later, we restarted the detailed monitoring of the sheep population initiated by Jewell and Boyd, using modern techniques of demographic and genetic analysis, to investigate the ecological causes and evolutionary consequences of this unusual instability.

1.2 Individuals and populations

So what is the point – as visitors to St Kilda commonly ask – of expending so much effort counting sheep over so many years? Can we – as the more sceptical visitors enquire – really hope to discover anything

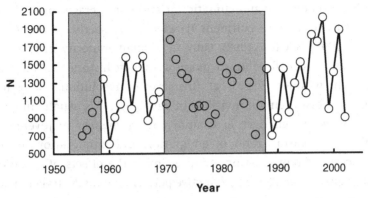

FIG. 1.5. Total numbers of sheep counted on Hirta 1955–2000. Between 1959 and 1969 annual counts were carried out systematically by Boyd, Jewell and Grubb (see Grubb 1974c) while between 1985 and 2001 a similar system was used by members of the current research team. During the other periods of the time series (shown shaded in the figure) counts were carried out less systematically by a single observer and several values are surprising given numbers in the preceding or succeeding year. We consequently believe that analyses of changes in population size are best confined to the periods 1959–69 and 1985–2001.

new? And, if we do, is it likely that it will shed light on the regulation of other animal populations?

Let us start on firm ground. The successful conservation, management and control of animals needs to be based on an understanding of the impact of environmental factors on survival and breeding success and their relative contributions to changes in population size (Riney 1982; Caughley and Sinclair 1994). Without this knowledge, too much effort can easily be spent attempting to modify aspects of the environment or components of the life-history that have little impact on population size. But is further research really necessary to achieve this? Are there really still important gaps in our understanding? Such questions reflect a fundamental misconception of the current state of ecological knowledge. In fact, our knowledge of the ecological processes underlying changes in population size in wild animals is very limited (Murdoch 1994; Turchin 1995). Only in a small number of studies of vertebrates has it been possible to measure the effects of population density and climatic variation on average survival, breeding success and recruitment (Fowler 1987; Saether 1997; Gaillard *et al.* 1998; Newton 1998). Only in a small minority of these studies has it been feasible to monitor the life-histories of individuals though, without records at this level of resolution, many important processes remain invisible. In particular, it is usually impossible to distinguish changes in fecundity from changes in juvenile mortality, to measure rates of emigration and immigration, to assess the costs of reproduction, or to estimate the effects of variation in early development on reproductive performance and survival in adulthood.

Most vertebrate studies that have investigated population dynamics using populations of marked individuals have involved monogamous birds (Newton 1998). The list of classic studies includes research on blue and great tits (Kluijver 1951; Lack 1966; Perrins 1979); Galapagos finches (Grant and Grant 1989); collared flycatchers (Gustafsson 1988); song sparrows (Smith 1988); house martins (Bryant 1979); Seychelles warblers (Komdeur 1996); red grouse (Watson *et al.* 1994); sparrowhawks (Newton 1985); oystercatchers (Ens *et al.* 1992); kittiwakes and fulmar petrels (Coulson and Wooller 1976; Ollason and Dunnet 1988; Thomas and Coulson 1988), barnacle geese (Owen 1984) and flamingoes (Johnson *et al.* 1999). However, research on the population dynamics of monagamous birds cannot be reliably generalised to mammals where most species are polygynous and substantial sex differences in body size, growth, dispersal, survival and age-related breeding success are common (Clutton-Brock 1988b). Fewer studies of mammals have been able to investigate demographic changes using individual-based data: exceptions include some rodents (Cockburn 1988; Wolff 1992), a few carnivores (Schaller 1972; Packer *et al.* 1988; Caro 1995; Creel and Creel in press), one or two primates (Crockett and Rudran 1987a, b; Altmann *et al.* 1988; Pope 2000), and several ungulates, including roe deer (Gaillard *et al.* 1993; Langvatn and Loison 1999), black-tailed deer (McCullough 1979), red deer (Clutton-Brock *et al.* 1985b), kudu (Owen-Smith 1990) and bighorn sheep (Festa-Bianchet *et al.* 1998).

The relative stability of many populations of longer-lived verte-brates means that exceptions are of particular interest since they may offer important insights into the factors generating stability and instability. Over the last twenty years, an increasing number of studies of large mammals have documented substantial changes in population size (Sinclair and Norton-Griffiths 1979; Prins and Weyhauser 1987; Leader-Williams 1988; Milner-Gulland 1994; Sinclair and Arcese 1995). Several of the cases where numbers fluctuate persistently involve resource-limited populations of ungulates living on islands – including feral goats on the Isle of Rum (Boyd 1981), mouflon in the sub-Antarctic Kerguelen Islands (Boussès *et al.* 1991) and reindeer on Antarctic and Arctic islands (Leader-Williams 1988).

The existence of persistent fluctuations in these island populations raises questions about their causes and also provides an opportunity to investigate their consequences. Why do numbers not stabilise? To what extent are changes in population size caused by changes in sur-vival versus reproductive success? What effects do changing numbers have on growth and reproductive performance? To what extent does climatic variation interact with the effects of population density? And do neighbouring populations show synchronous fluctuations? Answers to these questions are relevant to more general issues. Is the magnitude or frequency of fluctuations in population size related to the potential growth rate of the population? Are fluctuations in pop-ulation size likely to be larger and more persistent in isolated popu-lations than in larger ones from which dispersal is possible? And are populations that lack predators particularly likely to fluctuate in size? Since many populations of large herbivores are now contained within circumscribed ranges and are no longer subject to natural predators, answers to these questions are of relevance to the management and conservation of many populations of large mammals.

The first part of the book focuses on the causes of changes in population size in the St Kilda sheep and the consequences of these changes for growth, breeding success and survival. Chapter 2 describes the historical background of the St Kilda sheep population and the factors affecting reproduction and survival. Chapter 3 outlines the dynamics of the population, reviews our attempts to predict population size and examines the effects of variation in population size on development. Chapter 4 investigates the effects of changes in sheep numbers on the production and diversity of the island's vegetation while Chapter 5 examines relationships between sheep numbers and parasite densities and impact. Finally, Chapter 6 describes the distribution of reproductive success in males and examines the extent to which this changes with population density.

1.3 Population density and selection

Stimulated by the demographic analyses of Malthus (1798), Darwin's description of natural selection in the *Origin of Species* (Darwin 1859) focuses on the evolutionary consequences of differences in survival between phenotypically different individuals in populations close to carrying capacity. However, as Haldane and his successors pointed out, selection operates through differences in breeding success and survival or longevity is only one of several components of fitness (Haldane 1956; Futuyma 1986).

For practical reasons, few studies of selection in natural populations have been able to measure individual differences in breeding success, or to compare changes in the relative contributions of survival and fecundity during periods of population increase, stasis and decline or between stages of the life-history (see Endler 1986; Clutton-Brock 1988a). As a result, we know relatively little about the frequency with which selection changes in intensity or direction. A number of examples are known where the intensity or direction of selection in natural populations changes either between stages of the life-history or between periods of population growth (Hall and Purser 1979; Grant 1986; Pemberton *et al.* 1991). In addition, there is more extensive evidence of variation in the direction of selection between stages of the life history in laboratory populations of *Drosophila* (Simmons *et al.* 1980; Rose and Charlesworth 1981; Rose 1984; Luckinbill *et al.* 1984). These results have fuelled a debate as to whether selection in natural populations is typically 'hard' (strong and consistent across environmental conditions) or 'soft' (intermittent or inconsistent in strength or direction as environmental conditions change) (Wallace

1975; Kreitman *et al.* 1992). This debate is directly relevant to understanding the maintenance of genetic diversity, which is likely to erode more rapidly if selection pressures are 'hard' than if they are 'soft'.

Studies of large, long-lived animals provide unusual opportunities for comparing selection pressures on different components of fitness and investigating their stability. Though selection through mortality can be easily identified in small organisms with rapid life-histories, it is usually difficult to measure individual differences in breeding success in natural populations. (See above). Moreover, survival is often strongly affected by short-term changes in climate in smaller animals, so that the effects of phenotype or genotype on breeding success or survival of individuals are often obscured by stochastic environmental variation. Studies of selection in iteroparous vertebrates have the advantage that it is possible to measure the reproductive success of the same individual in several breeding seasons, increasing the chances of recognising and controlling for stochastic effects and detecting the consequences of variation in phenotype or genotype (Clutton-Brock *et al.* 1988b).

We have investigated three groups of questions about selection in the Soay sheep population. First, how do changes in population density and climate affect the total opportunity for selection, the extent to which different components of fitness contribute to this, and their relationships between fitness components? Second, how do density and climate affect the intensity of selection on particular phenotypic and genetic traits? And, third, to what extent are genetic differences responsible for differences in breeding success and survival? Chapter 7 describes selection pressures on different phenotypic traits while Chapter 8 examines selection pressures on genetic characteristics.

1.4 Adaptation in a changing environment

Just as an understanding of selection needs to be set in the context of population dynamics so, too, does an understanding of adaptation. Over the last thirty years, research has led to a rapid development in our understanding of the costs and benefits of different reproductive strategies in natural populations (Wilson 1974; Stearns 1992; Krebs and Davies 1993). Several recent studies show that many (if not all) of the costs and benefits of particular strategies change with environmental conditions and vary between different categories of animals (Albon *et al.* 1987; Clutton-Brock *et al.* 1987a; McNamara and Houston 1992; Festa-Bianchet *et al.* 1998). For example, in red deer, the costs of breeding increase in years when autumn rainfall is heavy or population density is high (Albon *et al.* 1987; Gomendio *et al.* 1990) and vary with individual differences in age (Clutton-Brock *et al.* 1983, 1987a, b; Albon *et al.* 1987). These results emphasise the dangers of measuring the costs or benefits of different reproductive strategies in a single season or in a particular category of animal: if the season is favourable or the sample of individuals selected is in good condition, average costs are likely to be underestimated while costs measured under adverse circumstances or in animals with few reserves are likely to be overestimates. Variation in the costs of breeding in female and male sheep is described in Chapters 2 and 9 respectively while Chapter 9 uses these estimates to investigate whether the unusually low breeding age and high reproductive rate of Soay sheep on St Kilda is likely to be maintained by selection.

Where breeding costs fluctuate, individuals may be able to increase their fitness by adjusting their investment in reproduction in relation to changing conditions or to individual differences in reproductive costs (McNamara and Houston 1992). Several recent studies of invertebrates provide convincing evidence that individuals adapt their breeding strategies to the circumstances they encounter. For example, some parasitoid wasps are more likely to lay eggs in sub-optimal hosts under environmental conditions likely to cause high mortality among adults (Roitberg *et al.* 1993; Fletcher *et al.* 1994). Similarly, in dung flies, the duration of copulation is adjusted to variation in body size (Parker and Simmons 1994). Though density-dependent changes in fecundity or maternal investment are usually regarded as inevitable consequences of food shortage, evidence of adaptive adjustments of this kind raises the possibility that some density-dependent changes in vertebrate life-histories may also represent adaptive responses to changing optima (Clutton-Brock *et al.* 1996). In the second part of Chapter 9, we use estimates of the costs and benefits of breeding during alternating periods of high and low population density to examine whether the sheep adjust their reproductive strategies to variation in the likely costs of breeding.

1.5 Soay sheep as a model system

Our study uses the St Kilda sheep population as a model system where it is possible to investigate questions of relevance to the population dynamics, selection and adaptation of other species. The final chapter of the book synthesises our understanding of population dynamics, selection and adaptation in the Soay sheep and examines the implications of our results for our understanding of the same processes in other mammals. The unusual aspects of the St Kilda population (including the constraints on movement imposed by the size of the island and the lack of effective predators) mean that the relative importance of particular demographic parameters in the sheep will not necessarily reflect their importance in other ungulate populations. Nevertheless, periodic starvation is a feature of many animal populations and the processes that we study are likely to be widespread in other populations where it is not feasible to investigate their effects.

1.6 Methods of research

Since 1985, members of the project have monitored the breeding success, growth, habitat use and survival of virtually all the Soay sheep living in the Village Bay area of Hirta. For this whole period, over 95% of sheep using the area have been marked with colour-coded ear tags, so that they are recognisable as individuals. We organise three main expeditions to the islands each year. In late February or early March, three or four members of the project visit Hirta for the first time that year, usually staying until early May. Ten censuses of the whole study population are carried out and data are logged directly into hand-held computers (Psion II). Each sheep seen is identified from its ear tag number and its location, activity, the plant community it is on and the other members of its group are recorded. These censuses provide estimates of the distribution of different individuals, the extent to which they associate with each other, and their use of different plant communities. In addition, we search the whole study area for carcasses and ear tags, especially the dry-stone shelters (cleits) which the sheep often use for shelter in bad weather. To allow us to investigate selection of anatomical characters, skulls are removed and allowed to rot, before being cleaned and removed for storage, complete with their horns. Jaws (complete with teeth) and one foreleg are cut off, boiled, dried and stored. As further animals die, their dates of death are recorded. In total, we find the bodies of at least 85% of all animals tagged as lambs in the Village Bay study area and also collect a substantial number of sheep from other parts of the island that have died in Village Bay during the winter (Clutton-Brock et al. 1992).

In late March or early April, the first lambs are born. Mothers usually move away from the grazing groups and stay close to their lambs. Approximately 95% of all lambs born in the Village Bay area are caught each year. We try to avoid catching lambs within twenty-four hours of birth to avoid desertions but catch as many as possible during the next twenty-four hours. Lambs run when approached closely but, until they can run up-slope (at around ten days), they can usually be caught. When lambs are caught, we record their mother's identity, their sex, coat colour and weight. They are tagged with Dalton's Jumbo 'Rototags', colour-coded for their year of birth and carrying a unique number, which is entered into our records. We collect small skin samples from the tag hole in their ears for genetic analysis (see Chapters 6 and 8). In 1988 and 1989, samples of individually identifiable lambs were stalked when asleep, caught and reweighed at intervals of two to thirty-six days, providing direct measures of increase in body mass throughout the first six weeks of life (Robertson et al. 1992). We subsequently use growth rates calculated from these data to estimate birth weight, reducing the observed weight by 108 g for every twenty-four-hour period since lambs were born.

A second visit occurs in July and August. Initially, a small team carries out ten censuses of the study population, while a larger team of fifteen people subsequently attempts to catch, weigh and, where necessary, re-tag as many of the sheep in the Village Bay area as possible. In the early years of the project, we relied on catching sheep in cleits or darting them using an immobilising gun to provide estimates of the weight of adults. Subsequently, with the help of an experienced team from the Mammal Conservation Trust, we developed methods of catching large numbers of sheep by enclosing them in temporary corrals. By this method, we have caught over 50% of the animals in the Village Bay sub-population each year since 1989. After sheep have been caught, they are weighed, re-tagged if their tags are broken or worn, and checked for reproductive status and milk. Horn and leg length are measured, samples of wool, blood and faeces are collected, ectoparasites are counted and the sheep are subsequently released. In addition, a count of the sheep on the entire island is carried out, with three groups of observers counting different sectors (see Fig. 1.5). While these counts give an estimate of the relative size of the population, some sheep can easily be missed and they are likely to be less accurate than our estimates of the Village Bay population, which are based on identifiable individuals.

Finally, between October and December, a third team visits the island to collect data on rutting activities. In the early years of the study, a small team carried out repeated censuses each day, recording the location and activity of individual males. More recently, a larger team carries out ten censuses of the whole population, checks for oestrus in all females each day, and collects samples of focal watches on particular males (see Chapter 9). Untagged rams, which immigrate into the area for the rut, are immobilised, tagged, measured and sampled for genetic analysis. After this, the sheep are left alone until the next spring, when we return to count and identify which individuals are still alive.

Long-term field studies have a reputation for collecting large amounts of data that are jealously guarded by their originators and are never fully analysed. We have tried to avoid this pitfall in the St Kilda sheep study. Research on the sheep was restarted in 1985 by members of the Departments of Zoology and Physiology at Cambridge (T.H. Clutton-Brock, S.D. Albon and P.A. Jewell), but the project rapidly came to involve several other groups. Over the last fifteen years, the project has involved members of the Institute of Cell, Animal and Population Biology at Edinburgh, the Institute of Biological Sciences at Stirling, the Institute of Zoology, Department of Biological Sciences at Imperial College (London), the Centre for Ecology and Hydrology (Banchory) and the Macaulay

Institute (Aberdeen) in addition to staff from Cambridge. Though the work initially focussed on population demography and dynamics, specialists in related disciplines have extended the study to involve the genetics of the sheep, their impact on the plant populations they live on and the effects of parasite populations on their survival and breeding success. More recently, our data have helped demographic theoreticians from the University of Kent (UK) and from the Macquarie University (Australia) to test and extend new methods of analysing demographic change (Catchpole *et al.* 2000) and to test methods of estimating population viability (Chapman *et al.* 2001). Each additional specialist that has joined the Soay sheep study has contributed new insights into the pattern and distribution of changes in survival and reproduction. Each has, in turn, benefitted from access to a population of recognisable individuals of known age, genotype and reproductive history – a situation which would otherwise require many years of preliminary work.

2

The sheep of St Kilda

T. H. Clutton-Brock *University of Cambridge*

J. M. Pemberton *University of Edinburgh*

T. Coulson *University of Cambridge*

I. R. Stevenson *University of Stirling*

A. D. C. MacColl *University of Sheffield*

2.1 Introduction

Sheep were domesticated in the Near East around 10 000 years ago and spread into Western Europe from there (J. Clutton-Brock 1981). Sheep similar to Soays had reached the Orkneys by 4000 BC and the sheep population of St Kilda may have originated around that date. In many aspects of their anatomy and physiology, they appear to be intermediate between contemporary domestic sheep and wild sheep (Boyd and Jewell 1974; Jewell 1986).

To understand the unusual dynamics of Soay sheep and their consequences for selection and adaptation, it is important to know something of their history as well as of the human inhabitants of St Kilda. The first two sections of this chapter describe the islands of St Kilda (section 2.2) and their history (section 2.3). Subsequent sections describe the appearance and anatomy of Soay sheep (section 2.4), their feeding ecology (section 2.5) and their reproductive system (section 2.6). Since variation in fecundity and neonatal survival affect the growth rate of the population, we describe the factors affecting the early development of lambs (section 2.7) as well as the factors affecting winter survival in juveniles and yearlings (section 2.8). Finally section 2.9 reviews the costs of reproduction and other factors affecting mortality in adults.

17

Soay Sheep: Dynamics and Selection in an Island Population, ed. T. H. Clutton-Brock and J. M. Pemberton.
Published by Cambridge University Press. © T. H. Clutton-Brock and J. M. Pemberton 2003.

2.2 The islands of St Kilda

The four main islands of the St Kilda archipelago lie 160 km to the north-west of the Scottish mainland (Fig. 1.1). Hirta, the largest, has a total area of 638 ha and consists of a large, horseshoe bay facing south-east (Village Bay) surrounded by five of the island's main hills: Ruaival and Mullach Sgar to the west, Mullach Mór and Conachair to the north, and Oiseval to the north-east. Across the high saddle between Mullach Sgar and Mullach Mór, the ground slopes down to Gleann Bay, facing north. Only at Village Bay, where a sand beach is backed by a storm beach of large boulders, is there easy access to the sea for, elsewhere, the island is edged with steep grass slopes and precipitous cliffs. The southern side of Village Bay is formed by the island of Dun (32 ha), a narrow ridge of land rising to a crest on its western side, home to a large colony of puffins. Off the north-western tip of Hirta lies Soay (99 ha), the original home of the sheep, whose steep cliffs rise to a central plateau (Fig. 1.4). Finally, 7 km to the north-east of Hirta lies the steep, rocky island of Boreray (77 ha) which, with its stacks, carries one of the largest colonies of North Atlantic gannets in the world.

St Kilda has an oceanic climate, though its hills increase its annual rainfall. Annual rainfall is around 1100–1300 mm on the lower ground and snow occurs in winter, but rarely lies long. Gales are common and occur from every direction and throughout the year (Campbell 1974). Winters on St Kilda can either be wet and windy or drier and colder: these contrasts are associated with large-scale atmospheric fluctuations over the North Atlantic, called the North Atlantic Oscillation (NAO) (Rogers 1984). Fluctuations in NAO provide an index of weather conditions that exert effects on many marine and terrestrial ecosystems (Forchhammer *et al.* 1998a; Belgrano *et al.* 1999). When pressure is low over Iceland and high over the Azores (high NAO), strong westerly winds bring warm, wet weather north into Europe, and winter gales are common. In contrast, when pressure is high over Iceland and low over the Azores (low NAO), cold, dry weather spreads west from Siberia, and winters in northern Europe are calmer and colder. Fluctuations in NAO account for much of the variation in winter weather conditions in the region: for example, in the Outer Hebrides, annual changes in the NAO winter index account for 61% of the variance in

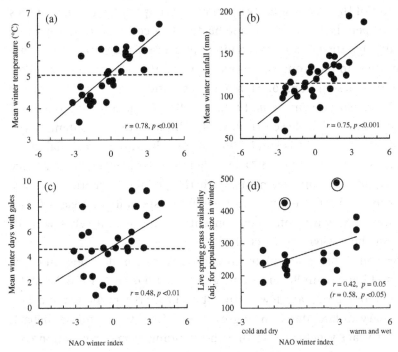

FIG. 2.1. Relationship between the North Atlantic Oscillation (NAO) winter index and (a) mean winter temperature, (b) total winter rainfall (months), (c) number of winter days with gales at Benbecula on the Outer Hebrides (Scotland) and (d) availability of live grass in spring on Hirta (adjusting for the effects of sheep numbers on overwinter offtake). Outlying circled points were omitted from the calculation of the correlation. Dashed lines indicate the thirty-two-year average. (From Forchhammer *et al.* 2001.)

winter temperature, 56% of variance in winter rainfall and around 70% of variance in average wind speed (Fig. 2.1a–c). High values of NAO in winter are associated with increased plant growth the following spring: in Norway, for example, 69% of plant species bloom earlier and 36% bloom longer after warm, wet winters (Post and Stenseth 1999). Similarly, on St Kilda, the standing crop of grass is higher after winters of high NAO than after winters of low NAO (Fig. 2.1d). However, high values of NAO in winter are correlated with reduced survival in wild ungulates because storms restrict movements and increase heat loss (Forchhammer *et al.* 1998b; Post and Stenseth 1998, 1999; Milner *et al.* 1999a, b; Post and Forchhammer 2002) and high NAO winters have adverse consequences for the sheep (see Chapter 3).

Most of Hirta is covered by seven main groups of plant communities (see Appendix 1). These include heaths, bog communities, wet grasslands, *Agrostis–Festuca* grasslands, biotic grasslands, maritime communities and a group of 'minor' communities (Gwynne *et al.* 1974). Detailed descriptions of St Kilda's vegetation are already available (McVean 1961; Gwynne *et al.* 1974), as well as a vegetation map (Ferreira *et al.* published in Jewell *et al.* 1974). The nutritional content of the pasture on St Kilda is similar to that of other highly seasonal grasslands (see Gwynne *et al.* 1974) and there is no evidence of a deficiency in any of the major elements, with the possible exception of phosphorus. Sodium levels are relatively high, especially in areas affected by sea spray, while potassium concentrations are relatively low in all except the low-lying grasslands.

Seasonal changes in plant growth on Hirta parallel those in other upland habitats in Britain. Some grass growth occurs throughout the winter but levels are low until February at the earliest. From late February to late March, new growth is offset by continuing dieback of last year's vegetation, so that the standing crop of vegetation continues to decline (Fig. 2.2a). In late March, growth accelerates, but offtake is still high, so that biomass remains low, and it is not until May that food becomes plentiful, leading to a rapid increase in standing crop between May and July.

The islands of St Kilda support large populations of seabirds, including around 134 000 fulmars, 121 000 gannets and 272 000 puffins (Murray 2002). They have their own indigenous subspecies of wren, wood mouse and, until recently, house mouse. The wood mouse has closer genetic affinities with populations in Norway than those in Scotland (Berry 1969) and the extinct house mouse resembled populations from the Shetlands and Faeroes, suggesting that both may have been introduced by the Norsemen (Berry 1970).

2.3 Human settlement on St Kilda and the history of St Kilda's sheep

The low-lying, grassy meadows of Village Bay and the slopes immediately above them (Fig. 1.2b) have been the site of the principal

FIG. 2.6. Variation in colour and horn growth in female Soay sheep. (a) Dark wild polled ewe. (Photograph by Tim Clutton-Brock.)

(b) Dark wild horned ewe. (Photograph by Tim Clutton-Brock.)

(c) Light wild scurred ewe. (Photograph by Tim Clutton-Brock.)

(d) Light wild horned ewe. (Photograph by Tim Clutton-Brock.)

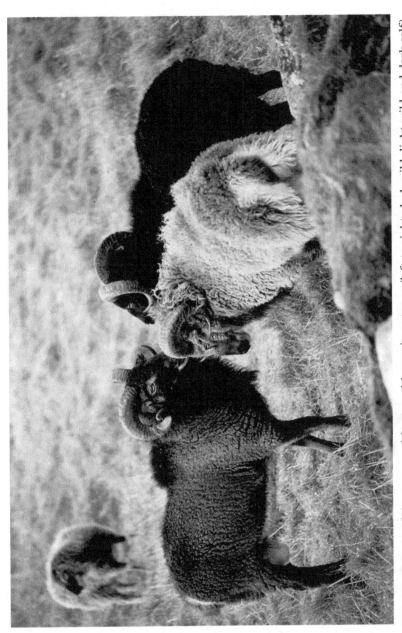

FIG. 2.7. (a) Party of three rams with normal horns in autumn (left to right: dark wild, light wild and dark self). (Photograph by Andrew MacColl.)

(b) Dark, wild, scurred ram. (Photograph by Ian Stevenson.)

(c) Dark wild yearling ram in summer with well-grown normal horns. The line delineating first- and second-year growth is clearly visible. (Photograph by Tim Clutton-Brock.)

(d) Sheep enclosed during the summer catch-up. (Photograph by Mick Crawley.)

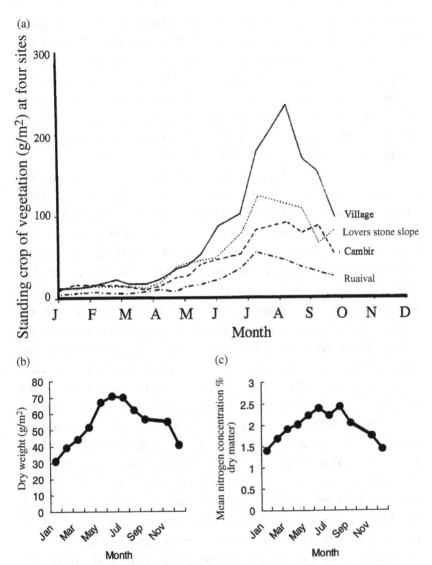

FIG. 2.2. Seasonal variation in availability and quality of food: (a) mean standing crop (dry weight g/m^2) in different months in the Village Bay area; (b) mean digestibility (*in vitro*) of organic matter (percentage) 'pinch' samples collected in the Village Bay area of Hirta in different months of 1964–5; (c) mean nitrogen concentration (percentage of dry matter) 'pinch' samples collected in the Village Bay area of Hirta over the same period. (From Milner and Gwynne 1974.)

human settlement on St Kilda since the islands were first occupied, probably in the Bronze Age (Campbell 1974; Harman 1995). St Kilda was subsequently settled by Norseman and then by Gaelic-speaking communities from the Western Isles. Much has been written of the people of St Kilda (see Williamson and Boyd 1960; Steel 1988; Harman 1997), from Martin Martin's description of their social organisation and economy in the eighteenth century (Martin 1753), to the present day (Buchanan 1995; Harman 1995). In 1727, the community was reduced to thirty by smallpox, but climbed to 100 and again remained at that level throughout the first half of the nineteenth century. In 1856, thirty-six emigrated to Australia and, from then until the 1914–18 war, the population remained around seventy. By 1928, numbers had fallen to forty, too few people remained to man the island's boat and till the fields and, in 1930, the entire community and their domestic animals were evacuated at their own request. The islands were sold to the Marquis of Bute, who transferred 107 Soay sheep from Soay to Hirta in 1932 and allowed the population to increase, largely unmanaged.

Many buildings used by the inhabitants of Hirta still stand. The fields and meadows that they cultivated are ringed by a continuous dry-stone wall, the Head Dyke, which used to exclude stock but now encircles the plant communities most heavily used by the sheep. The meadows and fields are subdivided by other dry-stone walls running away from the sea and are separated by the half-moon of the village street – a narrow path, backed by the derelict cottages (built in the 1860s) of the last inhabitants (Fig. 2.3a, b). Six of these cottages have been re-roofed and provide accommodation for visitors and scientists. The fields and meadows of Village Bay and the lower slopes of the surrounding hills are dotted with the dry-stone storage huts (cleits or cleitean) built by the St Kilda community to store food, which now are commonly used by the sheep for shelter in strong winds or heavy rain. From 1957, the island has also housed a military tracking station with a small garrison, which is part of the Hebrides missile range. In 1957, the National Trust for Scotland became the owner of St Kilda, leasing it to the Nature Conservancy and, later, to Scottish Natural Heritage, which manages the islands. The islands were designated as a National Nature Reserve in 1964 and a World Heritage Site in 1986

(a)

(b)

FIG. 2.3. (a) The village street with fields containing scattered cleits behind and the screes below Conachair in the background. (Photograph by Tim Clutton-Brock.) (b) The village, showing the encircled graveyard in the foreground and Oiseval in the background. (Photograph by Tim Clutton-Brock.)

but these changes have made little difference to the sheep which have not been fed, managed or culled throughout this whole period.

There are three naturally regulated populations of sheep on St Kilda today. Soay, off the north-west corner of Hirta, supports the original population of Soay sheep, which was the source of the animals introduced to Hirta in 1932 (see above). The name of Soay is probably derived from old Norse for 'sheep island' (Campbell 1974) and references to wild sheep on St Kilda can be traced back to the fourteenth century (Harman 1995). In the early sixteenth century, Boece mentions sheep populations on Hirta and then goes on to describe a separate population on a nearby island, presumably Soay: 'Another particular but uninhabitable island, lies near this. In it are animals by no means unlike sheep in shape, but wild and they cannot be caught except by surrounding them; they grow hair almost intermediate between sheep and goats, neither as soft as sheep's wool, nor as harsh as that of goats.' (M. Harman, unpublished data) In 1697, Martin estimated that there were 500 sheep on the island, but more recent estimates are lower (Campbell 1974). The highest of our counts suggests a total of at least 360 but, as on Hirta, numbers appear to vary considerably. There are historical records of the islanders visiting Soay to hunt sheep as well as to harvest their wool. During the eighteenth and nineteenth centuries, some sheep of later-developed breeds may have been introduced from Hirta, for the owner of St Kilda levied an annual tax on the people of Hirta of one lamb in seven, and he or his factor may have stored these lambs on Soay.

The crofters who occupied Hirta in historical times grazed the islands with cattle as well as sheep. Their sheep probably belonged to the four-horned Hebridean breed or dun-face, and were later replaced with Scottish black-face and Cheviot stock (Campbell 1974). The 1930 evacuation included all domestic stock and a subsequent expedition shot remaining stragglers so that few, if any, of the islanders' sheep were left on Hirta after 1932. The 107 Soay sheep (which included twenty-two castrated ram lambs and twenty intact males) introduced to Hirta increased rapidly. By 1939, it was estimated that the population had reached 500 and Fraser Darling judged that there were 650–700 in the following year (Boyd and Jewell 1974). The first organised census, in May 1952, gave a total of 1114 sheep. Between 1952 and 1960, when Boyd and Jewell began their intensive tagging programme, spring numbers ranged from 610 to 1344 (see Fig. 1.5). Hirta now carries a population of between 600 and 2000 Soays, the descendants of the animals introduced in 1932 and the focus of previous studies (Jewell *et al.* 1974), as well as of this one. Finally, Boreray carries a fluctuating population of around 400 black (or tan) faced sheep, a more recent and developed breed than Soays. These sheep are the descendants of sheep belonging to the community that left Hirta in 1930, which were abandoned on Boreray. Black-faces are still widely used to graze upland pastures on the Scottish mainland.

Population estimates of Soay sheep on Hirta fall into four main periods: from 1955 to 1960, when sheep were counted on the whole of Hirta each year, but limited research was carried out and separate numbers are not available for the Village Bay sub-population; from 1960 to 1968, when a team led by J. M. Boyd, P. A. Jewell and P. Grubb counted both Village Bay and the whole of Hirta each year; from 1968 to 1985, when detailed studies were abandoned but an attempt was made to count the sheep on the whole of Hirta by the resident warden; and from 1985 to the present (2002), when detailed work was restarted, most sheep in the Village Bay area were marked and the level of annual input was consistently high and included a whole-island count. In all four periods, sheep numbers fluctuated from around 600 to over 1600 (Fig. 1.5). Between 1968 and 1985 there were relatively few years of high mortality, and between 1973 and 1978, the annual growth rate was suspiciously low (Fig. 1.5), suggesting that the accuracy of counts over this period may have deteriorated.

Over the whole period of this study, the sheep population of Hirta was entirely unmanaged. However, between 1978 and 1980 Jewell castrated seventy-two male lambs in the Village Bay area and these castrates were still in the population during the early years of the current study (see Chapter 9).

2.4 Physical characteristics of Soay sheep on St Kilda

Soay sheep are small, relative both to wild sheep and to the earliest domestic sheep found in Mediterranean neolithic sites, which

FIG. 2.4. Dark wild polled ewe with twin lambs. (Photograph by Tim Clutton-Brock.)

resemble mouflon in size (J. Clutton-Brock, pers. comm.) (Fig. 2.4). Sheep of both sexes continue to gain weight until they are at least five years old (Fig. 2.5a). Adult female Soays of five years and above on St Kilda average around 24 kg in August and adult males average around 38 kg (Campbell 1974; Clutton-Brock *et al.* 1996; Milner *et al.* 1999a). Their leg lengths are around 91% of those of Scottish black-face sheep while hip width, body length and wither width are 75%, 77% and 68% respectively (Doney *et al.* 1974). We refer to sheep as lambs from birth to six months; as juveniles from six months to a year; as yearlings from one year to two years; and as adults from two onwards. In some analyses, we distinguish between adults and older animals which, depending on the analysis, may either be animals over six years or animals over ten years. In winter, the nutritional intake of the sheep from the swards in the Village Bay area is below the level necessary maintenance from the end of November until the end of January (Milner and Gwynne 1974; Clutton-Brock *et al.* 1997a; A. W. Illius, pers. comm.). They lose weight throughout the winter, reaching their lowest levels in February or March, when mortality shows a rapid increase, especially when

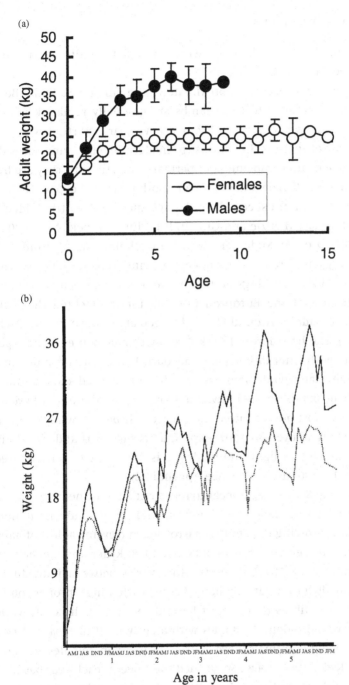

FIG. 2.5. (a) Mean weights of female and male sheep caught on Hirta in August 1990–2000. (b) Schematic plot of actual weight changes in females (dotted line) and males (solid line) in relation to season and age. (From Doney *et al.* 1974.)

population density is high or winter weather conditions are un-favourable (Fig. 2.5b).

Tooth eruption occurs (approximately) annually on St Kilda (Boyd *et al.* 1964). The milk teeth of lambs are gradually replaced by permanent teeth over a period of four or five years, starting with the central pair of incisors. In contrast to modern breeds, some Soays do not have their full complement of permanent teeth until they are five or six years old (Benzie and Gill 1974). Tooth wear of sheep on Hirta is similar to that of hill sheep of similar age and there is little evidence of dental disease (Benzie and Gill 1974; J. Clutton-Brock *et al.* 1990).

Soay sheep on St Kilda vary widely in colour, ranging from pale buff to very dark brown (Fig. 2.6, colour plate). Most dark sheep have contrasting light markings on the face and pale-coloured rumps and stomachs, resembling European mouflon though lacking the mouflon's white 'saddle'. Around 6% of all sheep are uniformly dark ('self-coloured'). The hair fibres of dark Soays with pale bellies ('wild-type') have hair with alternating pale and dark bands on the individual fibres, like many wild mammals, while self-coloured dark animals have no pale bands on the individual fibres and light animals have no dark bands. Most buff-coloured sheep have pale bellies (wild-type), but a few are buff all over. Breeding experiments suggest that dark colouring is dominant to light and wild-type is dominant to self-coloured (Ryder and Stephenson 1968; see Appendix 2).

The coats of Soay sheep appear to represent an intermediate stage of an evolutionary sequence running from wild sheep to advanced modern breeds, involving the gradual narrowing and eventual loss of hairs and their replacement with wool (Ryder 1968; Ryder and Stephenson 1968; Doney *et al.* 1974). Coat structure varies between individuals: some animals have relatively hairy fleeces with a high proportion of straight, hairy fibres that project beyond the rest of the coat, while others have more definite staples with a curly tip (Doney *et al.* 1974). Males and dark animals have hairier coats than females or light animals (Ryder 1966, 1968). Soays shed their fleeces each year between the end of April and early June, though males tend to shed earlier and more quickly than females. Fleece growth occurs mostly between June and August and is reduced between December and February.

Soay sheep are unusual in having an inherited polymorphism for horn development (Doney *et al.* 1974). Around 85% of males grow spiral horns while around 15% have small, deformed horns ('scurred') (Fig. 2.7, colour plate). Horns are used in conflicts between rams and rutting behaviour differs between horned and scurred males: scurred males spend less time in consort with oestrous females than normal-horned males and more time moving around investigating potential partners (see Chapter 9). The horns of females are smaller than those of males and a substantial proportion of females are either polled (around 30%) or scurred (35%) (Fig. 2.6, colour plate). Hornedness is inherited but the alleles involved have different expression in the two sexes: our current understanding of the inheritance of horn type in Soays is described in Appendix 2.

2.5 Feeding behaviour and habitat use

Like Scottish hill sheep (Martin 1962) as well as wild sheep (Geist 1971), the Soay sheep on St Kilda select younger and more succulent vegetation, feeding on narrow-leafed grasses throughout the year, using broad-leafed grasses heavily in late spring and early summer and eating heather most commonly in autumn and winter (Milner and Gwynne 1974). Depending on the weather, grass growth slows and then virtually ceases in late September or October and the sheep rapidly reduce the standing crop, which falls to low levels by mid December (see fig. 2.2b, Chapter 4). Changes in the quality of the vegetation follow this annual cycle: both digestibility and nitrogen levels are lowest between December and February and peak between May and July (Fig. 2.2c).

In all seasons, the sheep show a strong preference for grassland areas and, especially, for areas dominated by *Holcus* and *Agrostis* (Table 2.1) though this is weaker in the autumn when they spend relatively more time in areas of *Calluna* heath (Table 2.2). Though adult males and females use the same ground, they are usually segregated in single sex parties (Table 2.1) and castrates, too, form separate groups. Ewes spend more time on *Holcus–Agrostis* grassland than rams (Table 2.1). When sheep numbers are high, ewes spend slightly less time in areas of *Holcus–Agrostis* grassland and more time on

Table 2.1. *Proportion of groups of ewes and, rams seen on census day on different sward types in the Village Bay throughout the year (1985–2000)*

Habitat type[a]	Ewes	Rams	Habitat availability[b]
1	58.4	55.6	14.6
2	15.8	16.1	11.5
3	3.3	2.6	0.9
4	5.8	5	9.5
5	16.6	20.6	51.9
6	0.1	0.1	11.7

[a] 1, *Holcus–Agrostis* grassland; 2, *Agrostis–Festuca* grassland; 3, *Festuca* grassland; 4, *Molinia* grassland; 5, *Calluna* dry and wet heath; 6, rest of habitat.
[b] Percentage availability of different sward types.
Source: K. E. Ruckstuhl and T. H. Clutton-Brock, unpublished data.

Agrostis–Festuca swards and *Molinia* grassland. Differences in the grazing behaviour of rams between years of high and low density are generally small (Table 2.3).

Like bighorn sheep (Geist 1971; Festa-Bianchet 1991a), most female Soays spend their lives in the area where they were born. Three separate groupings or 'hefts' can be identified among sheep in the Village Bay area (Coulson *et al.* 1999a). These three groups use different parts of Village Bay: sheep belonging to the East Heft use Gun Meadow, Oiseval and An lag; those belonging to the Central Heft use the Old Village, the south side of Conachair, Mullach Mór and Signals Meadow; and members of the West Heft use the West Meadow and the coastline round to St Brianan's (Fig. 2.8). The East Heft is the largest, containing between 10% and 50% more females and males than either of the other hefts.

Of the total Village Bay sheep population, around 80% can be assigned to one of these three hefts while 20% are rarely seen and cannot be clearly assigned to any heft. Over 80% of female sheep remain in the same heft throughout their lives, while a larger proportion of males change hefts (Coltman *et al.* 2003). As a result of female

Table 2.2. *Proportion of groups of ewes and, rams seen on census days on different sward types in spring, summer and autumn (1985–2000)*

Season	Habitat type[a]	Ewes, %	Rams, %
Spring	1	70.2	63.6
	2	18.6	23.6
	3	0	0
	4	8.9	11.3
	5	2.3	1.6
	6	0	0
Summer	1	76	89
	2	1.7	0
	3	6.9	3.4
	4	7.9	3.6
	5	7.6	4.1
	6	0	0
Autumn	1	35.4	33.7
	2	18.2	16.2
	3	6.2	4.7
	4	0.8	0
	5	39.3	45.2
	6	0.2	0.3

[a] 1, *Holcus–Agrostis* grassland; 2, *Agrostis–Festuca* grassland; 3, *Festuca* grassland; 4, *Molinia* grassland; 5, *Calluna* dry and wet heath; 6, rest of habitat.
Source: K. E. Ruckstuhl and T. H. Clutton-Brock, unpublished data.

philopatry, there are consistent differences in phenotype and genotype frequency between hefts: more sheep belonging to the Central Heft are polled (31%) compared to the West Heft (16%) and the East Heft (18%). The frequencies of specific alleles also differ consistently between hefts and, within hefts, relatedness between individuals increases with the proximity of their individual ranges (Coltman et al. 2003). Pairs of females whose central 'focus' of ranging lies within 50 m of each other are, on average, related at the level of second cousins.

There are also consistent demographic differences between hefts. Though there are no significant differences in birth weight or adult

Table 2.3. *Proportion of groups of ewes and rams seen during census days (spring and summer) on six different sward types in the Village Bay during years of high and low population density*

Group type	Habitat type	Use in low population density years, %	Use in high population density years, %
Ewes	1	78.2	66.4
	2	11.8	16.5
	3	1.3	2.4
	4	6.3	10.5
	5	2.8	4.2
Rams	1	70.2	72.6
	2	13	17.4
	3	1.9	0.7
	4	13.9	6.3
	5	1	3.

[a]1, *Holcus–Agrostis* grassland; 2, *Agrostis–Festuca* grassland; 3, *Festuca* grassland; 4, *Molinia* grassland; 5, *Calluna* dry and wet heath; 6, rest of habitat.
Source: K. E. Ruckstuhl and T. H. Clutton-Brock, unpublished data.

weight, survival of lambs is around 5% higher and survival of adults around 20% higher in the Central Heft than in the East and West Hefts (Coulson *et al.* 1999a). Fecundity (the proportion of ewes breeding each year) is higher in the East Heft (82%) compared to the Central Heft (68%) and the West Heft (74%), though there are no significant differences in the frequency of twins between hefts.

Juvenile and yearling females associate with their mothers and usually graze within 20 m of them, though direct interactions are rare and there is no obvious social hierarchy among ewes (Grubb 1974c). As in wild sheep and goats (Schaller 1977; Festa-Bianchet 1991a), bonds between females and their daughters gradually weaken and, by the time daughters are twenty-four months old, they show no consistent tendency to associate more with their mothers than with other females with home ranges in the same area (Grubb 1974c; Jewell

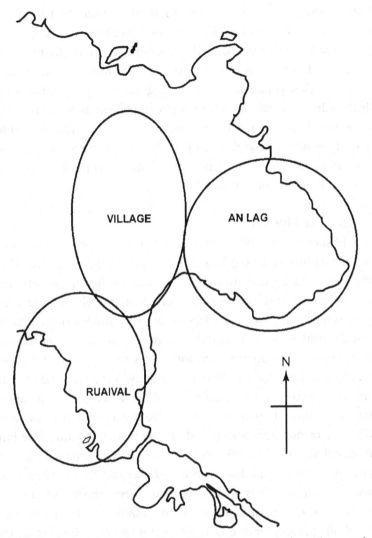

FIG. 2.8. Spatial divisions of the Village Bay population. Three main groupings or 'hefts' can be identified: the East Heft, the Central Heft and the West Heft.

and Grubb 1974). However, because females commonly adopt ranges overlapping those of their mothers and other sisters, the population continues to show genetic sub-structure (Coltman *et al.* 2003).

Male lambs typically leave their mothers in their first rut, when they are around six months old, and spend much of their time searching for females in oestrus. After the rut, most return to their mothers' home range, but no longer graze close to their mothers, often associating with other young males. By the age of two, young males have typically adopted separate home ranges and spend most of their time in close association with other males, forming groups that move independently of females (Grubb and Jewell 1974). As in wild sheep (Geist 1971), males show a well-defined dominance hierarchy, with larger and older males at the top (Grubb 1974a).

2.6 Reproduction

Like wild sheep (Geist 1971; Schaller 1977), Soays rut in autumn and bear their lambs in spring (Fig. 2.9a). Captive Soay sheep have been extensively used by Lincoln and his colleagues for investigations of physiological controls of reproduction in mammals, so the mechanisms underlying seasonal changes in reproductive behaviour are unusually well known. Seasonal changes in day length entrain annual cycles in reproductive hormones and testis size in Soay rams (Fig. 2.9b) which, in turn, control reproductive behaviour (Lincoln and Short 1980; Lincoln 1989; Lincoln *et al.* 1990; Lincoln and Richardson 1998). These annual changes in testis size are around 20% greater in Soays than in domestic sheep and closely resemble values for European mouflon (Lincoln 1989, 1998). Starting in September, Soay males engage in butting and horn-clashing fights which establish dominance relationships between males that affect priority of access to females in the mating season (Grubb and Jewell 1973; Grubb 1974a, b). Throughout September and October, males regularly inspect females for oestrus and, by the peak of the rut in November, they spend as little as 30–40% of daytime grazing, while females spend between 80% and 90% (Grubb 1974b, c) (see Chapter 9).

Except for some juveniles, virtually all females come into oestrus for one to four days during November. Individual females cycle at fifteen-day intervals until fertilisation is successful (Jewell and Grubb 1974; Jewell 1989) and, as in bighorn sheep (Festa-Bianchet *et al.* 1998), it is rare for adult ewes not to produce lambs. Mating seasons in domestic

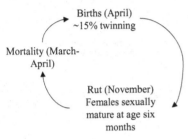

Births (April)
~15% twinning

Mortality (March–
April)

Rut (November)
Females sexually
mature at age six
months

Maximum longevity
Males: 11 years
Females: 16 years

(b)

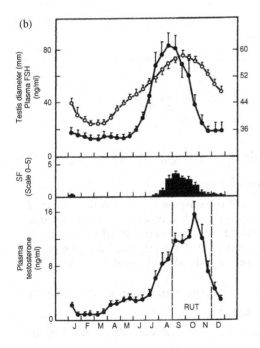

FIG. 2.9. (a) Summary of the life cycle in Soay sheep. (b) Seasonal testicular cycle in adult Soay rams living out of doors near Edinburgh (56° N); biweekly changes (mean ± SEM, $n = 7$) in the diameter of the testes (o), blood plasma concentrations of follicle-stimulating hormone (•) and testosterone and the intensity of the sexual skin coloration (SF, histogram). The period of most intense sexual and aggressive behaviour (rut) is also shown. (From Lincoln 1989.)

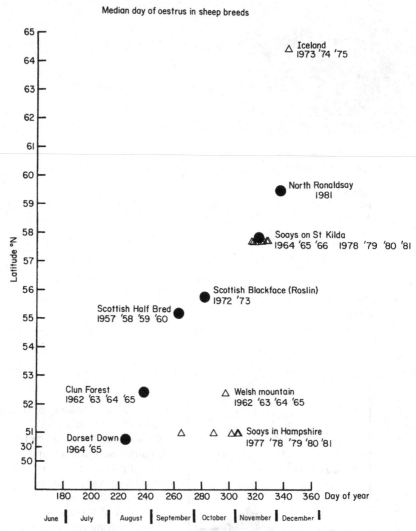

FIG. 2.10. The median dates of the onset of oestrus (plotted as day of year to encompass leap years) in different breeds of sheep adapted to a range of latitudes, calculated from the median date of births less 151 days of pregnancy (from Jewell 1989). Data are from flocks where rams or teasers were running continuously with the ewes.

breeds have adjusted to latitude and the conception dates of Soays on St Kilda are typical for their location (Fig. 2.10).

Once males find a female that is in oestrus, there are two outcomes. A single male may consort with her and guard her until he is supplanted by a more dominant male, or he gives up (Chapters 4 and 9). Consort pairs commonly last for several hours and consorting males mate repeatedly with the female they are guarding: one ram watched by Grubb (1974b) attempted to mate with six different ewes on thirteen occasions on the same day. Alternatively, oestrous ewes can be chased by multiple males for an hour or more. Chases end either when a dominant male regains control, or when the female stops running, at which point she may be mated by a number of the pursuing males. Females involved in chases commonly mate with several different males: in five-and-a-half hours of a single oestrus in 1996, a female was seen to be mounted 163 times by a total of seven males (K. Wilson, pers. comm.). Younger males and males with scurred horns often gain paternities by this method known as coursing, as in mouflon and wild sheep (Hogg 1988; Hogg and Forbes 1997). A large proportion of these matings probably involve smaller ewes, for larger males selectively consort larger females so that the degree of female promiscuity declines with female weight (see Chapter 9).

The first lambs are born in late March or early April. Mothers usually move away from grazing groups and give birth in sheltered sites or on steep slopes, as in mouflon and bighorn sheep (Welles and Welles 1961; Pfeffer 1967; Geist 1971). Unlike deer, the sheep make little effort to hide their lambs, who follow them closely from birth. Depending on population density, between 6% and 81% of Soay females conceive in the first November of their life, when they are around seven months old (see Chapter 3). In this respect, Soays resemble domestic sheep rather than wild sheep, which typically bear lambs for the first time when they are at least twenty-four months old (Schaller 1977). Whether or not different individuals on St Kilda conceive in their first year is related to their weight in late summer (see Chapter 3).

Twinning rates in adult females vary from 2% to 23% between years (Clutton-Brock *et al.* 1991). Females that conceive in their first year of life do not bear twins while, among older ewes, twinning rates range from 0–8% among ewes of 15–20 kg to 29–38% among ewes of over 25 kg in August. These rates fall within the range found in domestic sheep (Lindsay and Pierce 1984) and in wild Asiatic sheep (Schaller

1977), though twinning is rare in North American wild sheep (Geist 1971).

The sheep on Hirta lack predators. Greater black-backed gulls, lesser black-backed gulls, ravens, hooded crows and great skuas may sometimes attack and kill weak animals or newly born lambs, but this is not common. As a result, the ratio of lambs to females in autumn (typically 80–100 lambs per 100 females) is high compared to wild sheep, which typically show autumn lamb:female ratios of 20–60 lambs per 100 females (Schaller 1977; Douglas and Leslie 1986; Hass 1989).

As in wild sheep (Altmann 1970; Schaller 1977), development is rapid and Soay lambs will nibble grass by the time they are five days old. Weaning appears to be relatively early: the frequency and duration of sucking bouts starts to decline when lambs are two weeks old (Robertson *et al.* 1992) and, by mid June, sucking bouts are short and infrequent, though lambs suck occasionally throughout the rest of the summer. Though weaning is notoriously difficult to define, this suggests that it occurs substantially earlier than in bighorn sheep which, like red deer, continue to suckle their young until autumn or early winter (Geist 1971; Festa-Bianchet 1988a, 1998).

2.7 Early development and neonatal mortality

The birth date of lambs varies between years as well as between individual mothers (Jewell 1989; Clutton-Brock *et al.* 1992). Birth dates have ranged from 25 March to 12 August while median lambing dates have ranged from 15 April to 25 April (Fig. 2.11). Birth date is related to the mother's age, getting progressively earlier as a ewe gets older (Clutton-Brock *et al.* 1992). It also varies with litter size: twins are born on average two to three days earlier than singletons. Among twins (but not singletons), pairs of females are born earlier than pairs of males.

Birth weights of lambs range from 0.6 to 3.6 kg, while means for particular years vary from 1.65 to 2.21 kg (Clutton-Brock *et al.* 1992). Single lambs are born heavier than twins. There is also a small sex difference in birth weight: male lambs born as singletons average 2.02 ± 0.64 kg at birth compared with 1.94 ± 0.59 kg for females, a difference of around 4%. The sex difference in birth weight increases in twins to around 8%, despite their lower birth weight (males average

(a)   (b)

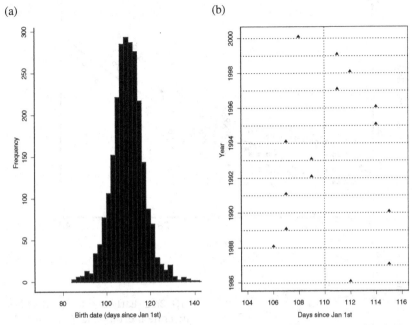

FIG. 2.11. (a) Distribution of birth dates in the Village Bay population, 1993. (b) Median birth dates in different years, 1986–2000.

1.64 ± 0.42 kg and females 1.52 ± 0.41 kg). Lambs born to young and old mothers tend to be born lighter than those born to middle-aged females and heavy mothers generally produce relatively heavy lambs (Clutton-Brock *et al.* 1996).

The effects of maternal characteristics on prenatal development in ungulates appear to vary. In some cases, birth dates or weights are consistently related to maternal age, rank or weight (Clutton-Brock *et al.* 1984; Festa-Bianchet 1988b; Verme 1989; Bon *et al.* 1993; Meikle *et al.* 1996; Byers 1997; Kojola 1997) while in others, they are not (Green and Rothstein 1991; Linnell and Andersen 1995; Côté and Festa-Bianchet 2001). The extent to which these differences are a result of genuine interspecific contrasts in early development, variation in environmental stringency or variation in the quality of data is unclear.

Lambs double in weight by the time they are three weeks old and quadruple their weight by six weeks (Fig. 2.12a). Growth rates are higher for singleton lambs (120 g/day) than twins (75 g/day) and do not differ significantly between the sexes, as in other caprids (Côté

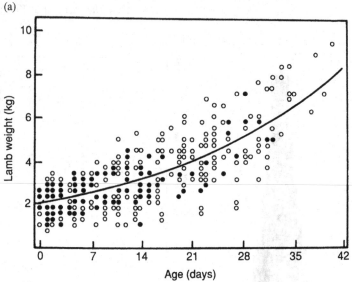

(a)

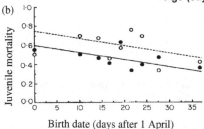

(b)

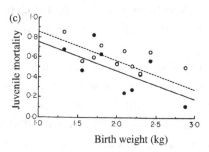

Birth date (days after 1 April)

(c)

Birth weight (kg)

(d)

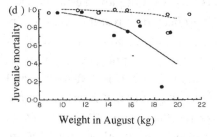

Weight in August (kg)

FIG. 2.12. Early development and juvenile mortality. (a) Generalised growth curve for lambs born in the Village Bay population in 1988 and 1989, fitted through data collected from lambs of known age (o, single observation; •, more than one observation). (From Robertson *et al.* 1992.) (b) Juvenile winter mortality plotted on birth date, controlling for the effects of population size (•, females; o, males); (c) juvenile winter mortality plotted on birth weight, allowing for effects of population size (•, females, o males); (d) juvenile winter mortality plotted on weight at four months, allowing for the effects of population size (•, females, o males); in all plots, lines are fitted equations through all data while displayed points are mean mortality for records in different categories. (From Clutton-Brock *et al.* 1992.)

and Festa-Bianchet 2001). Early-born lambs grow more slowly than late-born ones and growth rate increases with birth date at about 3 g/day. At four months, early-born lambs are heavier than late-born ones though, by the time lambs are sixteen months, the effects of variation in birth date are no longer obvious (Clutton-Brock *et al.* 1992). Differences in birth weight are positively correlated with weight at four months and persist until at least twenty-eight months and probably throughout the individual's life (Albon and Clutton-Brock 1988; Albon *et al.* 1991).

The survival of lambs through the weeks after birth is related to birth date and weight, as in several other northern ungulates (Bunnell 1980; Festa-Bianchet 1988b; Clutton-Brock and Albon 1989; Hass 1989; Birgersson and Ekvall 1997; Festa-Bianchet *et al.* 1987; Anderson and Linnett 1998; Keech *et al.* 2000: though see also Côté and Festa-Bianchet 2001). Early-born Soay lambs are more likely to die shortly after birth than lambs born later in the season (Clutton-Brock *et al.* 1992; Milner *et al.* 1999a). Lambs born below the average birth weight are more likely to die than heavy-born lambs: the probability that a lamb will die in its first month of life falls from 0.36 for lambs born under 1.0 kg to 0.04 for lambs born over 2.0 kg (Clutton-Brock *et al.* 1992). Since light mothers produce light lambs, these differences may, in part, be a consequence of variation in the mother's weight and ability to invest in her lambs. As in bighorn sheep, red deer and several other ungulates, there are no obvious sex differences in neonatal mortality overall (Clutton-Brock and Albon 1989; Fairbanks 1993; Bérubé 1997; Côté and Festa-Bianchet 2001) though females may show higher survival among the offspring of lighter mothers.

2.8 Juvenile and yearling mortality

Early development continues to influence survival during the first winter. Juvenile males are more likely to die than juvenile females, both absolutely and relative to their birth weight (Clutton-Brock *et al.* 1992). Early-born juveniles and those born below average weight are more likely to die than late- or heavy-born juveniles (Fig. 2.12b, c) and, as in bighorn sheep, the effects of low weight increase when population density is high (Festa-Bianchet *et al.* 1997; Milner *et al.*

1999b). Weight in August, at around four months, provides a better predictor of whether or not a juvenile will die in its first winter than birth weight (Fig. 2.12d), and whether or not they are pregnant affects the chance that juvenile females will survive their first winter. Both these effects vary between years when mortality is high and those when it is low (see section 2.9).

The age and weight of females influence the survival of their off-spring (Clutton-Brock *et al.* 1996). Relatively heavy mothers produce relatively heavy lambs which show high survival rates, especially when winter density is high (Clutton-Brock *et al.* 1996). This contrasts with results for bighorn sheep, where juvenile mortality does not vary with maternal weight (Festa-Bianchet *et al.* 1998). Juveniles produced by mothers less than a year old show relatively high mortality partly because young mothers produce relatively light lambs. However, lower birth weights are not responsible for all the effects of mother's age for the offspring of young mothers still show higher mortality through their first year than the offspring of older females when differences in birth weight are controlled (Clutton-Brock *et al.* 1996). In contrast, the lower survival of twins compared to singletons appears to be caused primarily by their lower birth weights, for this difference disappears when the effects of birth weight are controlled for (Clutton-Brock *et al.* 1992).

Mortality of yearlings varies between years from less than 10% to over 80% (see Chapter 3) and is higher in males than females, especially when population density is high (see Chapter 3). Yearlings born to young and old mothers are more likely to die in winter than those born to prime-aged mothers (Fig. 2.13). The mortality of yearlings is related to their weight at sixteen months (Clutton-Brock *et al.* 1992) and, males less than 20 kg show particularly high mortality.

Growth and survival are closely related to early development in other northern ungulates, including bighorn and Dall's sheep (Bunnell 1980; Festa-Bianchet 1988b), mountain goats (Côté and Festa-Bianchet 2001), pronghorn (Gaillard *et al.* 1993), red deer (Guinness *et al.* 1978a,b), moose (Keech *et al.* 2000) and bison (Green and Rothstein 1991). In some cases, relationships are only apparent in particular categories of individuals (Festa-Bianchet 1996; Côté and Festa-Bianchet

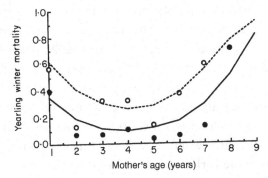

FIG. 2.13. Yearling mortality plotted on the mother's age for males and females, allowing for the effects of population size (●, females; ○, males). Lines are fitted equations through all data while displayed points are mean mortality for animals in different categories. (From Clutton-Brock *et al.* 1992.)

2001) or in years when population density is high (Festa-Bianchet *et al.* 1997). In most populations, early-born juveniles are heavier at the on-set of winter than late-born animals and show superior survival (Côté and Festa-Bianchet 2001). The tendency for early-born Soay lambs to be more likely to die in their first winter than those born later (see above) is unusual and suggests that adverse conditions during early development may have protracted effects that cannot be compensated for by longer growth periods. The positive correlation between weight and juvenile survival is found in many other ungulates, including bighorn sheep (Festa-Bianchet 1996), mountain goats (Côté and Festa-Bianchet 2001), red deer (Guinness *et al.* 1978b) and mule deer (Unsworth *et al.* 1999) though, in some cases, it is only apparent when population density is high or climatic conditions are adverse.

2.9 Adult mortality and the costs of reproduction

With the exception of a small number of females that die in parturition, virtually all deaths in Soay sheep older than six months occur in winter and early spring, mostly between January and April (Clutton-Brock *et al.* 1997a). Starvation is the usual cause of death, though resource shortage may also reduce the animals' immunity to parasites (see Chapter 5). In some years, body weight declines by as much as 30% between January and March (see Chapter 3). Autopsies show that sheep dying in late winter have virtually no fat reserves and that marrow fat is severely depleted (Gulland 1992). Juveniles, being smaller than adults, tend to die first – both because they have smaller body-fat reserves and because these are depleted at a higher rate through energy expenditure on thermoregulation (Clutton-Brock *et al.* 1997a). Adult males reach critical weights before mature females on account of heavy expenditure and reduced food intake during the autumn rut and usually die earlier in the winter (Chapter 3).

Both body weight of individuals and reproductive status affect survival in years when population density and winter mortality are high (Clutton-Brock *et al.* 1996; G. Tavecchia *et al.* unpublished data). Among juvenile females, survival is low among animals of less than 15 kg when overall mortality is high, especially if they have conceived lambs (Fig. 2.14a). Experiments that prevented juveniles from conceiving using progesterone implants confirmed that breeding reduces survival (Gulland 1991; G. Tavecchia, unpublished data). Among adult females, survival in years of high mortality is lowest among light females and those that have conceived twins (Fig. 2.14c). In contrast, when mortality is low, neither August weight nor breeding status have much effect on survival and females that fail to breed tend to show *lower* survival than those that have done so (Fig. 2.14b, d), probably because they are inferior phenotypes. A similar tendency for variation in phenotypic quality to be related to the costs of reproduction is found among older females in bighorn sheep (Festa-Bianchet *et al.* 1998).

While pregnancy has substantial costs to survival in years of high mortality, the number of offspring a female has raised in the previous year has no obvious effect on her weight in August, her probability of conception, her survival through the subsequent winter, or her chance of raising her subsequent offspring (Clutton-Brock *et al.* 1996). Differences in body weight in August among females that raise different numbers of offspring in the same year are relatively small and vary between age classes: among juvenile mothers, those that raise a lamb are slightly heavier in August than those that fail to do so; among yearlings, there are no significant differences in weight between those that breed successfully and those that fail to breed;

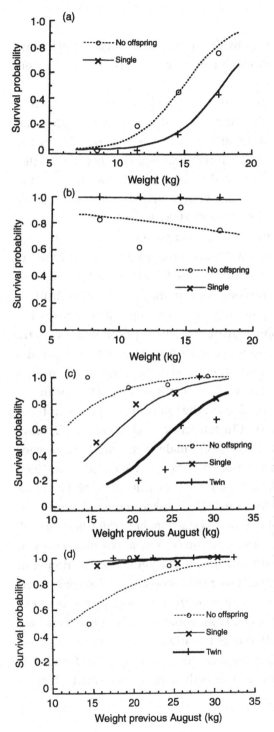

FIG. 2.14. Effects of breeding on survival for females of different body weight in the previous August. (a) Juveniles in years of high overall mortality; (b) juveniles in years of low overall mortality; (c) adult females in years of high overall mortality (crash years); (d) adult females in years of low overall mortality (non-crash years). Symbols show the proportion of animals in different categories that survived. (From Clutton-Brock *et al.* 1996.)

and, among older females, individuals that raise single offspring are lighter than those that either fail to breed or raise twins (Clutton-Brock *et al.* 1997a).

The relationship between reproduction and subsequent survival in Soay sheep resembles results for bighorn sheep, where females that have raised lambs successfully are no less likely to conceive in the autumn rut or to survive the winter than females that have failed to raise offspring, despite reduced weight gain following lactation and a positive correlation between body weight and longevity (Festa-Bianchet *et al.* 1998; Bérubé *et al.* 1999). However, it contrasts with the situation in red deer, where mothers that have raised calves through the summer months show reduced body weight and fecundity in autumn as well as lower survival through the following winter compared to females that have failed to breed successfully (see section 2.10).

In some sexually dimorphic ungulates, rearing sons depresses the mother's fitness more than rearing daughters (Clutton-Brock 1991). For example, in red deer, sons are born heavier and suck more frequently than daughters and mothers that have raised sons through the summer months are more likely to die the following winter and, if they survive, less likely to breed again the following year than mothers that have reared daughters (Clutton-Brock *et al.* 1981). Some similar differences exist in bighorn sheep: male lambs are born heavier than females and show faster growth rates (Festa-Bianchet *et al.* 1996) and mothers that have raised sons subsequently show higher parasite loads and are less likely to rear their subsequent offspring than mothers that have raised daughters (Festa-Bianchet 1989a; Bérubé *et al.* 1996). In contrast, in Soays, there are no consistent sex differences in sucking frequency or growth rate during the period of lactation and, unlike red deer, mothers that have raised sons do not show increased mortality in the following winter or depressed breeding success next spring (Clutton-Brock *et al.* 1992, 1996; Robertson *et al.* 1992). This may be the case because sex differences in early development in Soays are relatively small: male lambs are around 4% heavier than females at birth and sex differences in early growth rates are also small (Clutton-Brock *et al.* 1992).

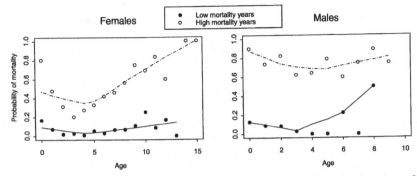

FIG. 2.15. Over-winter mortality in (a) females and (b) males in years of low mortality when population size increased by at least 10% versus years of high mortality when population size declined by over 23%. Data from 1985 to 1993.

As in many other ungulates, including bighorn sheep (Clutton-Brock and Albon 1989; Bérubé *et al.* 1999; Loison *et al.* 1999b), mortality increases in older females, especially in adverse years (Clutton-Brock *et al.* 1991, 1992; Catchpole *et al.* 2000) (Fig. 2.15a). Part of this effect may be due to increasing costs of reproduction: in years of high winter mortality, older females that have conceived twins show larger increase in mortality than prime-aged females (Fig. 2.16). However, in years of low mortality, much of the mortality in older animals occurs in females that have failed to conceive offspring, probably because these individuals are in relatively poor condition (Fig. 2.16a).

Reproductive performance also declines in older females though these changes occur later in the lifespan than changes in survival, as in red deer (Clutton-Brock and Albon 1989). While maternal age has no obvious effect on the birth dates of lambs (Fig. 2.17a, b), birth weights decline and neonatal mortality increases among offspring of the oldest females when winter mortality has been relatively low (Fig. 2.17c, e). These effects disappear after years of high winter mortality when few older females survive (Fig. 2.17d, f).

As in most other polygynous vertebrates (Clutton-Brock *et al.* 1991; Loison *et al.* 1999a), males have lower life expectancies and shorter lifespans than females (Fig. 2.15b). These differences are probably

(a)

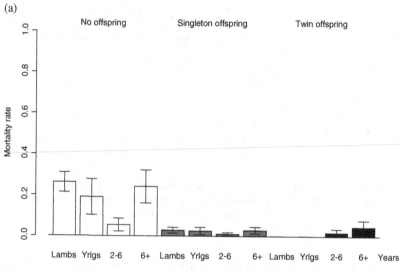

(b)

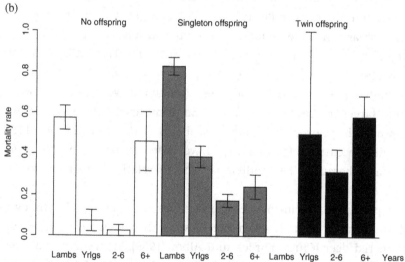

FIG. 2.16. Age-related mortality in females that have conceived different numbers of offspring in (a) years when winter mortality was low and population size increased by at least 10% and (b) years when winter mortality was high and population size declined by over 23%. Data from 1985 to 1993.

associated with the relative costs of reproduction, for castrates live considerably longer than intact adults of either sex (see Chapter 9). When overall mortality is high, the effects of age on male survival decline since few males survive.

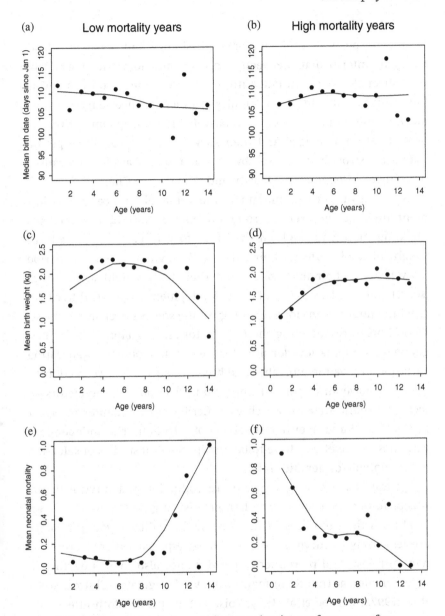

FIG. 2.17. Age-related changes in reproductive performance after winters of low versus high mortality. (a, b) Birth date of offspring; (c, d) median birth weight of offspring; (e, f) mean neonatal mortality.

2.10 Discussion

In many aspects of their anatomy, physiology and behaviour, Soay sheep are intermediate between wild sheep and domestic breeds. Like wild sheep, Soays rut in late autumn and bear their lambs in spring but, unlike most wild sheep, females commonly conceive in their first autumn and twinning rates are relatively high. Sheep populations on St Kilda are unmanaged and their survival and reproductive performance are strongly influenced by variation in food availability caused by changes in population density and climate (see Chapter 3).

As in many other wild mammals, individual differences in development are large and exert a strong influence on survival and eventual breeding success throughout the life-history. In particular, the birth weight of lambs affects their survival through the neonatal period as well as in their first winter and their body weight as yearlings (see Chapter 3). The effects of early development on survival have far-reaching implications for the ecology of the sheep as well as for the selection pressures affecting their life-histories. As Chapter 3 describes, changes in population density affect early development, generating variation in juvenile mortality and breeding success between cohorts. They also have an important impact on selection and the factors affecting juvenile survival which we describe in this chapter reappear in Chapters 7 and 8, either as targets of selection or as independent variables whose effects have to be incorporated in studies of selection on phenotypic or genetic traits.

Age exerts a strong influence on reproduction and survival in the sheep. Young females produce lighter offspring that show lower survival rates than mothers in their prime. In addition, young females are less likely to survive adverse winters, especially if they are pregnant. Survival and reproductive performance also show a decline in older females, as they do in many other wild ungulates (Clutton-Brock *et al.* 1997a; Bérubé *et al.* 1999; Loison *et al.* 1999b). Both effects are more apparent in years when overall mortality is relatively low, for adverse winter conditions affect all age categories, reducing the strength of age effects. The effects of age on survival and reproductive performance mean that fluctuations in the relative numbers of young and old animals in the population have an important influence on population growth rate and need to be incorporated in any attempts to model population dynamics (see Chapter 3). In addition, they affect the costs and benefits of breeding at different ages and the evolution of reproductive strategies (see Chapter 9).

Reproductive status also exerts important effects on survival and the costs of pregnancy are clearly high in adverse winters, especially among lighter females (Fig. 2.14). In years when overall mortality is relatively low, these differences disappear, presumably because the costs of pregnancy are obscured by positive correlations between breeding success and survival generated by a common dependence on weight or condition (Clutton-Brock *et al.* 1983; Festa-Bianchet *et al.* 1998). Unlike pregnancy, lactation has little effect on the survival of mothers (Clutton-Brock *et al.* 1996) because it occurs during midsummer when resources are plentiful and ends before food supplies decline at the end of the summer, allowing females to regain lost condition before the onset of winter (Clutton-Brock *et al.* 1997a). In this respect, Soays differ from red deer, where the costs of lactation persist until early winter and females that have reared calves are less likely to conceive in the autumn rut (Mitchell *et al.* 1976; Clutton-Brock *et al.* 1982a).

The high costs of reproduction in Soays raise important questions about the adaptedness of their life-histories. Why are Soay lambs born so early in the year? Is conception during the first year of life (which seldom occurs in wild sheep) really an adaptive strategy – or is it a by-product of artificial selection for fecundity during the initial domestication of Soay sheep? If so, why has it not been removed by selection? Are the benefits of early conception and twinning in years of low mortality large enough to offset the costs in years when mortality is high? Does the unusually high fecundity of Soay sheep leave them particularly prone to starvation? We return to these issues in Chapter 9, drawing on the relationships described in this chapter.

3
Population dynamics in Soay sheep

T. H. Clutton-Brock *University of Cambridge*

B. T. Grenfell *University of Cambridge*

T. Coulson *University of Cambridge*

A. D. C. MacColl *University of Sheffield*

A. W. Illius *University of Edinburgh*

M. C. Forchhammer *University of Copenhagen*

K. Wilson *University of Stirling*

J. Lindström *University of Glasgow*

M. J. Crawley *Imperial College London*

S. D. Albon *Centre for Ecology and Hydrology, Banchory, UK*

3.1 Introduction

A conspicuous feature of many naturally limited populations of long-lived vertebrates is their relative stability. Both in populations that are regulated by predation or culling and in food-limited populations, population size can persist at approximately the same level for decades or even centuries (Runyoro *et al.* 1995; Waser *et al.* 1995; Clutton-Brock *et al.* 1997a; Newton 1998). The persistent fluctuations shown by Soay sheep and by some other island populations of ungulates (Boyd 1981; Leader-Williams 1988; Boussès 1991) raise general questions about the causes and consequences of variation in the stability of populations (see section 1.2). How regular are they? How are they related to population density? What are their immediate causes? To what extent do fluctuations in food availability, parasite number or predator density contribute to them? And what are their effects on development and on the phenotypic quality of animals born at contrasting population densities? And how much do changes in phenotype contribute to changes in dynamics?

As yet, there are very few cases where we understand either the ecological causes or the demographic consequences of persistent

52

Soay Sheep: Dynamics and Selection in an Island Population, ed. T. H. Clutton-Brock and J. M. Pemberton.
Published by Cambridge University Press. © T. H. Clutton-Brock and J. M. Pemberton 2003.

fluctuations in the size of naturally regulated populations of mammals (Hanski 1987; Saether 1997). Since we are able to monitor the growth, movements, breeding success and survival of large samples of individuals as population density changes, the Soay sheep offer an opportunity to investigate the causes and consequences of changes in population size with unusual precision (see Chapter 1). While the absence of predators and the constraints on dispersal imposed by the habitat are likely to affect the relative contribution of different processes to changes in population size (see Chapter 10), the factors causing fluctuation in sheep numbers are likely to contribute to changes in population size in many other populations.

In this chapter, we start by describing the immediate causes of changes in the size and structure of the Soay sheep population on Hirta (section 3.2). Previous analyses of the population dynamics of the Soay sheep population of Hirta (Grubb and Jewell 1974) did not have access to a large enough time series to investigate the effects of variation in population density on reproduction and survival and this was an early objective of our study. Sections 3.3 to 3.5 examine how fecundity, development and survival change with population density and how climatic differences between years contribute to these effects. Variation in environmental conditions also affects early development with consequences for the subsequent growth, breeding success and survival of members of different cohorts (section 3.6). In sections 3.7 and 3.8, we describe our attempts to predict changes in the size of the population using progressively more complex models, while in section 3.9 we examine the degree of synchrony between sheep populations on the different islands of St Kilda.

3.2 Changes in population size

Numbers of Soay sheep on Hirta fluctuate widely as a result of variation in winter mortality (Fig. 1.5). Between 1959 and 1968, total numbers fluctuated from 610 to 1598 while numbers in the Village Bay population fluctuated from 174 to 404 (Grubb 1974a). Between 1985 and 2000, total numbers fluctuated from 663 to 2022 and numbers in the Village Bay population from 211 to 591. Increases in population size were associated with an increase in the number of animals using

(a)

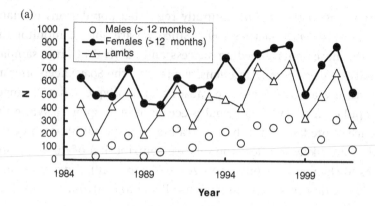

(b)

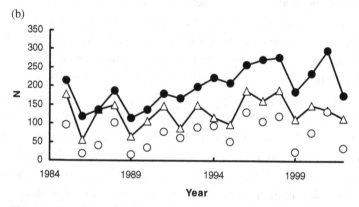

FIG. 3.1. Numbers of male and female sheep (over twelve months) and lambs (less than twelve months) (a) on the whole of Hirta (b) in the Village Bay sub-population 1985–2001. Estimates of population size for the whole island based on August counts; estimates for the Village Bay population based on censuses of marked individuals.

the most heavily populated areas rather than with an extension of the area used.

Fluctuations in numbers are more marked in some age and sex categories than others. Variation in male numbers is greater than in female numbers and variation in juvenile numbers is greater than in adult numbers (Fig. 3.1a), reflecting the increased susceptibility of males and juveniles to starvation (see sections 2.8 and 2.9).

While we counted the whole sheep population annually (see section 1.6), most of our work (like that of Jewell and his collaborators) was carried out on the Village Bay sub-population, which typically

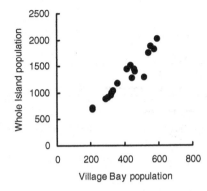

FIG. 3.2. Total number of sheep on Hirta plotted against totals for the Village Bay sub-population 1984–2002.

includes between a quarter and a third of all the animals on Hirta. Numbers of sheep using Village Bay show similar fluctuations to the whole island population (Fig. 3.1b) and their numbers are closely correlated with population size for the whole island (Fig. 3.2).

Both across the whole island and within Village Bay, year-to-year changes in population size show a pronounced threshold effect (Fig. 3.3a, b, Fig. 3.18). When there are fewer than 1100–1200 animals on the island, the population increases by about 1.27 per animal per year but, when there are more than this number, it declines by an average of around 0.2 per animal per year. Though juvenile mortality is more variable than adult mortality, variation in adult mortality has a greater impact on changes in population size since there are more adults in the population. Adult and juvenile mortality, together with the positive covariance between them, are the key factors responsible for changes in population size (Clutton-Brock *et al.* 1991; Coulson *et al.* 1999a) (Fig. 3.4).

Years of high mortality when numbers decline are usually separated by two or three years when population size increases (Fig. 1.5), generating an alternation of boom years followed by sudden crashes. This pattern was most apparent between 1985 and 1995, when three separate years of high mortality (1985, 1988 and 1991) were each followed by two years of rapid population increase. In contrast, after the crash of 1995, numbers increased in 1996 and in 1997 and remained high in 1998 before crashing in 1999. After this, two years of increase in population size were once again followed by a crash

(a)

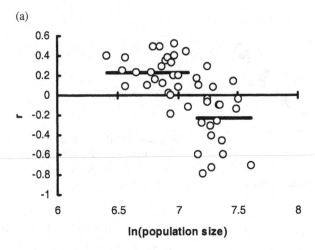

(b)

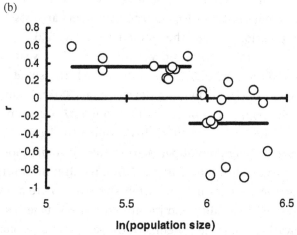

FIG. 3.3. (a) Changes in total population size on Hirta ($r = \ln [N(t + 1)/N(t)]$) as a function of log density, showing the threshold population density below which population tends to increase at a density-independent rate of 26.5% per year and above which population tends to decline at a density-independent rate of 19.2% per year (b) Changes in population size in the Village Bay sub-population (as above).

in 2002. Similar variability in the timing of years of high mortality occurred between 1960 and 1968. Cohorts born in different years have strikingly different demographies (see section 3.6 and Appendix 3).

The tendency for years of high mortality to be separated by two or three years of rapid population growth probably reflects the number

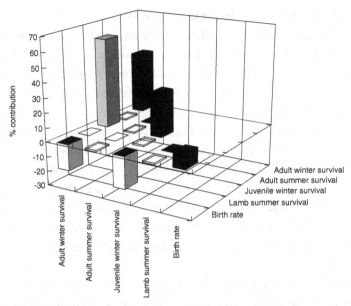

FIG. 3.4. Analysis of the key factors associated with relative changes in the size of the Village Bay population for 1986–1994 inclusive (from Coulson *et al.* 1999a). Structured demographic accounting of the variance of demographic change (Brown *et al.* 1993) was used to decompose the population fluctuations in the percentages due to changes in summer and winter mortality of lambs, juveniles and adults and to changes in the birth rate. The most important factors are juvenile and adult winter survival (the large black bars) and the positive co-variation between them (the large grey bar). Birth rate was not as important (the small black bar), explaining approximately 10% of the fluctuations in population size. However, birth rate did negatively co-vary with both juvenile and adult winter survival explaining 22% and 17% of the relative changes in population size respectively.

of years necessary for a basal population of less than 1000 sheep to regain a size at which heavy winter mortality is likely to recur (Clutton-Brock *et al.* 1997a). It is possible that longer intervals between crashes may have occurred between 1968 and 1985 (Fig. 1.5), but the reliability of these estimates is uncertain. Several aspects of the data from this period are surprising, including the progressive downward trend between 1971 and 1978, and we suspect that estimates of changes in population size over this period are unreliable.

3.3 Density-dependence in fecundity

Fecundity changes with population density in young and old ewes (Fig. 3.5a). The proportion of juveniles producing lambs when they are around twelve months old varies between years from less than 10% to over 80% and is low when autumn density is high (Fig. 3.5b) or winter weather conditions are wet and windy (G. Tavecchia *et al.* unpublished data). Among yearlings, the proportion of individuals producing lambs at around twenty-four months shows a weak tendency to decline with increasing density (Fig. 3.5c) while, among adults, around 80% of individuals produce lambs each year, irrespective of population size (Fig. 3.5d) or reproductive status in the previous year (Fig. 3.5e). However, twinning rates decline with increasing density (Fig. 3.5f) and the incidence of abortions during late pregnancy increases.

As in bighorn sheep (Festa-Bianchet *et al.* 1998), variation in fecundity in young animals is related to variation in body weight during

FIG. 3.5. Effects of population density on the fecundity of Soay sheep in the Village Bay population, 1986–98. Fecundity is the proportion of animals in the population that gave birth either to singleton or twin lambs in the spring while population density is the total size of the population in the Village Bay area the previous winter. As a result, after years when winter mortality is high, low fecundity is associated with high winter density but low density in the spring and summer after lambs are born. (a) Fecundity in relation to age, showing that the most parsimonious groupings of individuals are in four age classes: juveniles, yearlings, adults (two to ten years) and older animals (over ten years). (b) Changes in the proportion of juveniles giving birth at approximately twelve months. (c) Changes in the proportion of yearlings giving birth at approximately twenty-four months. (d) Changes in the proportion of adults (over twenty-four months) giving birth to lambs. (e) Changes in the proportion of females that had raised a lamb the previous year that gave birth. (f) Changes in the proportion of adults (over twenty-four months) giving birth to twin lambs. (g) Changes in neonatal mortality with population density the previous winter. (h) Recruits per adult. Fitted lines are from logistic regression models. Sloping lines represent a significant association between the fecundity measure and density, horizontal lines represent no significant association.

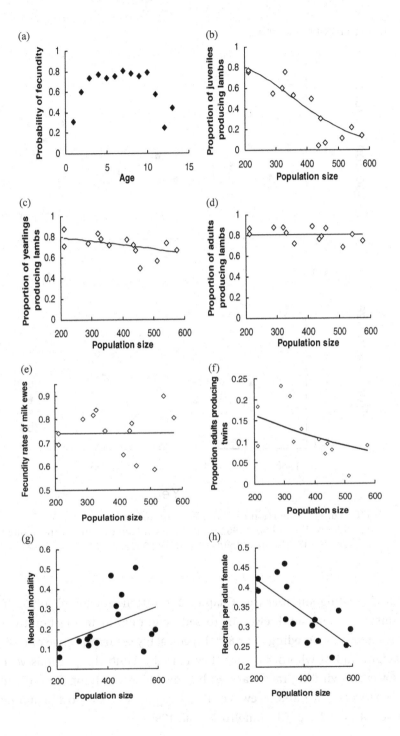

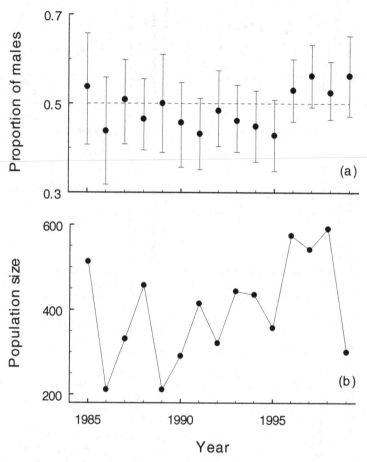

FIG. 3.6. (a) Proportion of males born in different years (±95% confidence limits (Zar 1996). (b) Population fluctuations in the Village Bay area 1985–98. (From Lindström *et al.* 2002.)

the preceding summer (see Chapter 2) (Clutton-Brock *et al.* 1996). In contrast, among adult ewes, increased weight has little effect on their probability of breeding: over 80% breed each year, irrespective of their weight and of whether or not they reared a lamb the previous year. However, twinning rates vary with body weight, ranging from 0% to 8% between years among ewes of 15–20 kg and from 23% to 38% among ewes of over 25 kg (Clutton-Brock *et al.* 1996).

In some mammals where adult males are larger than females, juvenile males grow faster than females before and after birth and males commonly show higher levels of mortality both before and after birth (Clutton-Brock *et al.* 1985a). As a result, female-biased birth sex ratios are sometimes found in populations where pregnant females are exposed to food shortage or harsh conditions (Clutton-Brock 1991) though these trends are by no means universal (Clutton-Brock and Iason 1986). For example, in red deer, the proportion of males born falls when winter density is high or winter weather is wet and windy (Kruuk *et al.* 1999a). In Soays, the proportion of male lambs born as varied from 43% to 56% between years (Fig. 3.6). Birth sex ratios are not consistently related to weather conditions or to the mother's age or weight though there is a weak positive correlation with population density the previous autumn (Lindström *et al.* 2002). A possible functional explanation is that it is advantageous for females to produce males after years when autumn density and winter mortality are high because these will join a relatively small cohort of lambs that are likely to show high survival and breeding success (West and Godfray 1997; see Chapter 6). However, the trend is weak and it is not yet clear whether it is consistent. In bighorn sheep, too, there is little evidence of consistent variation in birth sex ratios, apart from a tendency for old mothers to produce relatively few males (Bérubé *et al.* 1996; Festa-Bianchet 1996; Gallant *et al.* 2001).

3.4 Variation in birth weight and neonatal mortality

High population density and wet, windy weather depress the average birth weight of lambs and, in conjunction, population density and weather account for 76% of variation in mean birth weight between cohorts (Forchhammer *et al.* 2001). Birth weight of single lambs declines by around 200 g with every additional 100 sheep using the study area, falling from around 2.2 kg when population density the previous winter is low to 1.7 kg at high density (Clutton-Brock *et al.* 1992) (Fig. 3.7a). These results contrast with the absence of any relationship between density and birth weight in some other ungulates, including mountain goats (Côté & Festa-Bianchet 2001) and red deer

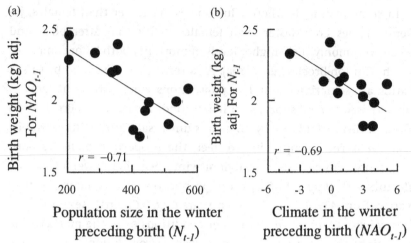

FIG. 3.7. Relationships between mean birth weight and (a) population size (N_{t-1}) and (b) climate severity (NAO_{t-1}) in the winter before birth. Correlation coefficients (r) are significant ($p < 0.05$). Through generalised linear models taking maternal age and weight into account, the combined variation in N_{t-1} and NAO_{t-1} was found to explain 28% and 14% of the variation in cohort birth date and birth weight respectively. (From Forchhammer *et al.* 2001.)

(Clutton-Brock and Albon 1989). When the effects of population density have been allowed for, birth weight also falls after wet, windy winters (Fig. 3.7b). While birth weight varies with litter size, lamb sex and mother's age (section 2.7), there is no evidence that changes in density either exaggerate or reduce these effects (Clutton-Brock *et al.* 1992).

Neonatal mortality, too, varies with population density the previous winter, increasing from less than 10% when population density the previous winter is low, to around 40% when it is high (Clutton-Brock *et al.* 1992; Forchhammer *et al.* 2001) (Fig. 3.8a, b). When the effects of population density have been allowed for, neonatal mortality rises after wet, stormy winters, when the North Atlantic Oscillation (NAO) index (see Chapter 2) is high (Fig. 3.8c, d). Changes in birth weight are responsible for much of this variation, but significant effects of density and winter climate in the year before birth remain when the effects of birth weight are controlled (Clutton-Brock *et al.* 1992). As in red deer, there is no evidence of any difference in the effects of

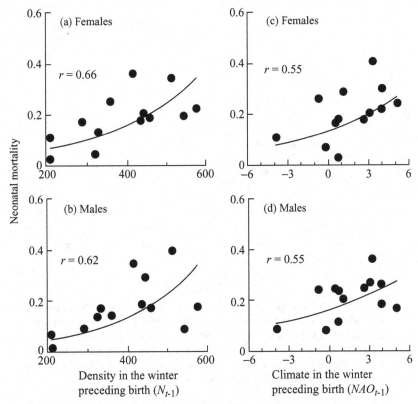

FIG. 3.8. Changes in neonatal mortality of female and male lambs with variation in density (N_{t-1}) and climate (NAO_{t-1}) in the winter preceding birth. (a) Mortality rates in female lambs plotted against N_{t-1}; (b) mortality rates in male lambs plotted against N_{t-1}; (c) mortality rates for female lambs plotted against NAO_{t-1}; (d) mortality rates for male lambs plotted against NAO_{t-1}. Correlation coefficients (r) are significant $(p < 0.05)$. Maternal age and weight, birth weight and date and twin birth were controlled for in all analyses as well as climatic variation (in a, b) or population size (in c, d). (From Forchhammer *et al.* 2001.)

density or weather on the relative survival of male and female lambs through the neonatal period.

As a result of density-dependent changes in juvenile fecundity and in neonatal mortality, rates of recruitment decline after winters when density is high (Fig. 3.9). However, this has little effect on the dynamics of the population since high winter density is associated with high

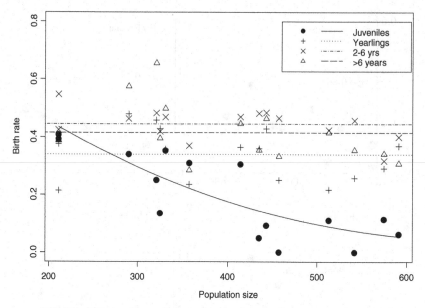

FIG. 3.9. Effects of population density in autumn on recruitment rates. Recruitment rate is the number of lambs produced that survived to six months of age per adult female. Solid points and line represent animals that were twelve months of age when they gave birth, while the three interrupted lines show recruitment rates for yearlings, two- to six-year-olds and older animals respectively.

mortality the following spring, so that reductions in birth rates occur after population density has fallen. As a result, density-dependence in fecundity may delay the rate at which the population recovers, but does little to slow the rate of increase once population size has started to rise (Clutton-Brock *et al.* 1997a).

3.5 Winter mortality

Winter mortality of juveniles and older adults varies with population density (Fig. 3.10a–d). Juveniles of both sexes are usually the first to die, followed by yearling and adult males and adult females while, among adults, older individuals typically die before animals in their prime (Clutton-Brock *et al.* 1997a). While high mortality is confined to years when summer density is high, the sheep do not always show high mortality when numbers are high. For example, sheep numbers

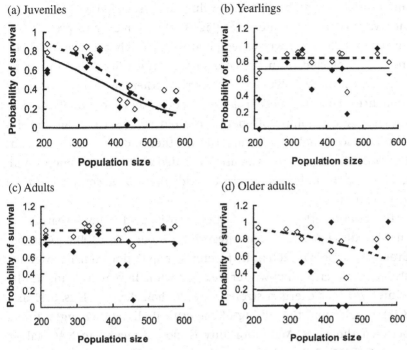

FIG. 3.10. Effects of population density on the survival of Soay sheep in the Village Bay population, 1986–98, showing survival plotted against total population density. Open symbols and dotted lines represent values for females, closed symbols and solid lines, values for males. The most parsimonious grouping of individuals for analyses of survival group animals in eight age classes: female juveniles, male juveniles, female yearlings, male yearlings, female adults (two to six years), male adults (one to six years), older females (over six years) and older males (over six years). (a) Changes in survival of juveniles; (b) changes in survival of yearlings; (c) changes in survival of female and male adults; (d) changes in survival of older females and males. (From Coulson *et al.* 2001.)

in the winter of 1997/8 were 542 in the study area and 1751 on the whole island but, despite the unusually large size of the winter population, winter mortality was below 10% (Fig. 1.5). However, in the following winter (1998/9), around 50% of sheep using the Village Bay area died between January and April. These two years differed in winter weather and in values of NAO: in the winter of 1997/8, NAO values were low while in 1998/9, NAO values were high, gales were frequent and rainfall was high. High NAO values are associated with depressed survival in most age categories (Fig. 3.11) and appear to predict survival better than low winter temperature, probably because the sheep are well protected from the cold by their thick fleeces (Milner *et al.* 1999b; Coulson *et al.* 2001). Wet, stormy weather has two separate consequences. First, the fleeces become sodden and individuals have to expend large amounts of energy to avoid hypothermia. Second, individuals seek shelter to escape the weather and are prevented from foraging. These two processes are associated with negative energy budgets which, if prolonged, cause individuals in poor condition to die of starvation.

The relative effects of population density and winter weather on survival differ between age categories (Milner *et al.* 1999b; Clutton-Brock and Coulson 2002). In juveniles, population density exerts a stronger influence than winter weather, while in yearlings and adults winter weather exerts a stronger effect than density (Figs. 3.10 and 3.11). In addition, the timing of the critical climatic changes differs between the sexes. Male mortality is closely related to NAO values between December and March as is the mortality of female juveniles and older females (Fig. 3.11a–d). In contrast, the mortality of female yearlings and prime-age females is more closely related to rainfall in February and March than to NAO between December and March (Fig. 3.11e, f).

Shortage of resources in winter is the principal cause of high mortality (see Chapter 4), though energy expenditure may also be affected by climatic variation and parasite load (Chapter 5). Plant biomass in spring is closely associated with population growth in the following year and is negatively correlated with sheep numbers the previous August (see Chapter 4). Estimates of the energy requirements of sheep in the Village Bay sub-population in winter, based on observed changes in plant biomass and estimated energy requirements of the sheep, suggest that the availability of vegetation per head during the winter is insufficient to cover the animals' requirements once female numbers in summer exceed 300 sheep, and predict a rapid decline in body weight, which is marked when total autumn numbers exceed 450 or female numbers exceed 200 (Clutton-Brock *et al.* 1997a) (Fig. 3.12). If animals die once they have exhausted their fat

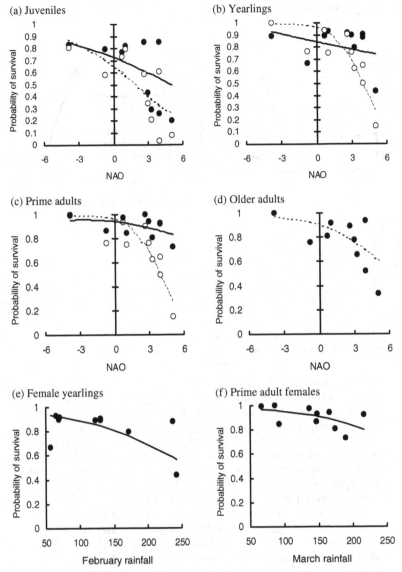

FIG. 3.11. Comparative effects of weather on the over-winter survival of different sex and age classes. Open symbols and solid lines represent females; solid symbols and broken lines represent males. Lines are from logistic regression mark–recapture–recovery models. (a) Changes in the survival of male and female juveniles with NAO between December and March. (b) Changes in survival of male and female yearlings with NAO between December and March. (c) Changes in survival of prime-aged adults (two to six years) with variation in NAO. (d) Changes in survival of older adults of both sexes with variation in NAO between December and March. (e) Changes in survival of female yearlings with February rainfall (mm). (f) Changes with survival of prime-aged females with March rainfall (mm).

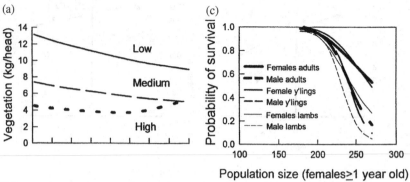

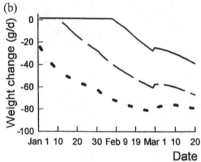

FIG. 3.12. Energetic predictions of food availability and survival.
(a) Predicted amount of vegetation available per head of sheep in
Village Bay, Hirta, between January and March under low (solid line),
medium (wide dashed line), and high (narrow dashed line) population
density (from Clutton-Brock *et al.* 1997a). The high rates of mortality at
high population density cause an increase in available vegetation per
head, slightly mitigating the rate of weight loss. (b) Predicted patterns
of weight loss in adult females. (c) Predicted proportions of each class
surviving winter plotted in relation to total population. Males are
unable to regain body reserves after rutting at high population
densities and are then more likely to die than females, while
juveniles/lambs of both sexes have lower body reserves than older
animals. The results were obtained by initialising the model to a
starting population of 400 sheep, in the ratio 0.5:0.2:1 females and
0.5:0.2:0.2 males (the ratios are for lambs: yearlings: adults) and then
running the simulation for 100 years.

reserves, these patterns predict an increase in mortality from starvation as total autumn population size rises above 400 or female numbers rise above 180 (Clutton-Brock *et al.* 1997a). Summer conditions,
in contrast, appear to have little effect on mortality. August standing
crop declines as sheep density rises (see Chapter 4) and the growth

rate of lambs and the body weight of females in August decline with increasing density (Clutton-Brock *et al.* 1992, 1997a). However, the magnitude of these changes is small and they principally affect growth in the year *after* a population crash.

Sex differences in energy requirements and expenditure probably account for much of the variation in mortality between age and sex categories. Juveniles and adult males would be expected to show higher mortality and earlier death dates than adult females as a result of their lower fat reserves at the onset of winter (Clutton-Brock *et al.* 1997a). They may consequently be more likely to be affected by conditions during the first half of winter while mature females, which are able to regain weight lost during lactation in the autumn months, may be more susceptible to variation in conditions in late winter. In addition, the energetic costs of gestation rise sharply in late winter (see Chapter 2). Another reason why population density may have less impact on mature females is that they are commonly the last animals to die, with the result that variation in population size in late winter is substantially lower than at the onset of winter and competition for resources is reduced (Clutton-Brock *et al.* 1997a).

Variation in the susceptibility of different age categories to high winter density has important consequences for population demography and dynamics. Though juvenile mortality is more variable than adult mortality, variation in adult mortality has a greater demographic impact, partly because adults constitute a larger proportion of the population and partly because their subsequent survival and breeding success is higher than that of juveniles (Fig. 3.4). Juvenile and adult mortality rates are positively correlated and the co-variation between them is also important in influencing the population dynamics (Fig. 3.4).

The age structure of the population in autumn also affects levels of winter mortality. The proportion of juveniles, prime adults and older animals in the population varies between years as a result of variation in mortality in previous years (Fig. 3.13). When the population includes relatively large numbers of juveniles and older adults, levels of winter mortality are likely to be high (see section 3.8). Since crashes remove many older animals, they also reduce natural mortality in

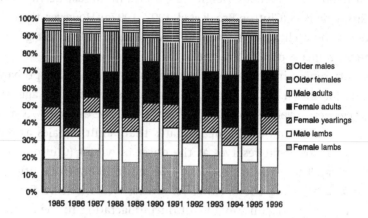

FIG. 3.13. Age structure in different years in spring. The figure shows the proportion of animals of different ages in different years. (From Clutton-Brock and Coulson 2002.)

the winter immediately following a crash, leading to an increase in age-related natural mortality as the number of years since a crash increases.

Higher mortality rates in males also have important consequences. When sheep numbers are low after crashes, there are up to eight or nine females per male among adults. In contrast, the adult sex ratio is close to parity in summers when population density is high (Fig. 3.14). Fluctuations in the adult sex ratio have repercussions for the rutting behaviour of males and the distribution of mating success across age classes (see Chapters 6 and 9).

3.6 Cohort variation in growth, survival and reproductive performance

As in many other vertebrates (Lindström 1999), conditions during early development generate differences in growth between cohorts in the sheep. Lambs born after winters of high density or high NAO are lighter as yearlings and adults (Clutton-Brock *et al.* 1992; Forch-hammer *et al.* 2001). Both density and climate appear to exert direct effects on growth. When the influence of birth weight is controlled for, cohorts born after winters when population density is high

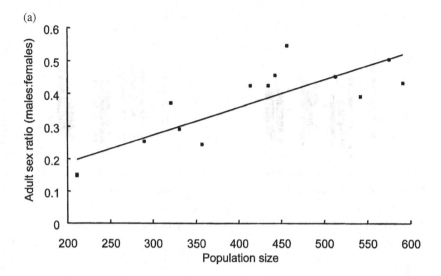

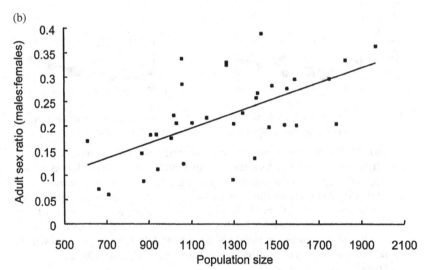

FIG. 3.14. Sex ratio (males:females) among adults (more than one year old) (a) in the Village Bay population in October 1985–98 and (b) the whole island population (1955–99) in relation to population size at the same time.

show lower body weights at four months in both sexes (Fig. 3.15a, b) and repeated weighing of lambs during the first weeks of life confirms that growth rate is depressed (Robertson *et al.* 1992). This is probably because population density depresses the biomass of live

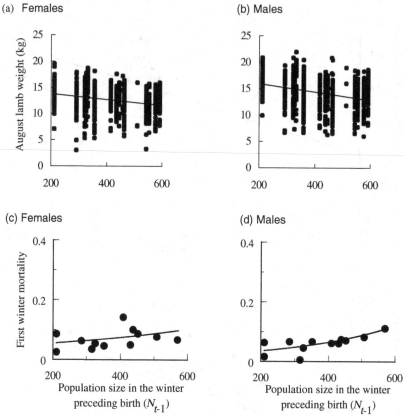

FIG. 3.15. Variation in weight and survival between cohorts. (a, b) Changes in weight at four months in females and males plotted on population size in the winter preceding birth. (c, d) Changes in mortality of lambs in their first winter, plotted on population size in the winter preceding birth. Both analyses control for the effects of maternal age and weight, birth weight and date, capture date, twin birth and NAO using generalised linear models. Correlations shown in (a), (b) and (d) were significant (see Clutton-Brock *et al.* 1992; Forchhammer *et al.* 2001.)

grass in spring (see Chapter 4) which is, in turn, closely related to the daily weight gain of lambs during their first four months of life (Robertson *et al.* 1992; see also Côté and Festa-Bianchet 2001). Light-born juveniles weigh less at subsequent ages than heavy-born ones (Chapter 2), and the average weight of cohorts at four months (as well as at sixteen and twenty-eight months) declines with increasing density in the winter before birth (Forchhammer *et al.* 2001) (Fig. 3.15a, b).

Variation in early development is also correlated with the growth and weight of adults (Forchhammer *et al.* 2001).

Differences in early development are associated with variation in mortality during the first winter of life. Cohorts that show relatively low birth weight suffer relatively high mortality during their first year of life (Forchhammer *et al.* 2001). After the effects of birth weight have been allowed for, cohorts exposed to high population density during the winter before birth tend to show increased mortality in their first winter, especially among males (Fig. 3.15c, d). In contrast, NAO values during the winter before birth are not consistently related to juvenile mortality. (These results differ from an earlier analysis (Clutton-Brock *et al.* 1992), which found no significant effect of population density during the winter preceding birth on subsequent mortality, but did not control for climatic effects.)

The effects of variation in early development extend into the breeding lifespan. Cohorts exposed to high population density *in utero* show relatively high mortality as yearlings and two-year-olds (Forchhammer *et al.* 2001). Subsequently, the effects of population density in the winter before birth disappear, but cohorts that experience high density in their first summer (N_t) show relatively high adult mortality while cohorts born after winters with high NAO indices show expected levels of mortality as juveniles but reduced mortality as adults (Forchhammer *et al.* 2001). Finally, conditions during early development are related to variation in fecundity. The proportion of juveniles that conceive during their first summer and give birth at around twelve months of age varies between cohorts from around 20% to over 80% (see section 2.6). Cohort differences in the age of females at first breeding are not consistently related to population size in the winter before their birth (Fig. 3.16a). However, females breed later when the winter preceding their birth is characterised by high NAO (Fig. 3.16b) or when population density is high in the summer that they are born (Fig. 3.16c). In addition, yearlings born in summers when population density is high produce relatively few lambs (Fig. 3.16d), though this effect disappears in adult females. In conjunction, population size in the first summer and NAO in the previous winter account for 81% of inter-cohort variation in age at first breeding (Forchhammer *et al.* 2001).

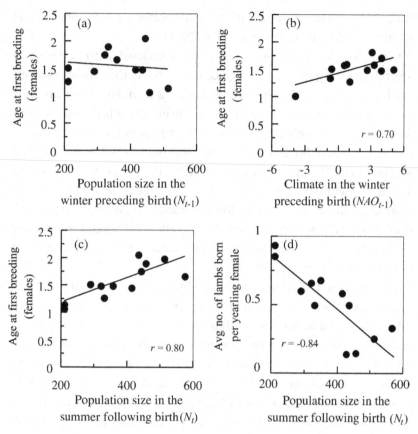

FIG. 3.16. Variation in fecundity between cohorts. (a) Age at first breeding in females plotted on population size in the winter preceding birth (N_{t-1}). (b) Relative age at first breeding plotted on NAO values in the winter preceding birth (NAO_{t-1}). (c) Relative age at first breeding plotted on population size following birth (N_t). (d) Average number of lambs born per yearling female plotted on population size in the summer following birth (NE). Correlation coefficients shown in (b), (e) and (d) are significant ($p < 0.05$). (From Forchhammer *et al.* 2001.)

What consequences do the differences in survival and fecundity between cohorts have for the dynamics of the population? Since weight at four months exerts a strong effect on survival through the first winter, density-dependent changes in growth could, in theory, play a major role in generating instability if cohorts of lambs produced at high density had insufficient reserves to survive their first winter (Clutton-Brock *et al.* 1992). However, in practice, the demographic effects of these changes in growth are slight. While changes in birth weight are probably responsible for part of the reduction in neonatal survival at high density, the latter has little effect on changes in population size since neonatal survival usually falls *after* winters when population density and mortality are both high (Clutton-Brock *et al.* 1991). As a result, increases in neonatal mortality may delay population recovery after years of high mortality but probably have little effect on the magnitude of population crashes (Clutton-Brock *et al.* 1997a). Similarly, birth weight and weight at four months are lowest in summers that *follow* crashes rather than those that *precede* them, so that density-related changes in growth affect survival in the first winter *following* a winter when mortality is high. Since mortality is generally low in these years, the principal effect of these changes in growth may again be to reduce the rate at which the population recovers.

3.7 Modelling the impact of density-dependence

Like most other attempts to predict animal numbers, we initially constructed a simple deterministic model of the dynamics of the sheep population based on our measures of the effects of density on fecundity and survival between 1985 and 1995 and structured for different age and sex categories (Grenfell *et al.* 1992). Over this period, there was a non-linear relationship between annual survival (S) and total population size and density-dependence was strongly *over-compensatory* (Grenfell *et al.* 1992). Over-compensatory mortality has been demonstrated in a number of relatively short-lived organisms (Hassell *et al.* 1976; Peterson *et al.* 1984) but is unusual in longer-lived species (Fowler 1981, 1987; Clutton-Brock and Coulson 2002). The dynamics of over-compensatory density-dependence in seasonally reproducing populations have already been explored (Hassell *et al.* 1976; May 1976; Bellows 1981) and we began by using a simple model from the literature on insects, expressing the density-dependent survival pattern by the following standard function (Bellows 1981).

$$S = \frac{d}{1 + (aN)^b}$$

(3.1)

Here, the parameters d, a and b respectively control the level of density-independent mortality, the threshold population size $(1/a)$ above which density-dependent mortality occurs, and the strength of density-dependence, which is overcompensating for $b > 1$ (Maynard Smith and Slatkin 1973; Bellows 1981).

We used the model to explore the demographic consequences of the effects of density on fecundity and survival in the sheep. Since lambs reproduce in their first year of life and the survival curves of males and females are qualitatively similar in shape (Fig. 2.15), the dynamics of the population can be captured by adding across sex and age classes to produce the following model of dynamics (Bellows 1981; Grenfell *et al.* 1992).

$$N_{t+1} = \lambda S_t N_t = \frac{\lambda d N_t}{1 + (a N_t)^b} \tag{3.2}$$

In Equation (3.2) a, b and d are survival parameters, as described above, averaged over the age and sex variations in survival. The other component of the model is the annual rate of increase of the population due to reproduction, λ. This parameter is calculated as $\lambda = (f + 1)$, where f is the average individual reproductive rate; the product λd therefore represents the maximum net annual increase in population, allowing for density-independent mortality.

This model illustrates the potential impact of over-compensatory density-dependence on the dynamics of the Hirta sheep population. The dynamics of the model depend on an interaction between the degree of over-compensatory mortality (b) and the net density-independent rate of increase of the population (λd) (Bellows 1981). As b increases above unity, the model moves from a stable equilibrium through periodic population cycles of increasing complexity, to an irregular, chaotic pattern of crashes (Grenfell *et al.* 1992). To compare the degree of over-compensation in the sheep population with values observed from previous studies, Fig. 3.17a plots b against λd for lambs yearlings and adults, in comparison with the equivalent figures from a range of insect studies (Bellows 1981). It shows that the degree of over-compensatory density-dependence found in the Soay population in some periods is unusually high: the values of b ob-

(a)

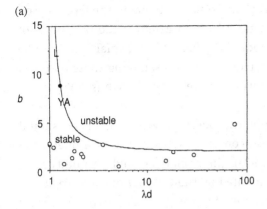

(b)

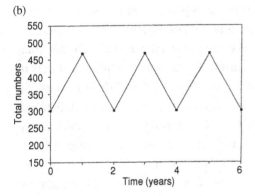

(c)

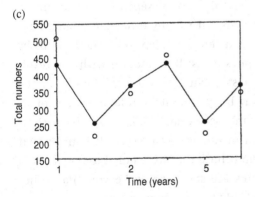

FIG. 3.17. (a) Local stability boundary for equilibria of the simple non-age-structured model (Equation (3.2)) (Bellows 1981) in terms of the degree of over-compensatory density-dependence (b) and the average net reproductive rate (λd). The b estimate for sheep (•) was calculated by averaging parameter estimates from the survival curves for each age class; the corresponding λd estimate refers to the average net reproductive rate of Soay sheep (Grenfell *et al.* 1992). L shows the value for lambs and juveniles, Y and A values for yearlings and adults. Equivalent estimates for insect populations (Bellows 1981) (o) are included for comparison. (b) Biennial fluctuations in population size predicted by a more complex age-structured model of the sheep population with moderate levels of b ($b = 10$) (Grenfell *et al.* 1992). (c) Triennial fluctuations predicted by the same model with high levels of b ($b = 50$). Observed changes in sheep numbers are shown by the open circles.

served for different age and sex categories of sheep (running from 6.08 for adult females to 16.31 for male lambs) is much higher than in most insect populations, where the observed upper limit is around $b = 5$ (Bellows 1981). Thus even though the reproductive rate of the sheep is low relative to insects and small vertebrates, the level of over-compensation is still sufficiently high to generate the possibility of unstable dynamics. With moderate values of b, biennial cycles can be expected (Fig. 3.17b) while triennial cycles resembling those observed in the sheep will occur if b values are extremely high (see Fig. 3.17c, where $b = 50$).

These calculations prompt the question why Soay sheep should show such high levels of over-compensatory mortality (Grenfell *et al.* 1992; Clutton-Brock *et al.* 1997a). Their relatively high fecundity for a medium-sized mammal (generated by their ability to conceive at less than a year old and the relatively high incidence of twins) is almost certainly important, for it permits sheep numbers to rise by over 60% in a single summer, producing a situation where, in some years, the number of animals entering the winter is substantially greater than the food supply can support (Clutton-Brock *et al.* 1997a). The effects of high fecundity are augmented by the lack of any significant decline in fecundity during periods of population increase. Density-dependent changes in the proportion of females lambing at twelve months, in twinning rate and in average recruitment per female (Fig. 3.5) are largely confined to breeding seasons following winters of high mortality and consequently have little effect on the rate of population growth. The reason underlying this lack of sensitivity of fecundity to rising density (which contrasts with many other large mammals: see Fowler 1987) is probably that female sheep wean their lambs by midsummer, when food supplies are still abundant, with the result that they can regain weight lost during lactation by the end of the summer and enter the rut at body weights similar to those of females that have not reared lambs (see Chapter 2). Comparisons show no differences in body weight in August between adult females that have raised singleton lambs and those that have failed to breed while females that have reared twins are significantly heavier than other categories (Clutton-Brock *et al.* 1997a; see Chapter 10).

3.8 Incorporating the effects of climate and age structure

It is clearly important to incorporate density-independent effects in any attempt to predict variation in sheep numbers for climatic variation has important effects on survival (see section 3.5). As in our previous model (see section 3.7), we started by relating population growth rate to population size for the whole population, in this case using data for all years between 1955 and 1999. Like Fig. 3.3, Fig. 3.18a plots instantaneous population growth rate ($r_t = x_{t+t} - x_t$) against the log of population size in the previous year $t(x_t)$ showing the close relationship between population size and population growth rate, as well as the presence of marked variation in survival at high population densities. A plot of x_{t+1} against x_t (Fig. 3.18b) reveals this pattern in more detail and shows that population growth rates are positive below a threshold population of around 1100 individuals ($x_t \cong 7$) while, above this level, population growth rate tends to be negative, though there is much variation (Fig. 3.3). We modified the balance of noise and density-dependence using a threshold auto-regressive model (Grenfell *et al.* 1998); Fig. 3.18b illustrates this fitted to separate linear auto-regressive models above and below the threshold. At low densities, the model indicates a variable increase in population size while, at high densities, the population can increase, remain constant or fall, depending on density-independent factors. Simulation of the model, incorporating the influence of density-independent factors and random noise, captures the essence of both the growth rate pattern (Fig. 3.18c) and the map of x_{t+1} on x_t (Fig. 3.18d). Like our previous model, this emphasises that the system sometimes shows strong over-compensatory density-dependence but, at other times, mild winter weather conditions allow the sheep to escape the consequences of high winter density and low food availability.

Since the effects of population density and climatic variation on mortality vary between age and sex categories (see section 3.5) it is also important to incorporate variation in the age structure of the population in predictive models (Coulson *et al.* 2001). Models which incorporate differences in population structure as well as variation in the responses of different sex and age categories to population density and winter weather predict changes in population size with considerably more accuracy than models that ignore these differences (Fig. 3.19). For example, stochastic, unstructured models of variation in sheep numbers account for little more than 20% of variation in population size (Grenfell *et al.* 1992, 1998), while the inclusion of variation in sex and age structure and in sex and age-specific

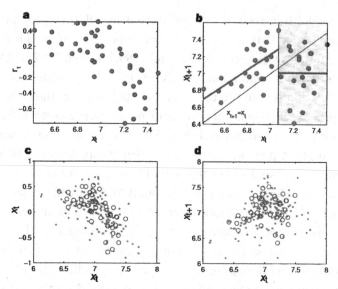

FIG. 3.18. Modelling population size for the whole of Hirta. (a) Plot of annual population growth rate $r_t = x_{t+1} - x_t$ against log population size (x_t). (b) Fit of an univariate SETAR threshold model to the scatter plot of x_{t+1} against x_t. The threshold, $x_t = C = 7.066$, and the linear relationship above and below it, were estimated by a least-squares fit to the data; see Grenfell *et al.* (1998) for more details. Shaded area denotes the regime above the threshold, heavy lines show the best-fit model and the thinner diagonal line is at $x_{t+1} = x_t$. (c) Comparing the observed r_t versus x_t plot (circles) with 150 iterates of the best-fit model with added noise (dots); a transient of 250 years was run off before recording the points. (d) The same comparison as (c), but plotted as x_{t+1} against x_t.

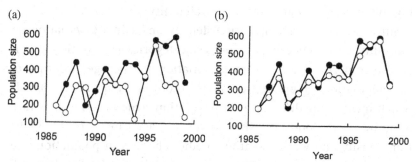

FIG. 3.19. Observed (solid points) and predicted (open points) values of population size in different years using (a) a time series model fitted to count data and winter weather and (b) age-structured Markov models incorporating the effects of variation in age structure as well as in population density and weather.

responses to density and weather (Fig. 3.19) raises the proportion of variation accounted for to nearly 90% (Coulson *et al.* 2001).

3.9 Synchrony in dynamics between populations

Where density-independent factors exert a strong influence on population dynamics, separate populations in the same area should show synchronous changes (Moran 1953). Numbers of black-faced sheep on Boreray have been counted approximately every other year between 1955 and 2000 and, as expected, their numbers are closely correlated with the number of Soay sheep on Hirta (Fig. 3.20). Moran (1953) showed that the expected correlation between populations in the absence of density-dependence equals the correlation between environmental perturbations so, with no density-dependence, the expected environmental correlation would equal the observed correlation between populations ($r = 0.685$). In fact, our analysis shows that, because of relatively strong effects of density on the system (Royama 1992), closer environmental correlation than that observed would be necessary to generate the level of correlation between the sheep populations on the two islands (Grenfell *et al.* 1998).

To identify the specific climatic changes responsible for generating synchronous dynamics in the two populations, we investigated the effects of including a range of different weather variables in our model, including mean monthly temperature, mean monthly precipitation and the total number of hours with wind over 34 knots during the months of March and April (Grenfell *et al.* 1998). Our eventual model (which included a negative effect of March gales on population growth rate above and below the density threshold and a positive effect of April temperature on population growth rate below the threshold) suggested that known climatic effects could account for around 30% of the observed inter-island correlation (Grenfell *et al.* 1998). Correlations with sheep numbers on the smaller island of Soay are less close, possibly because of the difficulties of censusing sheep numbers accurately there. Recent studies of other northern ungulates also suggest that variation in winter weather associated with the NAO can synchronise fluctuations in separate populations. For example, fluctuations in caribou and musk ox populations on opposite coasts of

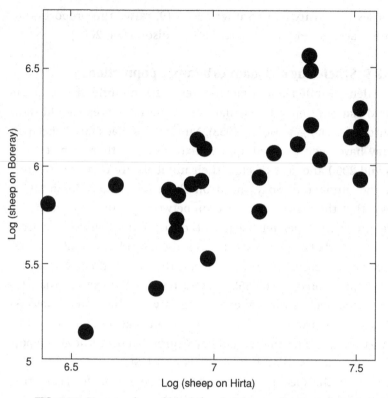

FIG. 3.20. Log number of black-faced sheep on Boreray plotted on estimates of log number of Soay sheep on Hirta. The association between the logged population sizes is significant (Pearson $r = 0.685$, 95% bootstrap confidence limits: 0.447–0.838).

Greenland are synchronous and are correlated with changes in NAO (Post and Forchhammer 2002).

3.10 Discussion

As in many other northern ungulates (Fowler 1987; Clutton-Brock and Albon 1989; Saether 1997; Gaillard *et al.* 1998; Post and Stenseth 1998), late winter mortality is the most important cause of changes in population size in the sheep. Young of the year, males and old females are particularly likely to die (Saether 1997; Loison *et al.* 1999b). Though mortality is less variable in adults than juveniles, changes in adult mortality contribute more to changes in population growth since adults form a larger portion of the population (Fig. 3.4).

Variation in mortality is affected both by population density and by density-independent factors: high mortality of adults is largely restricted to years when summer numbers exceed 1100–1200, although favourable winter weather conditions can allow a large summer population to escape high winter mortality. The immediate cause of fluctuations in winter mortality is starvation combined with variation in energy requirements resulting from climatic differences. Chapter 4 examines the effects of sheep density on primary production and food availability. In addition, density-dependent changes in parasite numbers exacerbate the effects of food shortage on the sheep and accentuate fluctuation in their growth and survival (see Chapter 5).

While there is no inherent cyclicity in sheep numbers, there is some regularity in the rise and fall of the population. It usually requires at least three breeding seasons for a population that has suffered heavy mortality to regain a level where the next population size reduction is likely to occur (1100–1200). As a result, years of high winter mortality are usually separated by at least two intervening years of low mortality (Fig. 1.5). This pattern is most clearly seen in changes in the number of adult females, and is least obvious in the changes in the number of juveniles and males (Fig. 3.1), presumably because juveniles and males are more susceptible to starvation than females (Figs. 3.10 and 3.11), so that fluctuations in climate can generate substantial mortality even when population density is relatively low. Interactions between population density and climatic factors have been recorded in other northern ungulates (Douglas and Leslie 1986; Loison and Langvatn 1998; Post and Stenseth 1998, 1999; Post and Forchhammer 2002) though they do not always lead to persistent fluctuations in population size. Climatic factors in winter interact with population density to affect survival in Scottish red deer populations, while summer weather conditions have relatively little effect (Albon and Clutton-Brock 1988). In coastal populations, wet, stormy weather has a greater effect on mortality than winter temperatures (Albon and Clutton-Brock 1988) and values of NAO predict changes in winter mortality (Albon *et al.* 2000). In mainland red deer populations, where winter temperature varies more widely, mid-winter temperatures exert stronger effects (Albon *et al.* 1992) and similar effects occur in continental populations (Loison and Langvatn 1998). Like young sheep, juvenile red deer are more susceptible than adults to variation in population density combined with adverse winter weather and males are more susceptible than females, leading to consistent correlations between population density and the adult sex ratio (Clutton-Brock *et al.* 1997a) though the extent of sex differences varies between populations (Loison and Langvatn 1998).

So why does the fecundity of female sheep not decline with increasing density as it does in many other ungulates, including red deer (Clutton-Brock *et al.* 1997a)? The likely answer is that Soay lambs develop rapidly and, by the time they are six weeks old, they obtain most of their food themselves (Jewell *et al.* 1974; Robertson *et al.* 1992). In combination, the relatively early breeding season of the sheep and the rapid development of their offspring allow female sheep that have reared lambs successfully to regain condition between June and August, when food supplies are super-abundant so that, by the rut, successful breeders do not differ in body weight from animals that have failed to rear lambs (Clutton-Brock *et al.* 1997a). As a result, the link between population density, food availability, body condition and fecundity is broken, and female sheep are almost as fecund in years when population density is high as in other years (Fig. 3.5 and 3.9). In ungulates with longer lactation periods, like bighorn sheep and red deer, mothers are unable to regain condition before the October rut so that high density or harsh weather are more likely to affect weight in the rut (Clutton-Brock *et al.* 1997a; Festa-Bianchet *et al.* 1998). Where these differences in weight at the onset of the breeding season are relatively small (as in bighorn sheep), they may have little effect on subsequent fecundity or survival (Festa-Bianchet *et al.* 1998). In contrast, where they are large (as in red deer) they can affect the probability that females will conceive in the autumn rut as well as their survival through the following winter (Clutton-Brock *et al.* 1983).

Extrinsic factors too, may be important. By preventing dispersal, the isolation of the St Kilda sheep population may lead to higher local densities and increased opportunities for instability. Rodent populations in fenced enclosures commonly attain artificially high population densities and may also show high mortality (Finnerty 1980; Cockburn 1998). Dispersal is necessarily reduced in island populations, preventing the gradual leakage of subordinate animals into inferior habitat and leading to increased levels of mortality *in situ*. However, the contrast in population stability between the red deer populations of Rum and the Soay sheep of Hirta shows that the constraints on dispersal imposed on the sheep do not, on their own, account for the persistent instability of population size (Clutton-Brock *et al.* 1997a).

The relatively fluid social organisation of the sheep may also contribute to their unusual dynamics. Dominance interactions and displacements are rare among females and there are no obvious social divisions within the population, apart from those between hefts (see Chapter 2). In other ungulates, including red deer, dominance relationships are often well defined among females as well as males, and subordinates or intruders are commonly displaced from resources by dominant members of the group (Thouless 1986). Consistent dominance relationships may focus the effects of rising population density on particular animals or groups, generating a more graded response to rising density and greater variance in dates of death (Clutton-Brock *et al.* 1997a). Since early deaths reduce competition for limited resources, increased variance in death date may increase the food available to survivors, contributing to a reduction in late-winter mortality and tending to increase the stability of population size.

Early development appears to exert an important influence on many aspects of growth, development, survival and breeding success of the sheep as it does in many other mammals (Lindström 1999). High population density in autumn is associated with relatively early birth dates and light birth weights the following spring and cohorts born after wet, stormy winters (high NAO) tend to be born lighter than those born after cold, dry winters (Fig. 3.7). Population density and climatic factors in the year before birth affect juvenile survival and growth during the first year of life partly through their influence on birth weight (see Chapter 2) and similar effects are found in other northern ungulates (Anderson and Linnett 1998; Portier *et al.* 1998) though they are not universal (see Côté and Festa-Bianchet 2001). Both density and climatic factors also appear to have independent effects on growth and survival for, after the influence of birth weight is controlled for, cohorts born after winters when population density is high show low survival through their first summer, low body weights at four months and, in males, relatively low survival during the first winter (section 3.6). In addition, cohorts of females that experience high density in the first year of life are less likely to breed in their first year.

Our analysis of cohort effects sheds some light on the causal pathways underlying the effects of density and climate on development. Sheep density in winter depresses the availability of live grass in late winter and spring (Chapter 4); this may constrain prenatal growth rates, leading to low birth weights which, in turn, contribute to neonatal mortality. In some animal populations, high NAO values in winter are associated with improved performance or survival the following spring, possibly because relatively mild, wet winters are associated with reduced snow cover or with an earlier onset of plant growth in spring (Douglas and Leslie 1986; Saether *et al.* 1996; Forchhammer *et al.* 1998a, b; Post and Stenseth 1999). The contrasting tendency for wet, windy weather in winter to depress birth weight, neonatal survival and juvenile fecundity in the sheep probably occurs because the energetic costs of adverse conditions depress fat reserves and prevent conception. Food availability during the first summer may be the principal factor affecting juvenile development and reductions in food availability caused by high sheep density in the first year of life probably account for the reduction in the survival and fecundity of juveniles when density is high.

Our results suggest that delays in the effects of rising population density on fecundity and survival in the sheep are responsible for allowing summer numbers to rise above winter carrying capacity. Combined with scarcity of shelter, high parasite load, and no possibility of dispersal, competition for limited resources during late winter leads to intermittent, savage reductions in population size. This explanation of the dynamics of Soay sheep resembles the conclusions of recent studies of population dynamics in other resource-limited herbivore populations, which emphasise the capacity of lags or delays in density to generate fluctuations in numbers (Saether 1997).

It is sometimes suggested that lagged responses to rising density and persistent fluctuations in population size are an inevitable consequence of the absence of predators (Messier 1994; Saether *et al.* 1996; Saether 1997). In contrast, our work suggests that delays in density-dependent changes in the sheep are a consequence of their unusually rapid development and of the timing of reproductive events. Comparisons with the dynamics of the resource-limited red deer population on Rum over the same period suggest that fecundity and recruitment may respond more quickly to changes in density in the deer because critical stages of growth extend into periods of the year when resources are limiting (see Chapter 10). While the presence of predators may often reduce fluctuations in the density of their prey (Messier 1994), stability may vary widely among resource-limited populations of herbivores as a result of contrasting responses to population density and weather generated by differences in reproductive timing and life-histories (Clutton-Brock and Coulson 2002). Indeed, it would not be surprising if similar differences in life-history were also important in affecting the population dynamics of species subject to high rates of predation.

While this chapter has been principally concerned with the causes of fluctuations in mortality and in population size, the unusual dynamics of the sheep have many important consequences. Changes in sheep numbers affect the structure of plant communities (see Chapter 4) as well as the numbers of parasites (Chapter 5). The sex differences in mortality that occur in years when mortality is high generate consistent fluctuations in the adult sex ratio (Fig. 3.14) which affect the distribution of mating success in males and the intensity of sexual selection (see Chapter 6). Intense competition for resources in crash years also exposes effects of variation in phenotype and genotype which are difficult to detect in more clement circumstances and increases the intensity of natural selection on a wide range of phenotypic and genotypic characters (see Chapters 7 and 8). Finally, fluctuations in mortality increase the costs of reproduction (Fig. 2.14) and affect the relative fitness of different reproductive strategies (see Chapter 9).

4
Vegetation and sheep population dynamics

M. J. Crawley *Imperial College London*

S. D. Albon *Centre Ecology and Hydrology, Banchory, UK*

D. R. Bazely *York University, Canada*

J. M. Milner *Scottish Agricultural College, Crianlarich, UK*

J. G. Pilkington *University of Edinburgh*

and

A. L. Tuke *Imperial College London*

4.1 Introduction

The relationship between the sheep and their food supply is a key element in understanding the population dynamics of Soay sheep on Hirta. This island population of Soay sheep provides an ideal model system for the study of plant–herbivore dynamics: there are no vertebrate predators like foxes or buzzards, and no competitors like rabbits or voles. The vegetation is relatively unpolluted by atmospheric nutrient inputs, there are no confounding management operations, and the population is closed to immigration or emigration. Because the sheep population is evidently food-limited, we expect that grazing will have a major impact on the biomass, spatial structure and botanical composition of the vegetation. In this chapter, we describe the relationship between the sheep and their food supply, and discuss the consequences of sheep grazing for plant performance and longer-term vegetation dynamics. In a plant–herbivore interaction where there are no competing herbivores and no vertebrate predators, we expect that herbivore numbers will be determined by the food supply available to the sheep during winter (Crawley 1983). Our study follows a long tradition of monitoring the response of vegetation to changes in the numbers of vertebrate herbivores: e.g. relaxation of rabbit grazing on chalk grasslands following the myxoma epidemic

89

Soay Sheep: Dynamics and Selection in an Island Population, ed. T. H. Clutton-Brock and J. M. Pemberton.
Published by Cambridge University Press. © T. H. Clutton-Brock and J. M. Pemberton 2003.

(Thomas 1960), African elephants (Cumming 1981), ungulate guilds in Serengeti (McNaughton 1985), introduced reindeer on South Georgia (Leader-Williams *et al.* 1987; Leader-Williams 1998), livestock in the New Forest (Putman *et al.* 1989), desert rodents in the USA (Brown and Heske 1990), moose on Isle Royale (McLaren and Peterson 1994), sheep on heather moorland (Welch and Scott 1995), lemmings in arctic tundra (Virtanen *et al.* 1997), whitetailed deer in North American forests (Cornett *et al.* 2000), kangaroos in Australia (Newsome *et al.* 2001) red deer on Rum (Virtanen *et al.* 2002) and many more. One of the great advantages of the model system on Hirta is that there are no other vertebrate herbivores on the island, so grazing effects on vegetation can be attributed unequivocally to the Soay sheep.

The key plant parameters affecting sheep performance are likely to be the annual productivity of the sward, the timing of the onset of growth in spring and the quality of the plant production (its digestibility, nitrogen content, toxin levels and so on; Crawley 1983). In particular we are concerned with the net rate of biomass production by the plant communities and with the rate of dry matter offtake by the animals (Fig. 4.1). The productivity of the sward is intricately bound up with grazing intensity, because of the well-documented trade-off between biomass and the net rate of carbon fixation (McNaughton 1993). There is also a correlation between sward height and utilisation efficiency, because tall swards allow large bite sizes and hence high intake rates, but tall swards lose a substantial amount of primary productivity through leaf turnover, i.e. the death and decay of leaves that might otherwise have been ingested by the sheep (Parsons *et al.* 1988; Newman *et al.*, 1995; Penning *et al.* 1995). In the short term, grazing typically increases food quality by encouraging regrowth, which tends to be high in nitrogen, and reducing the amount of material that accumulates in the sward as indigestible dead organic matter. In the longer term, selective grazing alters the competitive balance between plant species, and could lead to a decline in the mean abundance of palatable species, and an increase in the relative abundance of less palatable plants (Crawley 1997).

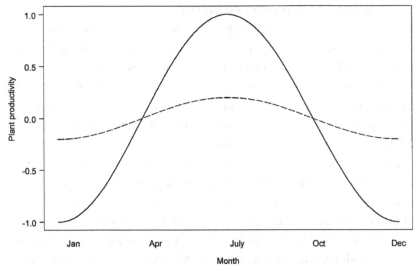

FIG. 4.1. Schematic representation of plant productivity (solid line) and offtake (dotted line) by the sheep (both in kg dry matter per hectare per week; after Crawley 1983). From October until March, offtake exceeds productivity and plant biomass declines. From April to September, productivity exceeds offtake and plant biomass increases (0 on the *y*-axis marks the point in spring where net productivity of the plants becomes positive; offtake is positive at all seasons). When sheep numbers are low, proportionately more of the summer productivity becomes standing dead organic matter in the sward in autumn and winter. Intake by the sheep is lower in winter because of their functional responses to reduced sward height, and higher in summer because of higher food availability, greater metabolic demands of lactation, and the need to regain body weight lost during winter.

To understand the impact of selective grazing on plant population dynamics and community structure, we need to address four sets of questions. (1) How does plant productivity respond to changes in sheep numbers and grazing pressure? (2) To what extent is the competitive ability of the different plant species affected by selective grazing by sheep? (3) How is plant recruitment from seed affected by reductions in seed production caused by sheep grazing? (4) And do changes in botanical composition under selective grazing lead to reductions in diet quality that could influence the dynamics of the sheep?

4.2 Grazing in theory and practice

Sheep numbers are affected by vegetation through the impacts of changes in diet quality and quantity on condition, mortality and fecundity (Chapter 3). Conversely, the sheep are expected to influence the vegetation through selective reduction of highly preferred species, and alteration of the outcome of interspecific competition. Plant species avoided by the sheep are likely to increase in abundance relative to those species that are preferred forage. This hypothesis is underpinned by a well-documented positive correlation between palatability and competitive ability (Pacala and Crawley 1992): those plant species that do best in the absence of herbivory (perhaps because they devote a large fraction of their resources to growth and a small fraction to defence), do less well in the presence of herbivory because well-defended, unpalatable plants become relatively more competitive following the selective defoliation of their more palatable neighbours.

The simplest theoretical models embodying these ideas assume that there is a fundamental symmetry to the plant–herbivore interaction: sheep dynamics are affected by vegetation, and sheep grazing affects vegetation dynamics. Because sheep numbers change rather abruptly, with most deaths occurring in February–March and most births in April–May, we adopt a difference equation approach, even though the impact of grazing on plant performance is obviously continuous. The model is of the form:

$$V_{t+1} = a_t \cdot V_t \cdot f(V_t) \cdot g(V_t, N_t) \tag{4.1}$$

$$N_{t+1} = \lambda \cdot b_t \cdot N_t \cdot h(N_t, V_t) \tag{4.2}$$

where V_{t+1} is the biomass of palatable vegetation in year $t + 1$, a_t is a random variable reflecting the 'goodness of the year' as it influences the intrinsic growth rate of palatable plant species, the function f describes competition between the plants, and would determine the equilibrium plant community in the absence of grazing by sheep. The function g describes the functional responses of the sheep, and the impact of the resulting selective defoliation on the growth of the component plant species.

Sheep numbers next year, N_{t+1}, are given by multiplying this year's population N_t by the maximum net reproductive rate, λ, by a random variable b_t reflecting the 'harshness of the year' (mainly as it affects late-winter mortality and lambing rate), and by a density-dependent function h which describes the numerical responses of the sheep, as determined by the way in which foraging success and food quality affect the rates of birth and death. Our research is aimed at understanding the nature of the three functions f, g and h. In particular, we want to find out how competition for food affects survival and fecundity for animals of different age, sex and body condition, and how interspecific competition amongst the component plant species is influenced by differential rates of defoliation.

The first step in our analysis of the dynamics of this system, therefore, is to determine the relationship between year-to-year sheep population change and food availability. We measure sheep population change as delta ($\Delta = \ln(N_{t+1}/N_t)$), which is positive when the population is increasing, zero when the population is constant, and negative when the population is declining (Fig. 4.2). The simplest measure of vegetation that combines elements of both the quantity and quality of food is the total biomass of green leaf for all palatable grass species. The question of which sample season to use in the analysis (spring, summer or autumn) is resolved statistically; all of the measures are entered simultaneously into a multiple regression, and the importance of each term is assessed by step-wise deletion (Crawley 1993). The statistical model looks like this:

$$\Delta = \beta_0 + \beta_1 N_t + \beta_2 V_a + \beta_3 V_b + \dots \tag{4.3}$$

where the parameters $\beta_0 \dots \beta_n$ are estimated by maximum likelihood, and the vegetation availabilities on different sample occasions, $V_a, V_b \dots V_n$, are assessed by deleting them from the model, one at a time. Note that a term for sheep population (N_t) appears in the maximal model because of the importance of density-dependence already demonstrated by Fig. 3.3. It turns out that the minimal adequate model contains no terms for food availability at all:

$$\Delta = \beta_0 + \beta_1 N_1 \tag{4.4}$$

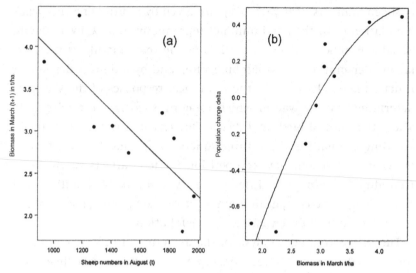

FIG. 4.2. Grass biomass in March drives fluctuations in sheep numbers. (a) Dependence of March grass biomass on sheep numbers in the previous August ($r^2 = 0.70$, $n = 9$); (b) relationship between population change $\Delta = \ln(N_{t+1}/N_t)$ and March grass biomass (second-order polynomial, $r^2 = 0.89$). All of the increasing populations were associated with total biomass in March of more than 3 tonnes per ha (t/ha). In the years of steepest population decline (1999 and 2002) the standing crop in March was much lower than this (2.23 and 1.80 t/ha respectively). In those high-density years when the population did not crash (e.g. 1996 and 1997) the total biomass in March was not depressed below 2.8 t/ha. In a multiple regression model, March biomass is a much better predictor of Δ (deletion test, $p = 0.036$) than is population size ($p = 0.347$), presumably because it incorporates annual variation in plant productivity (especially, over-winter plant growth) as well as the number of mouths. There is a much lower correlation between August sheep numbers and grass biomass in August ($r^2 = 0.12$; not shown), highlighting the resilience of the swards on Hirta to grazing by sheep, and suggesting that variations in summer grass biomass have relatively little impact on sheep population dynamics.

where $\beta_1 < 0$. We observe the expected pattern, in which sheep population change is positively correlated with the availability of palatable grass; the more food available per sheep, the faster the population grows. But we know already that there is a strong relationship between Δ and sheep population (the negative step-function of Fig. 3.3). Thus, dividing food availability by sheep numbers could introduce a spurious correlation between Δ and food availability. The correct analysis is to plot the residuals from a model that contains sheep population and other significant terms (but not vegetation) against vegetation. Once the effects of competition for food are taken into account by the sheep density term, there is no relationship at all between Δ and the availability of palatable biomass on any sampling occasion. The model of sheep dynamics that is suggested by this analysis is therefore extremely simple. With densities written in logs, we have:

$$N_{t+1} = N_t + \beta_{low} + \varepsilon_{low} \qquad (N < N_T) \qquad (4.5)$$

$$N_{t+1} = N_t + \beta_{high} + \varepsilon_{high} \qquad (N < N_T) \qquad (4.6)$$

where the five parameters are the threshold density (N_T), the constant mean increase at low density (β_{low}), the constant mean decline at high density (β_{high}), and the two variances associated with the two zero-mean stochastic terms ε_{low} and ε_{high} (Fig. 3.3 shows that $\sigma_{low}^2 < \sigma_{high}^2$ (Grenfell *et al.* 1998)). In the long run, we hope to discover how vegetation affects each of these parameters.

A measure of the resilience of the vegetation to grazing is obtained by comparing the year-to-year variability of grass biomass in March (when grazing impacts are at their greatest, and starvation can cause significant sheep mortality) with variability in August (after the rapid growth phase of the sward, when there is little or no sheep mortality). The much lower variation in August biomass than in March biomass indicates that the vegetation shows extraordinary resilience. High rates of defoliation over winter can be compensated by changes in the duration or the rate of the rapid growth phase in spring and early summer. Because of this resilience, sheep numbers have relatively little effect on summer production or on standing crop at the beginning of winter, and this is presumably the main reason why we see no strong, time-lagged density-dependence on sheep population dynamics acting through grazing pressure or impaired plant productivity.

Population density is not the only factor affecting sheep population growth, and variation in climate also plays an important role. As Chapter 3 describes, winter mortality is high when high sheep density coincides with wet, windy winters characterised by high NAO indices. However, even though the standing crop of live green matter in spring is relatively high after winters when NAO is high (Fig. 2.1), variation in plant biomass explains little of the variation in Δ after the simple, first-order density-dependence has been taken into account (Fig. 4.2b). This supports the evidence that fluctuations in winter weather affect the survival of sheep principally through their direct effects on heat loss and foraging opportunities rather than through any effects on their food supply. Summer weather conditions have little or no direct effect on sheep survival because food is superabundant in summer (Fig. 4.1).

4.3 Methods

Vegetation data are gathered twice per year: once in March when food availability for the sheep is at its lowest, and again in August when plant biomass is close to its annual peak. The study area is traversed by five permanent transects (see Fig 1.2b), with six permanent (but initially randomised) sampling stations on each transect. There are two sampling stations between the beach and The Street, two between The Street and the Head Dyke, and two on the hillside outside the Head Dyke. On each sampling occasion, 25 sward heights are measured, 10 cm apart in a straight line, at each of the 30 locations. Next, 20-cm square quadrats are placed contiguously along a 2-m randomly located sampling-pole, and each of the 10 quadrats is scored as being a gap (close-grazed turf containing much *Trifolium repens*) or a tussock (taller sward containing substantial standing dead organic matter and typically lacking *T. repens*). Quadrats that do not fall into either category are scored as 'indeterminate'. The position of the sampling-pole is randomised three times at each location to give a total of 30 gap or tussock scores. These counts are used as statistical weights in calculating biomass (see below). Finally, eight gaps and eight tussocks in the vicinity of the last position of the sampling-pole are selected and one gap and one tussock is selected at random (by generating two random numbers between 1 and 8, with replacement). For the selected gap and tussock, a 20 × 20-cm biomass sample is cut at ground level and bagged. Later, the biomass samples are sorted to species, oven-dried at 80 °C for 24 h then weighed. Bryophytes are lumped into a single category, while lichens are sorted into *Cladonia* spp. and *Peltigera* spp. Dead organic matter from all species is lumped into a single category. Any heather in the biomass sample is cut using fine scissors into two categories (green leaf material and wood) and these are dried and weighed separately. Biomass is estimated from the tussock and gap dry weights at each of the 30 sampling positions, using the tussock and gap cover scores to calculate a weighted average (indeterminate quadrats are taken as having a dry mass equal to the mean of the appropriate gap and tussock dry masses). Two variables are measured in August but not in March. Flower stem densities are counted for all flowering species in 90 quadrats, each measuring 40 × 40 cm, placed randomly in three strata along each of the five transects A–E (450 quadrats in all); 30 quadrats are taken from the Meadows between the shore and The Street, 30 from the Fields between The Street and the Head Dyke, and 30 outside the Head Dyke. Finally, 30 permanent 1 × 1-m quadrats (one at each sampling station) are searched and the numbers of seedlings of all vascular plant species are recorded.

4.4 Spatial distribution of grazing

Other things being equal, the sheep might be expected to distribute themselves over a heterogeneous matrix of plant communities in such a way as to maximise their fitness (Fretwell and Lucas 1970). At low population densities, therefore, the animals would be expected to aggregate in the best habitat patches. As sheep density increases, then the extra gain obtained by foraging in the best patches is likely to be offset by increased intraspecific competition, and by a reduction in the effective rate of resource supply (because of exploitation competition) (Hunter 1964). We should expect this process to continue, with inferior habitat patches being added successively, as population density increases. In practice, however, it is not always obvious what aspects of habitat quality to measure, nor what constitute sensible boundaries for the patches.

On Hirta we are fortunate in that habitat quality is relatively easy to assess. Apart from arthropod and gut parasites, the sheep have no predators, so we do not need to be concerned with the assessment of 'enemy-free space' that hinders the measurement of habitat quality in so many studies. Again, the plant communities of Hirta have clear and distinct boundaries (Gwynne *et al.* 1974), so census work, carried out at lambing time (March–April) in mid summer (July–August) and during the rut (October–November) can locate individual marked sheep within plant communities at precise grid references. Habitat quality is estimated using data from the sorted biomass samples (see above), from which we can calculate positive elements (like the biomass of green material of palatable species) and negative elements (like the biomass of bryophytes, or the mass of standing dead organic matter in the sward).

The distribution of sheep across habitats is non-random at all seasons. Not surprisingly perhaps, the best predictor of spatial variation in sheep density is the total biomass of green grass of all palatable species, and the habitat selection index (calculated as the proportion of sheep sightings divided by the proportion of the total area covered by the habitat in question) increases exponentially as the ranked biomass of palatable grasses (*Festuca, Poa, Holcus, Anthoxanthum* and *Agrostis*; from most to least palatable) increases from heather-dominated hillsides or *Sphagnum* bogs (ranked 6, with index <0.2) (Fig. 4.3) to productive pastures within the Head Dyke (ranked 1, with index >10).

Within communities, small-scale structure of the vegetation has important effects on grazing behaviour. The productive grasslands inside the Head Dyke are made up of a dynamic mosaic of tussocks and gaps: gaps are close-cropped, lawn-like swards containing an abundance of prostrate plants (notably *Trifolium repens*) and a luxuriant growth of mosses (principally *Rhytidiadelphus squarrosus, Pseudoscleropodium purum* and *Pleurozium schreberi*), while tussocks are patches of taller grass, woven into a matrix of standing dead grass stems (mainly *Agrostis capillaris* and *Holcus lanatus*). The sheep clearly prefer to graze from gaps rather than tussocks (J. M. Milner and A. L. Tuke, unpublished data), but grazing behaviour varies from season to season. In summer, food quality in gaps is high (because of the high proportion of palatable species, and low proportion of standing dead organic matter), but bite

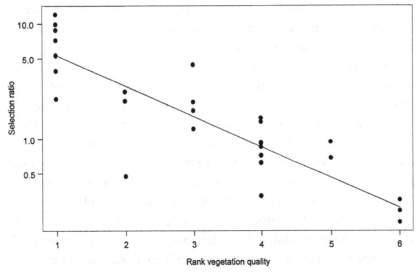

FIG. 4.3. Spatial variation in the use of plant communities of different botanical composition by the sheep. As food quality declines, the proportion of sightings of sheep within a plant community declines rapidly (note the log scale of the habitat selection index). Habitat selection index is calculated as the proportion of all sightings of sheep in a plant community divided by the proportion of the total study area made up by that plant community. Rank food quality index is determined by assigning each plant community a rank between 1 (*Festuca rubra/Poa* community) to 6 (*Sphagnum/Eriophorum* community) on the basis of published information on palatability and average digestibility of forage for sheep (Hunter 1964; King and Nicholson 1964).

size, and hence the intake rate, is low. Potential bite sizes are high in tussocks, but food quality is lower, due to the abundance of standing dead organic matter. It is not surprising, therefore, that sheep prefer to graze in summer on the 'indeterminate' swards (neither gap nor tussock) which combine relatively high biomass with relatively high food quality. In winter, the picture is somewhat different because the availability of palatable green grass in gaps is diluted by an (often copious) growth of *Rhytidiadelphus squarrosus* and other bryophytes, which the sheep would rather avoid. In March, the tussocks often contain substantial quantities of new young shoots of highly desir-able species like *Poa humilis* and *Festuca rubra*, but these are diluted by a dense weave of standing dead material and are essentially unavailable to the sheep. As the green shoots extend above the top of the tussock, they are rapidly clipped back by the sheep.

The structure of gaps and tussocks varies throughout the year and is affected by sheep numbers. In April, the rate of grass production first begins to exceed the rate of consumption, and green biomass starts to accumulate. After grass growth slows or stops in late summer, biomass begins its slow decline through autumn and winter into the new year. Some plant material becomes standing dead organic matter (particularly in tussocks), some decomposes, and some is consumed by the sheep. The major effects of grazing on plant biomass are seen at the end of winter (Table 4.1). In the high year, 1996, there were 1832 sheep in total while in the low year, 1995, there were 1177 (representing a 1.56-fold change from 1995 to 1996, one of the highest increases observed during the entire forty-one-year study period). Note that, contrary to expectation, the total biomass is higher in the year when sheep numbers were 50% higher. There were two causes of this; the proportional cover of tussocks was greater in the high-density year, and the average mass of a tussock was greater. Species' live biomass showed subtle shifts. In gaps, the proportion of *Holcus* and *Agrostis* went up while *Festuca* and *Anthoxanthum* went down with higher sheep numbers, consistent with what we know about their relative palatabilities. There were no pronounced patterns for the herbs. In tussocks, all live grass biomass was lower in the high-density year but it was *Festuca* and *Holcus* that decreased most. One marked pattern (even more striking in the winter samples) is the effect of grazing pressure on the abundance of bryophtyes in gaps; much of the green matter may be less available or unavailable because it is mixed with a mass of *Rhytidiadelphus squarrosus* and other mosses. A similar case can be made for the high levels of dead organic matter in tussocks in the high-density year; this will protect green matter of palatable plants from grazing. Recall that there is probably little competition for food in August, even in years with the highest sheep densities.

Total standing crop (live and dead) in August is a function of sheep numbers, but there is the merest hint of a negative correlation. In March, however, the relationship is very strong; the higher sheep

Table 4.1. *Botanical composition in August for years of high and low sheep density. Data are mean dry mass in g dry matter per 20 × 20 cm sample for transects A, B and C inside the Head Dyke (n = 24)[a]*

Species	Tussock		Gap	
	Low-density Year	High-density Year	Low-density Year	High-density Year
Agrostis capillaris	9.59	8.41	1.08	1.23
Festuca rubra	3.58	2.08	0.73	0.42
Holcus lanatus	3.49	2.03	0.73	1.06
Anthoxanthum odoratum	2.57	2.07	0.77	0.45
Poa humilis	1.80	1.32	0.17	0.21
Lolium perenne	0.69	0.11	0.003	0
Ranunculus acris	0.66	0.38	0.25	0.11
Trifolium repens	0.53	0.53	0.80	0.69
Cerastium fontanum	0.15	0.25	0.04	0.05
Leontodon autumnalis	0.03	0.04	0.02	0.07
Plantago lanceolata	0	0.005	0.09	0.02
Achillea millefolium	0	0	0	0.014
Bryophyte	0.18	0.16	1.57	2.37
Dead organic matter	4.54	11.16	0.61	1.69
Total	28.68	30.96	7.44	10.49
Proportional cover of gaps or tussocks[a]	0.27	0.35	0.14	0.18
Total biomass (gaps and tussocks)	21.42	24.01		

[a]Gap and tussock cover do not sum to 1.0 because of the presence of 'indeterminate' (i.e. unclassifiable) quadrats.

densities, the lower the total biomass (the negative correlation is highly significant: $r^2 = 0.70$, $n = 14$, $p < 0.01$). The March effect is caused principally by a reduction in the cover of tussocks in years when sheep numbers are high. Grazing at this level appears to do no long-term harm to the vegetation, presumably because the plants are able to replenish their reserves during the relatively lower grazing pressures that prevail during summer. The resilience of the vegetation to heavy winter grazing is shown by the fact that total biomass was higher in August 1997 than in any preceding year, despite the fact that sheep numbers during the winter of 1996–7 were the highest recorded up to that point, and had depressed tussock cover in March 1997 to its lowest ever level.

We have so far been able to measure the effects of a crash on vegetation structure only twice, and the response can be exemplified by August 1999 data. Between August 1998 and August 1999, the island-wide population declined from 1968 animals to 933, and most of the 1000 or so deaths occurred between February and April 1999. The crash represented a 53% decline in numbers ($\Delta = -0.746$ on a log scale, see p. 93), and was the second biggest crash since our records began in 1955 (the biggest crash was in 1958–9 when the population declined from 1344 to 610; $\Delta = -0.790$). The 1998–9 crash was different from earlier crashes, however, because the pre-crash population was so high (nearly 2000 animals), that even after heavy mortality, the residual population was still relatively high.

The crash affected all components of the vegetation: total biomass, the relative cover of gaps and tussocks, flowering rates, pasture quality (the ratio of green grass to standing dead organic matter), and heather condition (the ratio of green to woody tissues). Figure 4.4a shows the proportion of the ground covered by tussocks in March of the pre-crash, crash and post-crash years. In March 2000, tussock cover was roughly double the value measured during the crash year of 1999. The size of individual tussocks (dry mass per 400 cm^2) also varied with sheep numbers: individual tussocks were significantly smaller in the crash year than in either the pre- or post-crash years (Fig. 4.4b), presumably because sheep were forced to consume dead organic matter when high-quality food availability was so low. Individual gaps supported significantly higher biomass in the post-crash year, but gap biomass had been reduced to its lowest level the year *before* the crash occurred (Fig. 4.4c). The combined effects of altered tussock density and tussock size on biomass and botanical composition averaged across the entire grassland are shown Fig. 4.4d. In March of the pre-crash year (1998) green grass biomass was at very low levels, as a result of several successive years of high sheep numbers, but there was a substantial amount of dead organic matter (DOM) in the

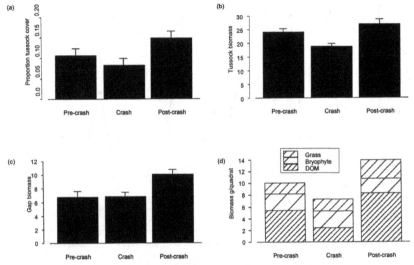

FIG. 4.4. Botanical changes associated with a crash in sheep populations: a comparison of March sward data from within the Head Dyke in pre-crash (1998), crash (1999) and post-crash (2000) years. (a) The cover of tussocks (proportion of ground area) recovers immediately after the crash to levels greater than the long-term average; (b) the biomass of individual tussock samples (g dry matter per 20 × 20 cm) is significantly lower in the crash year than in either pre- or post-crash years; (c) the biomass of individual gap samples (g dry matter per 20 × 20 cm) is significantly higher in the post-crash year than in either pre-crash or crash years; (d) overall botanical composition of the sward (weighted average of tussocks and gaps) was different in each of the three years (bryophyte dry mass was roughly constant, palatable green biomass was significantly higher in the post-crash year, but similar in the crash and pre-crash years, while dead organic matter (DOM) was lowest in the crash year and highest in the post-crash year).

sward. By March of the crash year (1999) both green biomass and DOM were at very low levels. Presumably, the low DOM was the result of two processes: (1) high grass consumption in summer/autumn 1998 left little grass to turn into DOM; and (2) sheep numbers were so high, and good-quality food availability so low, that sheep were forced to eat DOM, thereby further worsening body condition. In March of the post-crash year (2000), the amount of palatable grass was significantly higher than in either of the two preceding years, and the amount of DOM was substantially higher, reflecting the greatly reduced overall grazing pressure.

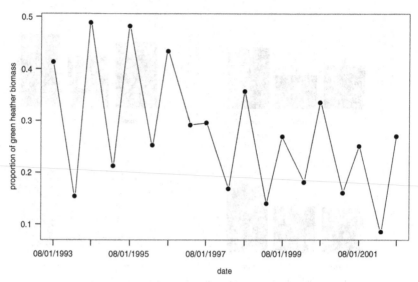

FIG. 4.5. The proportion of heather biomass made up of green shoots (green shoot dry weight divided by total dry weight of green shoots plus woody tissues) in August and March from 1993 to 2002. Green heather proportion shows strong seasonal fluctuations (low in March, high in August), and was higher in the period 1993–6 than more recently, i.e. 1997–2002 ($p < 0.05$). Green heather proportion was particularly low during the crash of March 2002.

Judged by the percentage of DOM in the sward, diet quality appears to be relatively low in the post-crash year, but this crude measure may be misleading. At low population densities the sheep are able to maintain a quality diet by foraging preferentially in the gaps, where there is much less DOM. Thus, as sheep numbers increase, grazing pressure on the gaps remains relatively high, and the proportional ground cover of gaps increases while tussock cover declines (and vice versa). At low populations (e.g. post-crash) a much smaller proportion of total sward area is heavily grazed, and the cover of unpalatable tussock vegetation reaches its maximum.

Outside the Head Dyke, the sheep feed extensively on heather. A simple measure of heather condition is the proportion of *Calluna* biomass represented by green shoots in March. This index is much higher in gaps (long-term average 28% of dry matter) than in tussocks (average 13%). In the crash year (March 1999) there was little difference from the long-term average green matter in heather tussocks, but the proportion of green shoots in gaps was significantly reduced, presumably as a result of unusually heavy grazing. By March 2000, however, the proportion green shoots in gaps had recovered almost to pre-crash levels, and was higher than the long-term average (33% green). There is a strong seasonal pattern in total *Calluna* biomass, but there have been no significant downward trends in heather abundance in recent years (data not shown).

4.5 Effects of grazing on flowering, botanical composition and plant species richness

The number of flower stems per unit area varies greatly from year to year (Fig. 4.6). Some species are so heavily grazed that they do not produce any flowering stems at all (*Festuca rubra* and *Rumex acetosa*). We know that the weather is suitable for flowering of these species, because they flower profusely in places that are inaccessible to the sheep (cliffs, cleit-tops and smaller, ungrazed islands; see below). Other plants produce flowers every year, but the rate of flower-stem production is correlated with grazing pressure (e.g. the less palatable grasses like *Agrostis capillaris* and *Anthoxanthum odoratum*).

The consequences of reduced flower production for recruitment from seed remain to be investigated. For the duration of the study so far, detailed monitoring of permanent quadrats has not detected any seedling recruitment at all for the longer-lived species, even for those that have produced copious seed. It appears that recruitment is microsite-limited for many if not all of these plants (Crawley 1997). If that is the case, then a combination of high soil disturbance during the winter of a crash year, followed by relaxed grazing pressure in the summer, might lead to a substantial bout of seedling recruitment during the immediate post-crash year. We await this with interest. The expected post-crash mass flowering did not occur in summer 1999 or 2002. This could be because the residual sheep population was so high that grazing pressure was sufficient to inhibit flowering, or it might be that the plants need more than a few months to recover from prolonged heavy grazing before flowering is possible. The crash was associated with little or no soil disturbance, so that even if a large seed

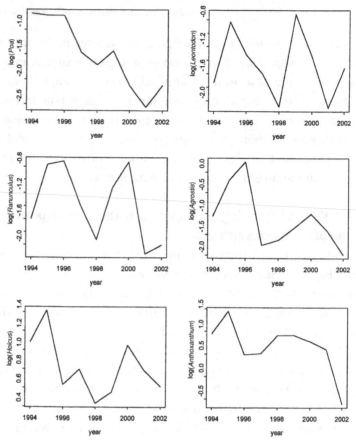

FIG. 4.6. Time series (1994–2002) in log flowering stem density (per 0.16 m^2) in August for the six most frequent flowering species (note the different scales), based on 450 random quadrats per year: *Poa humilis, Leontodon autumnalis, Ranunculus acris, Agrostis capillaris, Holcus lanatus, Anthoxanthum odoratum. Agrostis, Ranunculus* and *Leontodon* show very low flowering in the peak sheep years of 1997 and 1998, but flowering in *Anthoxanthum* appears to be highly grazing tolerant. The anticipated post-crash mass flowering did not occur in either August 1999 or 2002. Only *Leontodon autumnalis* ($r = -0.70$, $n = 9$) showed a significant negative correlation between flower stem density and current sheep numbers. Only *Ranunculus acris* showed a negative correlation between this year's flowering and last year's sheep numbers ($r = -0.52$). Groups of species showed significant correlations in their cross-year patterns of flowering (e.g. *Ranunculus* and *Leontodon* had $r = 0.62$), but no two species showed exactly the same pattern of flowering in their time series. Evidently, sheep grazing pressure is not the principal determinant of flowering stem density in most of these species. Note, however, that the two most palatable species (*Festuca rubra* and *Rumex acetosa*) did not flower at all in grazed communities during the nine-year study period, despite flowering freely in all years in places that were inaccessible to the sheep (e.g. cliffs and certain cleit roofs).

crop had been produced, it is likely that seedling recruitment would have remained very low as a result of microsite limitation (Crawley 1997).

There are several species on St Kilda that decrease in abundance under sheep grazing ('decreasers'). Some species are simply never found except on the most inaccessible and hence sheep-free places (*Angelica sylvestris*, *Tripleurospermum maritimum*, *Oxyria digyna* and *Sedum rosea*); these might be thought of as extreme decreasers. Other plants are so palatable to the sheep that they are selectively grazed down to very small size, and although they are not eliminated from plant communities, they seldom produce flowers; the best examples of these decreasers are *Festuca rubra* and *Rumex acetosa*. The classic unpalatable species that increase in relative abundance under sheep grazing are *Nardus stricta* and *Molinia caerulea*. Both are present on Hirta (Appendix 1) but neither is particularly abundant in the study area. On the more productive grasslands within the Head Dyke, the principal increasers are *Agrostis capillaris* and *Holcus lanatus*. This is interesting, because both of these species are thought of as being relatively palatable in other Scottish grasslands (King and Nicholson 1964).

Plant species richness is low on Hirta as a result of its small size and extreme isolation. Given this background, it is interesting to ask whether sheep grazing has an effect on species richness. The currently ungrazed parts of the archipelago, like the island of Dun and ungrazed microhabitats within Hirta and the turf roofs of the tallest and most steep-sided cleits, show the effects of freedom from sheep grazing on plant diversity. On Dun, species richness is relatively low because of the tall growth of nitrophilous plants like *Angelica sylvestris* and *Tripleurospermum maritimum* in places where bird droppings are most densely deposited. On cleit roofs, species richness is low because of the development of a virtual monoculture of *Festuca rubra*. Both of these patterns support the trade-off between competitive ability and palatability mentioned earlier in which tall, rapidly growing but highly palatable species outcompete the slower-growing, but less preferred plants when sheep are excluded.

Amongst the grazed communities, species richness is greatest on the heathy grasslands inside the Head Dyke to the west of the road, where species like *Gentianella campestris* and *Euphrasia* spp. are to be found. This is consistent with a mass of data from other grasslands that shows a trend of declining species richness with increasing biomass (Crawley 1997). This is due principally to the fact that individual plants grow larger at higher levels of soil fertility, hence reducing the opportunity of small plant species to respond to diversity-enhancing, small-scale heterogeneity. Against this trend, however, species richness is slightly lower (fifteen species) in grasslands above The Street than below it (seventeen species) where biomass is higher.

To investigate the effects of grazing on plant species composition, we excluded the sheep from small patches of vegetation by fencing them off. The effects on plant diversity of excluding the sheep proved to depend on soil fertility. On fertile soils, areas fenced in 1991 showed rapid and profound changes in biomass and botanical composition. Freed from grazing, preferred grasses (chiefly *Festuca rubra*) grew quickly to form a dense mat, which over the course of three or four years, completely excluded all low-growing grasses and herbs. By five years, plant diversity had collapsed, and a relatively broad spectrum of species (see Appendix 1) had been replaced by a virtual monoculture of *Festuca rubra*. A few plant species, however, were virtually confined to these temporary exclosures; these included highly palatable species like *Angelica sylvestris* and *Arrhenatherum elatius*, and some more puzzling examples like ragwort *Senecio jacobaea*. On nutrient-poor, or very wet soils, the response of the vegetation to sheep exclusion was much less pronounced, and much slower to develop. This could be a consequence of low plant growth rates, coupled with local seed limitation of preferred plant species that would be competitive under sheep-free circumstances. The sheep exclosures were taken down in August 1996, and there was little trace of where they had been by August 1997; the protected crop of *Festuca* was quickly eaten down and replaced by less palatable grasses (mainly *Agrostis capillaris* and *Holcus lanatus*).

4.6 Grazing and infection by fungal endophytes

One of the features of grasses which may act as a deterrent to herbivory is the presence of fungal endophytes. These members of the tribe Balansieae (Clavicipitaceae: Ascomycetes) and their anamorphic derivatives, *Acremonium* spp. (Deuteromycetes), occur in sedges, rushes and several hundred graminoids (White 1987; Clay 1990; Vicari and Bazely 1993). They cause no disease symptoms in the vegetative plant, but they may prevent or reduce flowering (Clay 1990). Many endophytes produce secondary compounds such as ergot alkaloids which are toxic to both vertebrate and invertebrate herbivores (Powell and Petroski 1992). By altering the rate of defoliation of different plant phenotypes, fungal endophytes may influence the outcome of inter- and intraspecific plant competition (Ball *et al.* 1993). Since the consumption of endophyte-infected grasses can be a major cause of animal illness (Powell and Petroski 1992), it is possible that endophytes play a role in the sheep population crashes on St Kilda, because the sheep would be unable to avoid heavily infected, and hence potentially toxic, pastures at critical times.

Festuca rubra is a major component of the diet of the sheep on Hirta (Milner and Gwynne 1974), and we can ask whether *F. rubra* populations on Hirta have a higher proportion of endophyte-infected plants, or higher densities of fungal hyphae within grass tissues, than those on the ungrazed island of Dun. Within Hirta, we can ask whether ungrazed habitats like cleit roofs have lower rates of infection than adjacent, grazed communities. Bazely *et al.* (1997) found that a fungal endophyte, which is probably the anamorph of *Epichloe festucae* sp. nov. (Leuchtmann *et al.* 1994) was widespread on Hirta, where 65–100% of sampled tillers were infected (Table 4.2), consistent with the levels found in several other studies (Bacon *et al.* 1977; Shelby and Dalrymple, 1987; Clay and Leuchtmann 1989). In contrast, on Dun, ungrazed for more than 70 years, significantly fewer *F. rubra* tillers were infected (Table 4.2), supporting the hypothesis that the fungal endophyte defends its host, *F. rubra*, from herbivory. Alternatively, grazing may increase susceptibility to endophyte infection. The pattern is far from clear-cut, however, because endophyte infection rates of *F. rubra* on grazed and ungrazed cleit roofs were the same, and on the nearby islands of Rum and Benbecula there were no differences in endophyte infection between sites with different grazing pressures (Table 4.2).

Table 4.2. *Proportions of endophyte-infected* Festuca rubra *shoots on St Brianan's Fanks, Hirta and on Dun in the St Kilda group, on Rum and on Benbecula, August–September 1991*

	Percent infection	n
Grazed:		
St Brianan's	83.9	56
Ungrazed:		
Dun	24.0	50
Grazed:		
Cleits	94.0	65
Ungrazed:		
Cleits	100	55

	Percent infection	n	Percent	n	Significance	Mean
Rum	Outside exclosure		Deer Exclosure			
1	37.5	16	71.4	14	$p = 0.081$	53.3
2	0	11	0	8	$p = 1.000$	0
3	18.2	13	40.0	10	$p = 0.341$	26.1
Benbecula	Sheep field		'Ungrazed' Monastery			
	40.8	49	44.4	45	n.s.	42.5

4.7 Grazing and long-term changes in plant communities

The pattern of plant communities on Hirta reflects underlying heterogeneity in geology, soil type, drainage, exposure and human influence. Sheep grazing modifies the relative abundance of plants within these communities, but appears to have rather little influence in determining the precise location of the boundaries between communities. Early accounts (Barrington 1866; Petch 1933; Poore and Robertson 1949; McVean 1961) suggested that there had been major changes in the vegetation on Hirta during the seventeen years immediately following abandonment by the islanders in 1930, when a managed flock of black-face sheep was replaced by a food-limited population of feral Soay sheep. Subsequent detailed studies have failed to substantiate these anecdotal reports. Gwynne *et al.* (1974) made detailed observations on fifty-four plant communities, and thirty-six of these were reanalysed in 1992 by D. R. Bazely and T. A. Watt (unpublished data). They found no new species and no apparent loss of species. The general paucity of significant changes in plant community composition over this thirty-year period suggests that the sheep population had reached some form of dynamic equilibrium by 1963, and that a relatively stable interaction between sheep and vegetation has persisted to the present day. *Calluna* had increased somewhat in the maritime communities, but not in the heaths. *Molinia* was found to have decreased slightly in one community (wet grassland) but increased in three others. The *Agrostis–Festuca* grasslands showed some shifts towards species which are less palatable, such as *Molinia*, but the greatest changes occurred in the former cultivated areas in Village Bay, where formerly common ruderals and arable weeds were replaced by *Holcus–Agrostis* grassland (McVean 1961; Gwynne *et al.* 1974) (Appendix 1).

4.8 Discussion

Our main finding is that the relationship between sheep numbers and primary production is a simple one. The plant communities on Hirta appear to be extremely resilient to the levels of grazing imposed by the food-limited population of Soay sheep. Peak grazing intensities occur during February and March of crash years, at which time there is heavy sheep mortality. Although plant biomass is reduced to very low levels during these periods, rapid regrowth during spring and summer restores the competitive position of even the most palatable plant species. Post-recovery grass biomass in August is correlated with current (i.e. post-crash) sheep numbers, but not with sheep densities during the previous winter, suggesting that time-lagged effects of grazing on plant production are not of major importance.

Similarly, there is no reason to believe that Soay sheep are responsible for 'overgrazing' of the plant communities of Hirta. Typically, overgrazing is associated with managed ungulate populations, where the numbers of animals are kept artificially high during the summer time (Crawley 1997). Overgrazing is usually the result of supplementary winter feeding, or (in places like North Wales) due to the practice of overwintering the sheep in the lowlands, and importing animals at exceptionally high densities for the summer. Soay numbers on Hirta are regulated by intraspecific competition for food during the winter, and this means that the plant populations are able to recover during the summer, because offtake by sheep is relatively low compared to primary productivity in the summer (Fig. 4.1).

In general, the effects on plant biodiversity of grazing by Soay sheep appear to be positive: species richness is lower on ungrazed parts of the island, and falls rapidly when fences are erected to exclude the sheep. There are a few species that are restricted to cliffs and other inaccessible ungrazed places, but some of the rarest plants on Hirta, like *Gentianella campestris* and *Saxifraga oppositifolia* thrive in the grazed sward. Spatial variation in the productivity of preferred plant species is the major determinant of the spatial distribution of sheep across habitats within Hirta (Fig. 4.3). Results from grazing exclosures and ungrazed cleit roofs show how the competitive abilities of the different plant species are affected by selective grazing by sheep, and lend support to the hypothesis that there is a positive correlation between palatability and competitive ability (Pacala and Crawley 1992). Differential rates of defoliation interact with growth rates and differential degrees of grazing tolerance to determine relative abundance (Crawley 1997).

At this stage we have no information on the way that plant recruitment from seed is affected by reductions in seed production caused by sheep grazing. Some of the most palatable species (e.g. *Festuca rubra* and *Rumex acetosa*) did not set any seed at all in grazed areas of Hirta, even in the summers of 1999 and 2002 following crashes in sheep numbers.

Although many details remain to be understood, it is reasonably clear that the sheep and the vegetation of Hirta are in a stable equilibrium. The equilibrium is the product of a dynamic interaction between the animals and the plants, in which the productivity and botanical composition of the vegetation determine the average population density of the sheep. In turn, selective grazing pressure determines the productivity of the vegetation and the relative abundance of the different plant species.

5
Parasites and their impact

K. Wilson *University of Stirling*

B. T. Grenfell *University of Cambridge*

J. G. Pilkington *University of Edinburgh*

H. E. G. Boyd *University of Cambridge*

and

F. M. D. Gulland *Marine Mammal Center, Sausalito, USA*

5.1 Introduction

Highly pathogenic *epidemic* disease agents, like the rinderpest and myxomatosis viruses, have obvious and often dramatic consequences for the dynamics and evolution of their host populations (Osterhaus and Vedder 1988; Roelke-Parker *et al.* 1996; Vogel and Heyne 1996; Hudson 1997; Hochachka and Dhondt 2000). In contrast, the effects of *endemic* diseases are often more subtle and, until recently, had largely been overlooked (Grenfell and Dobson 1995; Hudson 1997; Hudson *et al.* 2002). The mathematical models of Anderson and May in the late 1970s (Anderson and May 1978, 1979, 1982a; May and Anderson 1978, 1979) were the first to highlight the potential of endemic parasites and pathogens to regulate host populations, and more recent theoretical studies have also implicated parasites as potentially important driving forces in the evolution of their hosts. Their impact is believed to extend to the evolution of secondary sexual characters (Hamilton and Zuk 1982; Read 1987, 1988; McLennan and Brooks 1991; Hamilton and Poulin 1997) and optimal life-history strategies (Michalakis and Hochberg 1994; Sheldon and Verhulst 1996; Richner 1998); the manipulation of host behaviour (Moore 1984; Poulin 1994; Moore and Gotelli 1996); the maintenance of genetic diversity and even to the evolution of sex (Van Valen 1973; Hamilton 1980; Hamilton *et al.* 1990; Moritz *et al.* 1991; Howard and Lively 1994).

Empirical support for these models comes largely from laboratory studies (Grenfell and Dobson 1995; Clayton and Moore 1997; Hudson

113

Soay Sheep: Dynamics and Selection in an Island Population, ed. T. H. Clutton-Brock and J. M. Pemberton. Published by Cambridge University Press. © T. H. Clutton-Brock and J. M. Pemberton 2003.

et al. 2002). Data from natural systems are much thinner on the ground, particularly for long-lived vertebrates, due mainly to the logistical difficulties of collecting long-term demographic data. These problems have largely been overcome on St Kilda, where demographic data have been collected continuously since 1985. Other problems associated with most other natural host–parasite systems include the complication of competing herbivores and predators (e.g. Hudson *et al.* 1992a; Stenseth 1995) and the possibility of migration into and out of the study population. These problems are much reduced on St Kilda, where the system can be characterized as a plant–herbivore–parasite interaction with zero dispersal and restricted sheep movement about the island (Grenfell *et al.* 1992, 1998; Coulson *et al.* 2001; Chapter 3).

In this chapter, we assess the importance of parasites for the evolution and population dynamics of their hosts, focussing on the Soay sheep of St Kilda. Specifically, we address the following questions: What effect does island life have on the parasite community? Which parasites are likely to cause most damage to the health of their hosts and how can we measure their abundance? What factors are likely to generate variation in parasite loads? What are the costs of parasitism for the host, in terms of growth, survival and reproduction? And, do parasites cause changes in the population density of their hosts or merely reflect them?

One of the advantages of studying the parasitology and epidemiology of Soay sheep is that there is a large complementary literature for their domestic relatives. Clearly, there will be similarities and differences between the domestic and wild situation, and throughout this chapter emphasis is placed on making these comparisons, in an attempt to understand both systems better. We also attempt to use the Soay sheep system to draw some insights into the impact of parasites on other vertebrate host populations in the wild.

5.2 The parasite community

THE PARASITES AND THEIR PATHOLOGICAL CONSEQUENCES
Island life potentially has a number of important implications for parasites. First, parasite communities on islands are likely to be

influenced by 'founder effects' (e.g. Grant 1998). These will determine not only which parasite species are carried to the island, but also the genetic make-up of the founding host population, which may affect their relative susceptibility to particular parasite species. A second consequence of island life is that the host population is typically more isolated from neighbouring populations. We can therefore expect some parasite species to be absent from islands merely as a consequence of the fact that they lack the capacity to reach, and subsequently infect, their hosts. Third, islands tend to harbour fewer species in general, and this will affect the transmission potential of those parasite species that rely on one or more intermediate host species in order to complete their life cycles. Thus, we can expect hosts living on islands typically to be infected with relatively fewer indirectly transmitted parasite species (i.e. those species that require intermediate hosts or vectors). A final consequence of island life is that host populations are generally smaller than those on the mainland. As a consequence, the number or density of hosts may be below the 'critical threshold level' for the persistence of some parasite and pathogen species (e.g. measles is endemic only in human populations of greater than 5 million people; Anderson 1993). This is because the parasite or pathogen runs out of susceptible hosts to infect (i.e. uninfected immigrants or newborns). This effect is likely to be particularly important for highly virulent viral or bacterial pathogens that either kill their hosts or induce long-lasting effective immunity (see Swinton *et al.* (2002) for a full discussion of the persistence of microparasite infections of wildlife). Conversely, immunity against protozoans and macroparasites (i.e. nematodes, tapeworms, flukes, etc.) tends to be short-lived and infections can persist in individual hosts for long periods of time. As a consequence, most protozoan and macroparasite populations can survive endemically in small host populations with low inputs of susceptibles (Anderson 1993). Thus, we expect rather few macroparasite and protozoan species to be absent from the parasite fauna on islands. So, what do we observe on St Kilda?

The parasites of the Soay sheep on St Kilda were first investigated in the early 1960s (Cheyne *et al.* 1974). Although these studies were rather limited in duration, most of the parasites now known from

St Kilda were first identified during this period (Table 5.1), and the conclusions drawn from these initial investigations remain largely true today (Cheyne *et al.* 1974). The parasite community is fairly typical of domestic hill sheep in Scotland (Soulsby 1982) with one or two interesting differences. For example, the sheep tick, *Ixodes ricinus*, and larval stages of the blow fly *Lucilia sericata*, are notable absentees from the ectoparasite faunae (Boyd *et al.* 1964; Cheyne *et al.* 1974), and the highly pathogenic and widespread nematode *Haemonchus contortus* is also conspicuous by its absence (Soulsby 1982). As expected, many common bacterial and viral infections of sheep also appear to be absent, presumably because the host population size is too small to sustain them (Grenfell and Dobson 1995; Hudson *et al.* 2002). These include Johne's disease (*Mycobacterium paratuberculosis*), enzootic abortion of ewe (*Chlamydia psittaci*), *Mycoplasma ovipneumoniae*, Border disease virus and Maedi–Visna virus (St Kilda Soay Sheep Project Annual Report 2001, unpublished). Conversely, there is a rich community of protozoans present on the island (Table 5.1).

The tapeworm *Taenia hydatigena* is a surprise inclusion to the parasite fauna. This parasite was first recorded on St Kilda in the 1960s, when 49% of adult sheep were infected (Cheyne *et al.* 1974). The prevalence of infection has remained high ever since, with 31% of adults harbouring this parasite in 1989 (Gulland 1992) and 30–53% in 1992 (Torgerson *et al.* 1992, 1995). The presence of *T. hydatigena* on St Kilda is noteworthy because its cystercercal stage, which encysts in the linings of the sheep's rumen and abomasum, must be ingested by a carnivore intermediate host before its life cycle can be completed. The absence of carnivores on St Kilda since the removal of dogs in the 1930 human evacuation (Chapter 2) makes the presence of this parasite on the island paradoxical (see above). It seems most likely that eggs of this parasite are being brought over to the islands repeatedly by birds from the mainland (Torgerson *et al.* 1995). Despite its relatively high prevalence, individuals rarely harbour more than two cystercerci (Torgerson *et al.* 1992) and so it is unlikely that this parasite is an important source of morbidity or mortality on St Kilda. The same can be said of the only other tapeworm on the island, *Moniezia expansa*. Adults of this species live in the small intestine, and

Table 5.1. *List of parasite species associated with Soay sheep on St Kilda*

Taxon	Specific name	Location
Protozoa	*Cryptosporidium parvum*	Small intestine
	Giardia duodenalis	Small intestine
	Eimeria ahsata	Small intestine
	Eimeria bakuensis	Small intestine
	Eimeria crandallis	Small intestine/large intestine
	Eimeria faurei	Small intestine/large intestine
	Eimeria granulosa	Unknown
	Eimeria intricata	Small intestine/large intestine
	Eimeria marsica	Unknown
	Eimeria ovinoidalis	Small intestine/large intestine
	Eimeria pallida	Unknown
	Eimeria parva	Small intestine/large intestine
	Eimeria weybridgensis	Small intestine
Bacteria	*Dermatophilus congolensis*	Skin
Flies	*Melophagus ovinus*	Wool
Lice	*Damalinia ovis*[a]	Wool
Tapeworms	*Moniezia expansa*	Small intestine
	Taenia hydatigena	Abdominal cavity
Nematodes	*Dictyocaulus filaria*	Lungs
	Muellerius capillaris	Lungs
	Teladorsagia circumcincta[b,c]	Abomasum
	Teladorsagia trifurcata[b,c]	Abomasum
	Teladorsagia davtiani[b,c]	Abomasum
	Trichostrongylus axei	Abomasum/small intestine
	Trichostrongylus vitrinus	Abomasum/small intestine
	Capillaria longipes	Small intestine
	Strongyloides papillosus	Small intestine
	Nematodirus battus	Small intestine
	Nematodirus filicollis	Small intestine
	Nematodirus helvetianus	Small intestine
	Bunostomum trigonocephalum	Small intestine
	Trichuris ovis	Large intestine
	Chabertia ovina	Large intestine

[a] Pseudonym *Bovicola* or *Trichodectes*.
[b] Pseudonym *Ostertagia*.
[c] Now believed to be a single species (see text).
Sources: Cheyne *et al.* (1974), Gulland (1992), J. G. Pilkington unpublished data, B. H. Craig, unpublished data.

infected individuals usually harbour just a single worm (K. Wilson, unpublished data).

Both lice and keds (wingless flies) are common on St Kilda. In 1964, the prevalence of these two ectoparasites was greater than 90% (Cheyne *et al.* 1974). Although absolute counts of keds and lice have not been made since, relative measures of ked abundance in August have been determined since 1988 (a count is made of the total number of keds observed during a 1-minute search of the wool on the sheep's belly). These indicate that keds are numerous only on young animals, particularly ram lambs. Sheep keds live in the wool of their hosts, but feed on blood. Heavy infestations can severely reduce host condition and cause anaemia; however their main damage results from the irritation they cause. Sheep lice also feed mainly on fragments of wool and other epidermal products, but occasionally feed on blood from open wounds. As with keds, the chief effects of lice on their hosts are due to the irritation they cause: infected sheep may become restless and reduce the amount of time they devote to sleeping and feeding (Soulsby 1982).

Two species of lungworm are found in sheep on St Kilda. *Dictyocaulus filaria* is a widespread, directly transmitted, parasite. The adult worms live in the tracheae and bronchi and cause a catarrhal parasitic bronchitis, which may develop into pneumonia (Soulsby 1982). Infected lambs may be observed coughing up parasite eggs, which are then swallowed and hatch as they pass through the alimentary tract, before being voided on to the pasture as larvae. This parasite predominantly affects lambs, whilst older animals are more commonly infected with *Muellerius capillaris*. This is probably the commonest lungworm of sheep in Europe and requires molluscs as intermediate hosts (Soulsby 1988). Adults of *M. capillaris* live in the alveoli and pulmonary parenchyma. Infected animals generally show no clinical signs, but heavy infections weaken the lungs and may lead to a reduction in the general fitness of the host (Soulsby 1982). Little is known about its prevalence or impact on St Kilda.

Soay sheep harbour a range of gastrointestinal nematodes, the most numerous of which belong to the genus *Teladorsagia*. During the population crash of 1989, these 'brown stomach-worms' comprised approximately 75% of the total number of worms living in the gut and, based on morphology, *T. circumcincta* (formerly referred to as *Ostertagia circumcincta*) represented about 85% of these, with *T. trifurcata* and *T. davtiani* making up the remainder (Gulland 1992). However, recent genetic studies on the worms from St Kilda could find no evidence to separate these three putative species based on nuclear or mitochondrial DNA sequences (Braisher 1999). It seems likely, therefore, that there is just a single *Teladorsagia* species on St Kilda and we will refer to this as *T. circumcincta*.

On St Kilda, individuals have been known to harbour more than 20 000 of these small (7–12 mm) parasites in their abomasum (Gulland 1992). On the mainland, *T. circumcincta* is extremely common, and pathogenic: symptoms of infection include poor weight gain or weight loss, loss of appetite and diarrhoea (reviewed by Holmes 1985; Symons 1985). Because the pathological consequences of parasitism by these trichostrongylids are dose-dependent (Downey *et al.* 1972; Coop *et al.* 1982), a range of control measures are commonly employed by farmers on the mainland to reduce levels of infection in domestic sheep, including treatment with anthelminthic drugs and transfer to clean pastures.

Post-mortems performed during the population crash of 1989 (Gulland 1992) found that the sheep were emaciated and had no fat reserves. The abomasal walls were reddened and lesions were present. These observations, and biochemical evidence (Gulland 1992), are consistent with the animals having died from protein-energy malnutrition, accentuated by damage caused by trichostrongylids (Cheyne *et al.* 1974; Gulland 1992). Thus, as well as being the numerically dominant parasite on St Kilda, for most of the sheep *T. circumcincta* also appears to be commonly the most pathogenic. For these reasons, the majority of the parasitological and epidemiological studies on St Kilda have centred on this parasite and *T. circumcincta* is the focus of the remainder of this chapter (for a discussion of some of the other parasites on St Kilda see Cheyne *et al.* 1974; Gulland 1992; Gulland and Fox 1992).

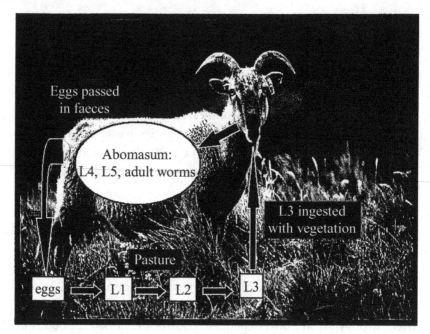

FIG. 5.1. Life cycle of *Teladorsagia circumcincta*. L1–L5 refer to first to fifth instar larvae. See text for further details.

LIFE-CYCLE OF *TELADORSAGIA CIRCUMCINCTA*

Teladorsagia circumcincta is a directly-transmitted, gastrointestinal trichostrongylid and its life cycle (Fig. 5.1) is typical of this family as a whole (for reviews see Dunn 1978; Soulsby 1982; Gulland 1991). Adult females produce eggs, which develop to the morula stage before being voided in the faeces. Hatching may take place in as little as 24 hours, but eggs can survive on the pasture for several months prior to hatching, depending on environmental conditions. The emerging first-stage larvae (L1) moult in to the second stage (L2), which undergo a second moult to become the infective third-stage larvae (L3). The cuticle of the L2 is retained as a loose sheath around the L3 and provides it with some protection against adverse climatic conditions.

Following ingestion by the host, the L3 larvae exsheathe in the rumen and pass to the abomasum, where they enter the gastric glands on the second or third day. In the glands, the larvae undergo a third moult, to become early fourth-stage larvae (EL4) followed by a fourth moult to the fifth stage, or immature adult. By day 12, most of the larvae have reached the mature-adult stage and, by day 16, they begin to emerge from the gastric glands as adult worms and attach themselves to the walls of the abomasum. These worms become sexually mature, copulate and the females lay eggs, which first appear in the faeces 17 or 18 days post-infection. The average lifespan of an adult worm is probably around 50 days, but mortality of adult worms may occur from 16 days post-infection. Although development within the host normally takes about three weeks, larvae may become arrested at the EL4 stage. This process, referred to as *hypobiosis*, occurs within the mucosa and may last for up to three months (Armour *et al.* 1969). The mechanisms determining whether, and for how long, a larva undergoes arrested development are not yet fully understood, and from an epidemiological point of view this is currently the largest gap in our knowledge (see below). However, genetic, climatic, density-dependent and immunological effects have all been implicated (Michel 1974; Gibbs 1986). In mainland Scottish sheep infected with *T. circumcincta*, the probability of an EL4 arresting development increases from low levels in late summer (August–September) to nearly 100% in late winter (December–January). De-arrestment occurs in early spring (April–May), coinciding with lambing (Reid and Armour 1972).

MEASURES OF PARASITISM

Since the sheep on St Kilda are protected, it is not possible to assess their worm burdens directly by sacrifice and post-mortem examination, and indirect measures have to be used, such as faecal egg counts (FEC). This involves collecting faecal samples from free-ranging, ear-tagged sheep and determining the density of parasite eggs per gram (wet weight) of faeces using a modification of the McMaster technique (Ministry of Agriculture Fisheries and Food 1971). Eggs are then classified in to several taxa: *Moniezia expansa*, *Capillaria longipes*, *Trichuris ovis*, *Nematodirus* spp. and strongyles (comprising *Teladorsagia* spp., *Trichostrongylus* spp., *Chabertia ovina*, *Bunostomum trigonocephalum*,

Strongyloides papillosus). In the remainder of this chapter, the term faecal egg count will be used to refer exclusively to counts of strongyle eggs. Counts of *Dictyocaulus filaria* larvae are made using similar methods (see Gulland and Fox 1992), but these will not be discussed in detail here. In addition to measuring parasitism rates, the densities of infective strongyle larvae on the pasture per kilogram of grass (wet weight) are also measured, using standard techniques (Gulland and Fox 1992).

Parasitism rates are frequently described in terms of *prevalence* (proportion of animals infected) and *intensity* (mean parasite burden per animal). On St Kilda, the prevalence of strongyle infection is high and shows relatively little systematic variation (Gulland and Fox 1992). Therefore, in this chapter we focus mainly on variation in parasite intensity, as measured by strongyle faecal egg counts (see Wilson *et al.* (2002) for a discussion of the biases associated with indirect measures of parasitism). A key assumption of these analyses is that FEC and strongyle worm burdens are positively correlated. However, because FEC is a function of both worm burden and worm fecundity, this assumption will not be true if there is strong non-linear density-dependence in worm fecundity (Cabaret *et al.* 1998). Fortunately, several lines of evidence suggest that FEC and worm burden are strongly correlated in Soay sheep. First, Gulland (1992) found that there was no density-dependence in the fecundity of female worms in sheep dying during the 1989 crash. Second, in Soay sheep dying naturally on St Kilda during the population crash of 1992, there was a strong positive linear relationship between FEC (measured several days prior to death) and the absolute number of *Teladorsagia* worms harboured in the abomasum (measured within 24 hours following death) (closed squares in Fig. 5.2) (Grenfell *et al.* 1995). Third, a similar relationship was observed in a feral population of Soay sheep culled on the island of Lundy in 1995 (open squares in Fig. 5.2) (Boyd 1999). Most importantly, despite spanning worm burdens that differed by nearly 100-fold, there was no significant difference between the two populations in the slopes or intercepts of the linear regressions describing the relationship between log-FEC and log-worm burden (Fig. 5.2). Thus, faecal egg count appears to be a useful measure of parasitism in this population.

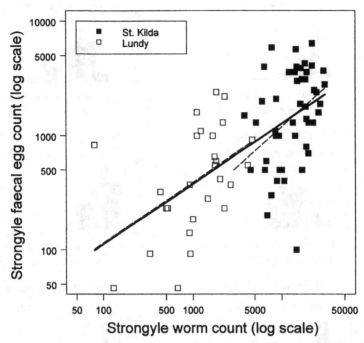

FIG. 5.2. Relationship between faecal egg count and strongyle worm burden. Filled symbols are for Soay sheep dying naturally on St Kilda (K. Wilson, unpublished data) and open symbols are for Soay sheep culled on Lundy (Braisher 1999; Boyd 1999). Strongyle faecal egg counts are number of strongyle eggs per gram of faeces. For the Lundy sheep, faecal egg counts were taken immediately prior to culling, whereas those for the St Kilda sheep were recorded approximately two weeks prior to death (K. Wilson, unpublished data). Worm counts are numbers of *Teladorsagia* adults in the abomasum. Lines represent least-squares fits of linear models; dashed lines are for the two separate data sets and the solid line is for the combined data set. The regressions for both St Kilda ($r^2 = 0.147$, $F_{1,46} = 7.899$, $p = 0.007$) and Lundy ($r^2 = 0.216$, $F_{1,25} = 6.91$, $p = 0.015$) were statistically significant. Neither the slopes ($t = 0.531$, $df = 73$, $p > 0.5$) nor intercepts ($t = 0.608$, $df = 73$, $p > 0.5$) differed significantly between studies. Overall regression: $\log_{10}$(faecal egg count) $= 0.992 + 0.528^*\log_{10}$(worm count), $r^2 = 0.392$, $F_{1,73} = 47.05$, $p < 0.0001$.

5.3 Variation in parasitism rates

QUANTIFYING VARIATION

Parasite distributions are usually highly aggregated, with some hosts having lots of parasites and most having just a few (Fig. 5.3) (Wilson

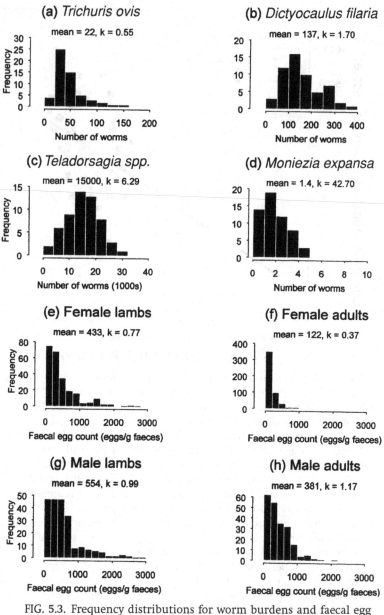

FIG. 5.3. Frequency distributions for worm burdens and faecal egg counts. (a)–(d): Data are post-mortem worm burdens for four taxa of parasites found in Soay juveniles that died on St Kilda during the winter of 1991–2: *Trichuris ovis* from the large intestine, *Dictyocaulus filaria* from the lungs, *Teladorsagia* spp. from the abomasum and *Moniezia expansa* from the small intestine. (e)–(h): Data are August faecal egg counts for four sex–age classes collected during the period 1988–93. For each distribution, its mean and exponent k (of the negative binomial) are shown. See Wilson *et al.* (1996) for details.

et al. 2002). Patterns like these are often best described by the negative binomial distribution (Fisher 1941; Pennycuick 1971), which is defined by its mean and an exponent, k, which determines the degree of skew in the data. When k is small, the distribution is highly aggregated with a long tail (e.g. Fig. 5.3a); whereas when k is large (>20), the distribution is often bell-shaped and approximates the Poisson distribution (e.g. Fig. 5.3d) – at this point, the parasites are randomly distributed across hosts.

The exponent k is a useful parameter to quantify: statisticians need to measure k because this determines the degree of skew in the parasite distribution that must be controlled for when performing statistical analyses (Wilson *et al.* 1996; Wilson and Grenfell 1997); while, for parasitologists, k is important because it identifies heterogeneities in parasitism and can lend some insights into the relative importance of different epidemiological processes, such as parasite-induced mortality and acquired immunity (Anderson and Gordon 1982; Pacala and Dobson 1988; Grenfell *et al.* 1995; Wilson *et al.* 2002). The shape of the parasite distribution is also important for evolutionary biologists and population dynamicists (Anderson and May 1978; May and Anderson 1978; Poulin and Vickery 1993): if parasites affect the health and survival only of those individuals with high parasite burdens (i.e. those in the tail of the parasite distribution), then the proportion of highly parasitised hosts will be determined not only by the *mean* of the parasite distribution, but also by its *shape*. Thus, if the parasites are randomly distributed across hosts (and follow the Poisson distribution), then a relatively larger proportion of individuals will be in the susceptible tail of the distribution, relative to when this distribution is highly skewed. In the former case, therefore, parasites are likely to be relatively more important as both a selection pressure (Poulin and Vickery 1993) and a regulatory influence (Anderson and May 1978; May and Anderson 1978). Knowing the shape of the parasite distribution is therefore important from both a statistical point of view and in terms of understanding the parasite–host interaction. However, only rarely have changes in the shape of the parasite distribution been tracked through time or across cohorts for a naturally infected host population (Grenfell *et al.* 1995).

In Soay sheep, the adult parasite distributions show variable degrees of aggregation (Gulland 1992; Wilson *et al.* 1996): *Trichuris ovis* is highly aggregated ($k = 0.55$), whereas *Dictyocaulis filaria* ($k = 1.70$) and *Teladorsagia* spp. ($k = 6.29$) are less so, and the tapeworm *Moniezia expansa* exhibits little aggregation ($k = 42.70$) and conforms to the Poisson distribution (Fig. 5.3a–d). In both Soay sheep (Gulland 1992) and domestic sheep (Hong *et al.* 1987; Grenfell *et al.* 1995), adult parasites tend to become increasingly aggregated with host age. These trends are probably due to the effects of genetic and developmental heterogeneities on host immunity (Grenfell *et al.* 1995).

Faecal eggs counts also conform to the negative binomial distribution (Fig. 5.3e–f) and k values on St Kilda generally range between about 0 and 2.5 (Grenfell *et al.* 1995). These values are much smaller than equivalent values for adult *Teladorsagia* worms (cf. Figs. 5.3c and e), but this partly reflects the much lower means for FEC than worm counts. The distribution of FEC tends to become increasingly aggregated with age in females (cf. Figs. 5.3e and f), but not in males (cf. Figs. 5.3g and h), indicating that different density-dependent processes are acting on the two sexes (see below). It is interesting that in Soay sheep and domestic sheep, parasite distributions tend to become increasingly aggregated as their means decline (i.e. the k of the distribution is positively correlated with its mean); Grenfell *et al.* (1995) discuss possible interpretations of this trend.

Because parasite distributions are aggregated, average parasite loads are often expressed in terms of their geometric means (back-transformed mean of the logged data) or are displayed on log-transformed axes. Such conventions are generally followed here.

TEMPORAL AND SPATIAL VARIATION IN EXPOSURE TO PARASITES

The density of infective (L3) larvae on the pasture shows two seasonal peaks (solid line in Fig. 5.4a). One peak in L3 counts occurs in the late spring (May–June), mainly due to the development of eggs deposited on the pasture by immunocompromised peri-parturient ewes (see below). A second peak in L3 density occurs in midsummer (around August), due to the development of eggs shed onto the pasture in the previous months predominantly by immunologically naive lambs

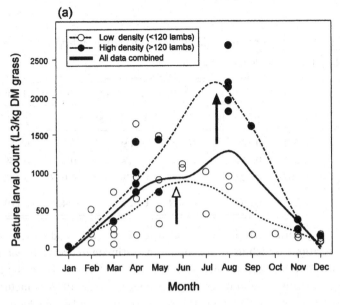

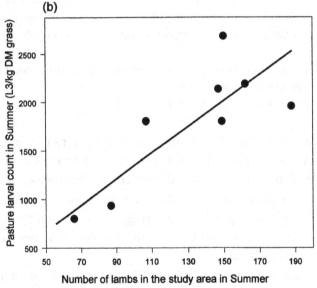

FIG. 5.4. Relationship between number of lambs using the study area and number of infective L3 larvae on the pasture (a) within years and (b) across years. In (a), data cover the period 1987–97 and are split into high lambing years (>120 lambs) and low lambing years (<120 lambs); the peak in L3 density is later in high density years than in low density years (lines are Loess fits and the peak densities are indicated by the arrows). In (b), data are for August in the years 1988–97 inclusive (excluding 1994 and 1995, for which there are no data). The linear regression equation is: Density of L3 on pasture in summer = $163.4 + 12.3 \times$ number of lambs in the study area in summer, $r^2 = 0.64$, $F_{1,6} = 10.61$, one-tailed $p = 0.009$.

Table 5.2. *Pearson's correlations between the density of infective L3 larvae on the pasture and the density of sheep*

Population estimate[a]	Spring L3 count		Summer L3 count		Autumn L3 count	
	r	$p^b(n=9)$	r^c	$p^{b,c}(n=8)$	r	$p^b(n=9)$
Lambs	0.391	0.150	**0.800**	**0.009**	0.571	0.054
Ewes	0.039	0.461	0.475	0.117	0.249	0.259
Rams	0.162	0.338	0.256	0.271	0.017	0.483
Total	0.214	0.290	**0.689**	**0.029**	0.439	0.119

[a]Estimated population size in same year as L3 count was made, except for spring correlations, which used the population estimate in previous year. For spring (April–May) and autumn (October–December), L3 counts cover the period 1989–97 inclusive. For summer (August), counts are for 1988–97 inclusive (excluding 1994 and 1995 for which there are no data). Lambs are all animals less than twelve months old; ewes are all females more than twelve months old; rams are all males more than twelve months old; total = lambs + ewes + rams.
[b]One-tailed *p* values, based on the prediction that L3 counts would be positively correlated with sheep density.
[c]Numbers in bold are significant at the 5% level.

(pasture counts decline in late summer partly due to the immunity acquired by these lambs). A closer look at Fig. 5.4a reveals that the magnitude of this second peak is dependent on the number of lambs in the population: in years when there is a small crop of lambs, there is generally a single L3 peak in late spring, whereas when the density of lambs is high, the second L3 peak in midsummer is revealed (cf. dotted lines in Fig. 5.4a).

Across years, the density of L3 in Village Bay in August increases in a linear fashion with the current size of the lamb population, but not with the density of adult males or females (Table 5.2; see also Fig. 5.4b). In both spring and autumn, the density of infective larvae on the pasture is again most strongly correlated with the number of lambs feeding on the pasture, but neither of these correlations is statistically significant (Table 5.2). Thus, although 'stocking density' influences the number of infective larvae on the pasture at some times of year (e.g. summer), as found for domestic sheep (Cameron and Gibbs 1966; Downey and Conway 1968; Thamsborg *et al.* 1996), other influences such as climate (temperature and rainfall) appear to predominate for much of the year (see also Ollerenshaw and Smith 1969; Paton *et al.* 1984; Besier and Dunsmore 1993).

The sheep on St Kilda segregate into hefts or social groups, which differ in their frequency of specific genotypes and in their survival and reproductive rates (see Section 2.5). Parasites could be partly responsible for generating these heterogeneities and so here we address spatial variation in the parasite distribution, both within and outside the host. There is consistent spatial variation in the density of infective larvae on the pasture both within (Fig. 5.5a) and between (Fig. 5.5b) years. Within years, the repeatability, r ($\pm$ approximate standard error; see Lessells and Boag 1986) is 0.30 ± 0.15 ($p = 0.018$) and across years, it is 0.74 ± 0.15 ($p = 0.007$). So, within years approximately 30% of the variation in the density of infective larvae on the pasture is due to variation between areas, whereas across years the explained variation increases to more than 70%. This spatial variation is partly maintained by variation in the density of sheep in the different areas (as indicated by the sizes of the different symbols in Fig. 5.5), but it is also influenced by variation in local topography and microclimate (e.g. some areas are particularly prone to flooding following heavy rainfall) (Suryahadiselim and Gruner 1985; Gulland and Fox 1992).

Since the density of infective larvae varies between areas, sheep feeding in these different areas face different parasitological threats, and we would expect this to be reflected in their faecal egg counts. Current analyses suggest that although there is consistent spatial variation in FEC for both males and females, this is only partially explained by variation in the local density of infective larvae (Fig. 5.5c and d). Other factors likely to influence spatial variation in parasitism rates include variation in the quality of the forage available in the different areas, stocking densities, and spatial differences in the 'quality' of the sheep themselves (i.e. genotype, body condition, previous parasitological history, etc.).

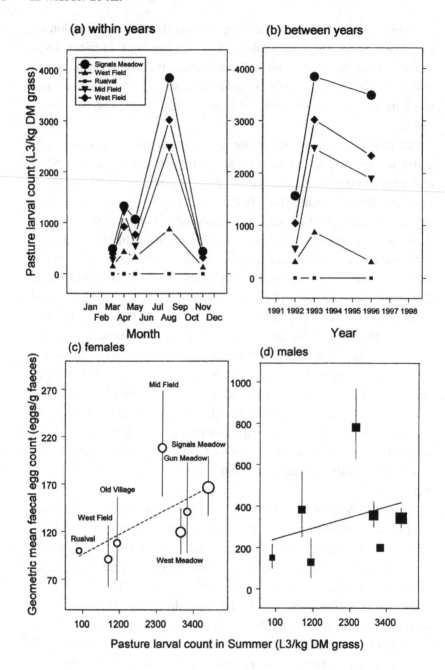

(a) within years

(b) between years

(c) females

(d) males

EARLY DEVELOPMENT OF PARASITISM

A recurrent theme throughout the remainder of this chapter is the striking difference between the two sexes, both in their resistance to parasite accumulation and in the impact of parasites on their survival and reproduction. As this next section shows, these sex differences become apparent early on in life.

The eggs of strongyle parasites first appear in the faeces of lambs when they are about 45 days old (Gulland and Fox 1992; Boyd 1999) (Fig. 5.6). Since the *prepatent period* (i.e. the time from infection until parasite eggs first appear in the faeces) for *T. circumcincta* is around 17 days, this means that infective larvae are first acquired when the lambs are less than one month old. Parasite loads (and FEC) gradually increase over the next few months, plateauing in midsummer (August–September). In most years, FEC then declines through the autumn and early winter, but in high-density years (like the one illustrated in Fig. 5.6) FEC may remain high thoughout this period (see also Gulland and Fox 1992). In lambs of both sexes, FEC rises again towards the end of winter when forage is in short supply, and again this trend is particularly marked in high-density years (see below). A sex difference in FEC first appears when animals are just ten weeks

---

FIG. 5.5. Spatial variation in parasitism. Spatial variation in the number of infective L3 larvae on the pasture is shown (a) within years and (b) between years; spatial variation in faecal egg counts is shown for (c) females and (d) males. In (a) and (b), the lines join L3 counts for one of five different areas within Village Bay. In (a), all of the data are for 1993; in (b), the data are for August L3 counts in 1992, 1993 and 1996 (in all other years, L3 counts were not separated into different areas). In (c) and (d), the symbols and bars are the geometric mean ± standard error faecal egg counts for animals occupying one of seven different areas within Village Bay in Summer 1993 (the five areas illustrated in (a) and (b), plus two others). Note that the y-axis scales differ in (c) and (d). In all four figures, symbol size reflects average sheep density in the different areas. These data show that there is consistent spatial variation in the density of L3 on the pasture both within and between years and although there is spatial variation in faecal egg counts, this is only partly explained by the density of infective L3 larvae on the pasture (neither regression line is statistically significant; $p > 0.1$).

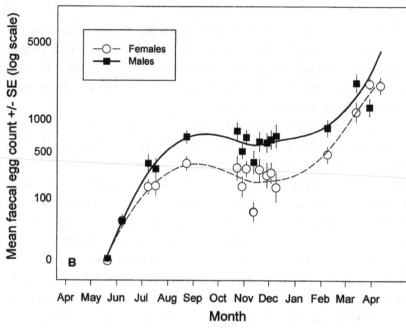

FIG. 5.6. Early development of parasitism in Soay sheep on St Kilda. The data are serial faecal egg counts collected from a single cohort of animals born in April 1994 (Boyd 1999). Females are the open symbols and males the closed symbols. Fitted lines are smoothing splines. B = mean birth date.

old (Fig. 5.6); at this time, the average FEC for ram lambs is more than 60% higher than that for ewe lambs (geometric means ± approx. standard errors: 487 ± 89 (males) versus 293 ± 51 (females); $t = 2.03$, df = 81, $p = 0.046$). By August this difference has increased (828 ± 106 versus 486 ± 72), and by the beginning of October, male FEC is more than double that of females (929 ± 152 versus 443 ± 114; $t = 2.44$, df = 39, $p = 0.019$).

The higher FEC of ram lambs could occur because parasites have a higher rate of establishment in males than females, or because established worms survive longer, or because parasites grow larger and hence females are more fecund. Our data do not allow us to distinguish between these possibilities, but experimental infections on the mainland suggest that a similar sex difference observed in domestic sheep is primarily due to a sex difference in the establishment rates of worms (Dobson 1964; Knight *et al.* 1972), especially for sheep on low-protein diets (Bawden 1969). The effects of host sex on parasite survival and fecundity are less clear (Dobson 1964; Frayha *et al.* 1971; Molan and James 1984).

The importance of sex differences in parasite establishment rates is indicated by the results of an experiment we conducted on St Kilda in August 1995 (K. Wilson *et al.* unpublished data). In this experiment, one group of lambs was given a short-acting anthelminthic 'drench' to remove its parasites, whereas a control group was given a placebo. The parasitism rates of the two groups of lambs were then monitored during the following autumn to determine how quickly the lambs re-acquired parasites. We found that by the first week of October, when faecal egg counts were first taken, the FEC of ram lambs in the treated group were not significantly different from those of animals in the control group (drenched: 293 ± 88, control: 536 ± 120; $t = 1.631$, df $= 26$, $p = 0.115$) and prevalences were approximately the same (Fig. 5.7). In contrast, the FEC of treated ewe lambs were substantially lower in early October than those of control animals (drenched: 60 ± 29, control: 392 ± 90; $t = 4.393$, df $= 30$, $p < 0.001$), and were not comparable until early December (week 9). A closer look at these data indicated that although the FEC of *infected* females were similar in both the drenched and control animals throughout the period October–December (data not shown), the prevalences were markedly different until December (weeks 9 and 10) (Fig. 5.7). This observation strongly suggests that it is the rate of establishment of new infections that differs most between the two sexes at this age, rather than the survival rates or fecundities of worms in existing infections. This result is consistent with those from a study conducted by Gulland (1991) using five-month-old captive Soay and blackface sheep which showed that, for a given larval intake rate, the percentage of larvae establishing was significantly higher for males than females ($F_{1,19}, = 5.40$, $p < 0.05$). This result was independent of breed, though Soays had proportionately lower establishment rates than blackface sheep ($F_{1,19}, = 6.17$, $p < 0.05$), suggesting that they may be genetically more resistant to *T. circumcincta* than blackface sheep.

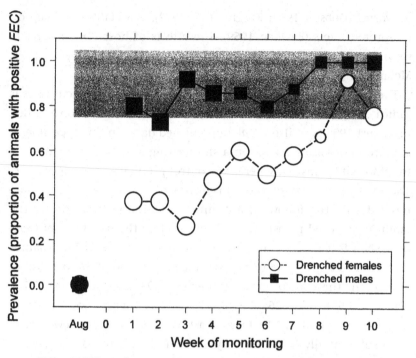

FIG. 5.7. Effect of sex on the rate of parasite establishment following drenching. Lambs were given an anthelminthic drench in August when they were approximately five months old and the prevalence of infection in both sexes was 100%. Drenching removed all parasites and faecal egg counts of all treated lambs were zero immediately following treatment. By week 1 of monitoring (first week of October), more than 80% of ram lambs had positive faecal egg counts, compared with just 40% of ewe lambs; by week 10 (second week of December), both sexes had prevalences of 80–100%. During the sampling period (October–December), the average prevalence of infection in control animals was 80–100% (indicated by the shading in the figure). Symbol size is positively correlated with sample size.

SEASONAL PATTERNS OF PARASITISM

Seasonal variation in FEC is determined by temporal trends in both the number of infective larvae available for ingestion on the pasture (as discussed above) and the immunological status of the sheep (determined by the sheep's age, nutritional plane and levels of immunologically depressive hormones). Both of these factors (as well as attributes of the parasites themselves) interact to determine temporal changes in parasite arrestment, development, mortality and

fecundity. Whilst we know little at present about how the immunological status of Soays varies through the year, we can measure seasonal variation in FEC and compare this with the variation observed
in domestic sheep, where the immunology is better understood.

We begin by comparing the seasonal patterns observed for ewes
on St Kilda with those seen in female Scottish hill sheep on the
mainland (Morgan *et al.* 1950, 1951) (we restrict the comparison to
females, because there is little comparable data available for males
on the mainland). As Fig. 5.8 illustrates, the temporal trends in the
two populations are broadly similar (Morgan *et al.* 1951; Gulland and

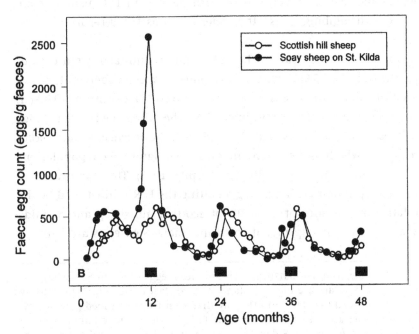

FIG. 5.8. Temporal variation in faecal egg counts in female Soay sheep
on St Kilda (closed symbols) and hill sheep on the mainland Scotland
(open symbols). The data are serial egg counts collected from animals
over a single year and joined together to indicate how the pattern
changes over the first four years of life (beginning at time B). The Soay
data are for animals sampled between August 1993 and April 1994
(except the first four points which are from ewe lambs May–July 1995).
The hill sheep data are from Morgan *et al.* (1950). For both
populations, lambing occurs in April (indicated by the solid boxes).
Soay ewes often lamb for the first time when they are twelve months
old, whereas Scottish hill sheep do not lamb until they are
twenty-four months old. Note the arithmetic scale.

Fox 1992). In both populations, lambs first acquire strongyle parasites within one or two months of birth and parasite loads continue to increase during the first five to six months of life. On St Kilda, faecal egg counts first peak in August, when the sheep are about five months old, and they generally remain high until late autumn or early winter (Gulland and Fox 1992; Boyd 1999) (see also Fig. 5.6). On the mainland, a similar pattern is observed but the peak in FEC is delayed by one to two months. In both populations, from the second calendar year onwards, temporal variation in ewe parasitism is dominated by a characteristic increase in FEC during April–June (Fig. 5.8). Because this spring rise in FEC usually coincides with lambing, it is often referred to as the *peri-parturient rise* (or PPR).

On St Kilda, the peak of the PPR occurs within about ten days of parturition (Fig. 5.9a), whereas in domestic sheep it generally peaks around two to four weeks later (e.g. Crofton 1954; Brunsdon 1970) (cf. the open and closed symbols in Fig. 5.8). The difference in the timings of these two peaks may be associated with the absence of *Haemonchus contortus*, which is prevalent in most mainland sheep populations (Procter and Gibbs 1968; Blitz and Gibbs 1972). The importance of pregnancy and/or lactation in generating the PPR is illustrated by the relative magnitudes of the PPRs of Scottish hill sheep and St Kilda Soays in their first spring: at this time, many Soay ewes are lambing

---

FIG. 5.9. The peri-parturient rise in faecal egg counts in Soay sheep on St Kilda. Data are mean ± standard error faecal egg counts for ten-day periods (a) in relation to the day that each female lambed (day 0) and (b) in relation to calendar date (note log scale). Average lambing day ± SD was 20 April ± 8 days (range 31 March – 18 May), and did not vary between the age classes ($F = 0.573$, df = 2,404, $p = 0.449$). In both figures, data are for the springs of 1989–95. In (a) data are only for females that gave birth to singletons and in (b) a comparison is made of the temporal patterns observed in females that lambed that year (open symbols) and those that did not (solid symbols). Note in (a) that for females in the two youngest age classes the peak FEC is during the week of lambing; whereas for older sheep the PPR peaks during the two to three weeks following lambing. Note in (b) that for yearlings, females that fail to lamb have faecal egg counts that tend to be higher and peak sooner than females that do lamb successfully.

(a)

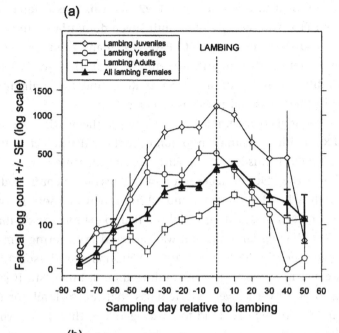

(b)

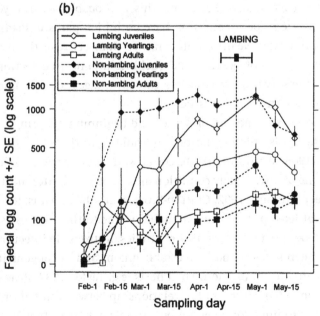

for the first time (Clutton-Brock *et al.* 1992) and this is associated with a very high PPR, whereas mainland hill sheep do not lamb until the following year (Morgan *et al.* 1950; Paver 1955) and their first PPR is considerably lower, though still evident (Fig. 5.8). Thus, variation in the fecundity schedules of sheep on St Kilda and the mainland is reflected in their seasonal patterns of parasitism.

Despite being of considerable applied interest, the proximate causes of the PPR remain unclear, even in domestic sheep (Parnell *et al.* 1954; Field *et al.* 1960; Brunsdon 1964). Early studies identified an association between the PPR and lactation and this was later confirmed experimentally: when lambs were removed from their mothers at a very young age, these non-suckling females failed to exhibit the dramatic rise in FEC following lambing that was observed in lactating females that retained their lambs (Connan 1968; Brunsdon and Vlassoff 1971). No such experiment has yet been conducted on St Kilda, but it is interesting to observe that yearling females that fail to lamb, or lose their lamb at a young age, tend to have *higher* FEC than those which raise a lamb successfully (Fig. 5.9b). This is probably because young females that are in poor condition have both high parasite loads and low conception and weaning rates, rather than because there is no parasitological cost of bearing and suckling a lamb (see section 5.4 below). This observation emphasises the advantages of experimental studies for examining these issues.

The PPR occurs at the end of winter and beginning of spring, when fresh vegetation is only just becoming available to the sheep (Procter and Gibbs 1968; Brunsdon 1970). On St Kilda, this period coincides with the time of lowest body condition and peak over-winter mortality for the sheep (Gulland 1992; Clutton-Brock *et al.* 1997a). Thus, it seems likely that at least part of the rise in FEC is due to the stresses associated with food shortage. A number of studies of strongyle infections of domestic sheep suggest that the establishment and pathogenicity of parasites is greater in malnourished hosts (e.g. Taylor 1934; Brunsdon 1962; Gordon 1964), although the relationship between nutrition and susceptibility to infection may not be a simple one (Abbott *et al.* 1985; Abbott and Holmes 1990). The importance of malnutrition, and other sex-independent mechanisms, in generating the PPR is highlighted by a comparison of the seasonal patterns in FEC during 1989 and 1990 (Gulland and Fox 1992). In spring 1989, which followed a high-density winter when nearly 60% of the sheep died due to malnutrition (Gulland 1992), both males and females exhibited a marked rise in FEC. Conversely, in spring 1990, when the winter population density was much lower, the spring rise in males was absent. Clearly, the male rise in 1989 was not due to any factors associated with lambing and must have been linked to the high population density and food shortage of that year (Gulland and Fox 1992).

In males, there is little seasonal variation in FEC except during late winter and early spring in high-density years, when FEC increases dramatically in response to food shortages (Gulland and Fox 1992).

PATTERNS IN PARASITISM ASSOCIATED WITH HOST DEMOGRAPHY

In this section, we examine variation in FEC (and prevalence) in relation to host age and population density, and how these patterns differ between the two sexes. We restrict our discussion to variation in summer FEC for which we have comparable data for both males and females.

In females, there is a striking decline in August FEC with age (Fig. 5.10a), and statistical models distinguish between four distinct age classes with successively lower faecal egg counts: lambs (four months old), yearlings (sixteen months), two-year olds (twenty-eight months) and older animals (over forty months). When the female population is characterised in the same way as in Coulson *et al.* (2001), the statistical model explains slightly less variation in FEC, but there is a similar monotonic decline with age (lambs > yearlings > prime-age females > seniles); there is no evidence for an increase in FEC in senile females that could contribute to the higher mortality of this age class. Age-related declines in the levels of parasitism are often interpreted as evidence for *acquired immunity*, i.e. resistance due to the development of protective immunity in response to the accumulation of exposure to parasite antigens. However, interpretation of such trends must be guarded because there are several alternative explanations for them (Anderson 1993; Wilson *et al.* 2002). For example, a decline in FEC with age might simply be due to the fact that as

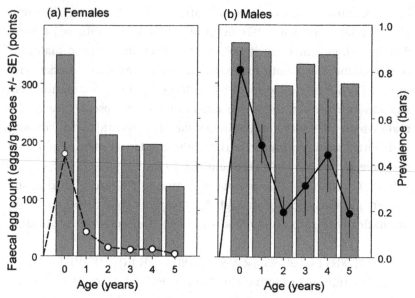

FIG. 5.10. Sex differences in faecal egg counts in Soay sheep on St Kilda. The symbols joined by lines are age-specific geometric mean ± SE summer faecal egg counts for (a) females and (b) males. The bars show the mean prevalence (proportion infected). Female faecal egg counts decline during each year of life for at least the first four years; male egg counts decline over the first year or two, but not subsequently. Females have lower prevalences than males, except when they are lambs.

animals age they get bigger, produce more faeces and their parasite eggs become 'diluted'. Alternatively, it might be that younger individuals are exposed to higher levels of infective larvae because of where, or how, they forage. A third possibility is that animals with high FEC are more likely to die from their parasites and so generate a decline in parasitism with age as the heavily infected individuals are selected out of the population. Finally, younger animals may have higher innate susceptibilities to infection regardless of their previous exposure. We have reviewed these competing hypotheses and conclude that although innate responses are likely to be important for Soay sheep during their first year of life, the subsequent decline in female FEC with age is primarily due to acquired immunity. Experimental infections with domestic sheep further support this assertion: protective immunity begins to develop when lambs are three to six months old (depending on genetic background) (Windon *et al.* 1980; Stear *et al.* 1996) and, subsequently, immunity increases with the animal's experience of infection.

In males, prevalence remains high throughout life, and any decline in August FEC with age generally extends only to the yearling stage (Fig. 5.10b); the faecal egg counts of animals aged one year and older are statistically indistinguishable from each other (unpublished analysis). The difference in FEC between males in the two youngest age classes is probably due to the fact that FEC is first measured when the animals are just four months old, before any protective immunity has developed; the data suggest that acquired immunity does not increase much beyond the yearling stage. The pattern for male Soays is similar to that observed in lungworm-infected bighorn ewes in North America, where lambs and adults differ in their faecal larval counts but age has no effect on the faecal larval counts of adult females (Festa-Bianchet 1991b).

Our analyses show that there is considerable year-to-year variation in parasitism rates in both sexes, much of which can be explained by adult population density (Fig. 5.11). This contrasts with the situation observed in bighorn sheep, where faecal larval counts of lungworm showed little variation from one year to the next, over an eight-year period (Festa-Bianchet 1991b). For female Soays, the density of infective larvae on the pasture explains the remainder of the variation in August FEC, whereas for males L3 density appears to have little impact of August faecal egg count. Thus, in females especially, year-to-year differences in parasitism can be attributed to the combined effects of sheep and infective larval density. This suggests that FEC is a function of the population's age-structure: the number of lambs in the population sets the upper limit to the density of L3 on the pasture, and the number of adult sheep determines food availability and immunocompetence (because adults consume most of the vegetation).

Despite the considerable variation in faecal egg counts between years, individual animals show consistent relative levels of parasitism. Across years, female faecal egg counts are highly repeatable: overall,

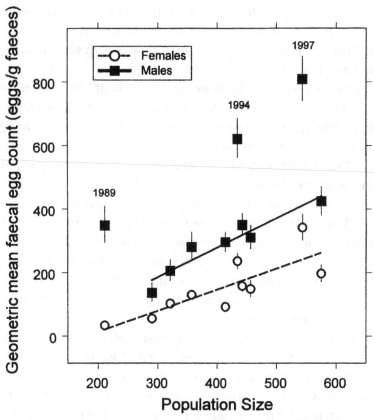

FIG. 5.11. The relationship between population density and faecal egg count in Soay sheep on St Kilda. The symbols and bars are mean ± SE faecal egg counts for all females (open circles) and males (solid squares). Overall, there is a significant positive relationship between population density and faecal egg count for females ($y = -126.87 + 0.6953x$, $r^2 = 0.631$, $F_{1,8} = 13.72$, $p = 0.006$), but not for males ($y = -133.66 + 1.3093x$, $r^2 = 0.369$, $F_{1,8} = 4.67$, $p = 0.062$), probably due to unusually high egg counts for males in 1989, 1994 and 1997. The regression for females is shown by the dashed line; the solid line is the regression for males, excluding 1989, 1994 and 1997.

the mean repeatability (± SE) is $r = 0.45 \pm 0.04$ ($p < 0.001$) and, after accounting for the decline in FEC with age and year-to-year variation in average FEC, the mean repeatability increases to $r = 0.58 \pm 0.03$ ($p < 0.001$). Thus, across years, approximately 45% of the variation in female FEC is due to differences between individual ewes, and once age and sampling year effects have been accounted for, the amount of variation explained increases to 58%. This compares with a value of $r = 0.25$ ($n = 238$ female–years, $p < 0.001$) for the repeatability of faecal lungworm counts from bighorn ewes (Festa-Bianchet 1991b). The repeatability of FEC for male Soays is much lower overall ($r = 0.14 \pm 0.09$; $p = 0.074$) but, after accounting for the decline in FEC with age and year-to-year variation in average FEC, the mean repeatability increases significantly ($r = 0.42 \pm 0.07$; $p < 0.001$). These results are important because the repeatability of a trait sets an upper limit to its heritability (Falconer and Mackay 1996).

GENETIC VARIATION AND THE HERITABILITY OF PARASITISM

Heterogeneities in parasitism rates are important because they may influence the stability of the host–parasite interaction; as a rule, the more extensive the heterogeneities, the more stable the interaction (Anderson and May 1978, 1979; May and Anderson 1978, 1979). Genetic heterogeneities are likely to be particularly important in this respect for two reasons: first, because they may evolve through time and, second, because (via antagonistic pleiotropy) they may be associated with variation in other important traits, such as growth rates (May and Anderson 1983; Read *et al.* 1995). Here, we examine the evidence for heritable variation in parasitism on St Kilda and associated genetic correlations with other important traits. Chapter 8 discusses the evidence for variation in parasite loads associated with specific genetic markers, such as particular alleles of adenosine deaminase and variation within the Major Histocompatibility Complex.

A number of studies of domestic sheep have estimated the heritability of FEC (Table 5.3), and published estimates range between 0.13 ± 0.07 (McEwan *et al.* 1992) and 0.53 ± 0.15 (Baker *et al.* 1991). Recently, the heritability of FEC in Soay sheep on St Kilda was estimated using pedigree information and multiple trait, restricted-estimate maximum-likelihood models implemented by the program VCE (Groeneveld and Kovac 1990; Groeneveld 1995). After removing sources of fixed effect variation (due to age class, sampling year, cohort, twin status and date), heritability estimates ranged between $h^2 = 0.11 \pm 0.02$, for male FEC in the summer, and $h^2 = 0.14 \pm 0.01$,

Table 5.3. *Heritability of nematode parasitism in domestic sheep breeds*

Sheep breed	Infection type	Worm species	$h^2 \pm SE$	Reference
Romney	Natural	Mixed	0.13 ± 0.07	McEwan et al. (1992)
Romney	Natural	Haemonchus contortus	0.21 ± 0.05	Bisset et al. (1992)
Merino	Experimental	H. contortus	0.23 ± 0.03	Woolaston and Piper (1996)
Romney	Natural	Nematodirus spp.	0.25 ± 0.09	McEwan et al. (1992)
Romney	Experimental	Mixed	0.27 ± 0.07	Bisset et al. (1994)
Red Maasai	Natural	Mixed	0.33 ± 0.10	Baker et al. (1994)
Scottish Black-face	Experimental	Teladorsagia circumcincta	0.33 ± 0.15	Bishop et al. (1996)
Merino	Experimental	H. contortus	0.34 ± 0.10	Albers et al. (1987)
Romney	Natural	Mixed	0.34 ± 0.19	Watson et al. (1986)
Romney	Natural	Mixed	0.39 ± 0.13	Morris et al. (1993)
Merino	Experimental	T. colubriformis	0.41 ± 0.04	Woolaston et al. (1991)
Merino	Experimental	Mixed	0.42 ± 0.14	Cummins et al. (1991)
Merino	Experimental	H. contortus	0.49 ± 0.17	Sreter et al. (1994)
Romney	Natural	Mixed	0.53 ± 0.15	Baker et al. (1991)

Source: For details, see Smith (1996).

Table 5.4. *Heritability of nematode parasitism[a] in Soay sheep on St Kilda*

Model	n^b	Mean $h^2 \pm$ SE	p
Males (summer)	687/493	0.11 ± 0.02	<0.001
Males (autumn)	836/306	0.13 ± 0.03	<0.001
Females (summer)	1250/576	0.13 ± 0.01	<0.001
Females (spring)	2294/348	0.14 ± 0.01	<0.001

[a]Nematode parasitism was measured as $\log_e$ faecal egg count.
[b]Number of observations/individuals.
Source: Coltman *et al.* (2001a).

for female FEC in the spring (Table 5.4) (Coltman *et al.* 2001a). Thus, although all heritability estimates for parasite resistance were significantly different from zero ($p < 0.001$), they were towards the lower end of the distribution of previously published estimates (Table 5.3), and approximately 50% lower than previous estimates from the same population (mean $h^2 = 0.26$), which used parent–offspring regression and sibling analyses (Smith *et al.* 1999).

There are several reasons why the heritability estimates generated by Coltman *et al.* (2001a) might differ from those generated in other studies of sheep. First, the method of analysis used by Coltman *et al.* (2001a) was far more robust than those employed in most other studies, in that it makes better use of the pedigree data, uses larger sample sizes, and generates much smaller standard errors. Second, many of the previously published heritability estimates are likely to be inflated by maternal effects, which were significant on St Kilda and were estimated separately by Coltman *et al.* (2001a). Finally, a direct comparison between the heritability estimates from St Kilda and elsewhere may be inappropriate because the sheep used in the mainland studies were generally experiencing regular drug treatment regimes whereas, on St Kilda, anthelminthics are rarely used and so sheep are usually exposed to continual mixed infections.

Theoretical considerations would lead us to predict that traits closely related to fitness (such as parasite resistance) should be subject to strong selection and that this should deplete levels of additive genetic variation (Gustafsson 1986). In line with this expectation, the heritability of parasite resistance in Soay sheep was significantly lower than most morphometric traits (e.g. in females, the heritabilities of body weight and hindleg length were $h^2 = 0.28 \pm 0.02$ and $h^2 = 0.35 \pm 0.21$, respectively, whereas the heritability of summer FEC was $h^2 = 0.13 \pm 0.01$). However, contrary to expectation, it does not appear that this was a consequence of the depletion of additive genetic variance due to selection, because parasite resistance traits had considerable additive genetic variance when measured by the coefficient of additive genetic variance. Instead the low heritability of parasite resistance was a consequence of high residual variance (Coltman *et al.* 2001a).

So, high levels of additive genetic variance for parasite resistance are maintained in the Soay sheep population, despite strong selection (see below). This leads to the obvious question of what is the mechanism maintaining this variation? One possibility is antagonistic pleiotropy, whereby negative genetic correlations across traits, such as parasite resistance and growth rate, result in a genetic trade-off between the two traits. However, there is no evidence, at present, for a genetic trade-off maintaining additive genetic variation in parasite resistance. Instead, Coltman *et al.* (2001a) found that there were *positive* genetic correlations among six of the eight pairwise comparisons of morphometric traits and parasite resistance traits (Table 5.5), indicating that selection on morphometric traits indirectly reinforces selection in favour of parasite resistance. This indicates that growth and parasite resistance are not traded off, but rather that genetically resistant individuals experience better growth. Other potential explanations for the maintenance of additive genetic variation in parasite resistance in this population are discussed by Coltman *et al.* (2001a) and in Chapter 8.

5.4 Costs of parasitism

Many studies have demonstrated a significant negative impact of parasites on host survival, reproduction and growth (see reviews by Grenfell and Gulland 1995; Gulland 1995). In Soay sheep, there is a consistent negative correlation between over-winter survival and faecal egg count (Illius *et al.* 1995; Coltman *et al.* 1999b; Milner *et al.* 1999b) and a number of experimental studies have shown that when parasites

Table 5.5. *Genetic correlations between nematode parasitism and morphometric traits*

Traits[a]	Hindleg length	Weight	FEC (summer)	FEC (spring/autumn)[b]
Hindleg length	–	$+0.78 \pm 0.05$ $p < 0.001$	-0.23 ± 0.08 $p < 0.01$	-0.31 ± 0.13 $p < 0.05$
Weight	$+0.80 \pm 0.02$ $p < 0.001$	–	-0.30 ± 0.25 ns	-0.39 ± 0.19 $p < 0.05$
FEC(summer)	-0.26 ± 0.02 $p < 0.001$	-0.05 ± 0.04 ns	–	$+0.71 \pm 0.09$ $p < 0.001$
FEC (spring/ autumn)	-0.22 ± 0.04 $p < 0.001$	-0.14 ± 0.04 $p < 0.001$	$+0.28 \pm 0.04$ $p < 0.001$	–

[a]Values above the diagonal are for males, those below the diagonal are for females.
[b]FEC spring/autumn refers to faecal egg count in spring (for females only) and autumn (for males only).
Source: For details see Coltman *et al.* (2001a).

are removed from Soay sheep survival rate is enhanced. Prior to the population crash of 1988–9, Gulland (1992) administered slow-release intra-rumenal anthelmintic boluses to 52 Soay sheep (19 male lambs, 12 female yearlings, 14 male yearlings and 7 male two-year-olds) to chemically remove their worm burdens. These boluses are designed to release 42 mg of albendazole per day for at least 100 days, but field observations suggest that their efficacy extends well beyond this time. The idea was to determine whether treated animals survived better than the controls. During the late winter of 1989, over 70% of the sheep died and there was no detectable difference in the proportional mortality of sheep in the treated (44/52–85%) and control (34/40–85%) groups. However, the daily survival rate of the treated animals was significantly higher than that of the controls (Gulland 1992).

A second bolus experiment was conducted prior to the winter of 1991–2 (Gulland *et al.* 1993), when the overall mortality rate was much lower (44%). This time, anthelminthic boluses were given to 55 sheep

(17 female lambs, 20 female yearlings and 18 male yearlings). In all three sex–age classes, the mortality rates of treated animals was lower than that of the controls and, for female lambs and male yearlings, this difference was statistically significant (Fig. 5.12a). The most interesting comparison here is between the male and female yearlings; in females there was a relatively small reduction in mortality, from 6/21 (29%) animals in the control group to 2/20 (10%) in the treated, whereas in males there was a massive reduction in the number of animals dying, from 8/17 (47%) in the controls to 0/18 (0%) in the treated. Thus, it appears that parasites are a much more important mortality factor for males than females, and parasites might explain at least part of the male-biased mortality observed in adult Soays (Chapter 3 and section 5.5; see also Wilson *et al.* 2002).

The impact of parasites on fecundity is less clear. Although the probability of females lambing in spring tends to decrease as their FEC in the previous August increases (logistic regression analysis: yearlings: $b = -1.081$, $\chi_1^2 = 7.30$, $p = 0.007$; two-year-olds: $b = -0.441$, $\chi_1^2 = 0.54$, $p = 0.462$; adults: $b = -0.777$, $\chi_1^2 = 4.33$, $p = 0.037$), it appears that the main impact of parasites on fecundity is probably via their effect on female survival, since parasites disappear as an important

---

FIG. 5.12. Experimental analysis of the effects of parasites on fitness. The fitness components examined are (a) over-winter mortality, (b) fecundity, (c) skeletal (hindleg length) growth, (d) weight gain and (e) horn growth. The figure shows the results of an experiment in which treated animals (closed bars) were given an anthelminthic bolus to remove their gastrointestinal parasites and control animals (open bars) were matched for age and sex. Sex–age classes included in the experiment were female lambs, female yearlings and male yearlings (Gulland *et al.* 1993). Over-winter mortality is defined as the percentage of animals failing to survive until the following year; fecundity is defined as the percentage of females that survived and produced a lamb that lived to weaning age. Growth rates (means ± SE) were calculated for animals which survived for at least a year following bolusing and were captured and measured in the summers of both 1991 and 1992 (growth is defined as the difference between these two measurements). Numbers at the base of each bar are the number of animals in each group (note that in (e), only measurements for normal-horned animals are included). Significance levels: – analysis not possible, NS $p > 0.05$, *$p < 0.05$, **$p < 0.01$, ***$p < 0.001$.

correlate of fecundity when we consider only those females that survived throughout the following year (yearlings: $b = -0.756$, $\chi^2_1 = 2.36$, $p = 0.124$; two-year-olds: $b = -0.009$, $\chi^2_1 = 0.001$, $p = 0.991$; adults: $b = -0.054$, $\chi^2_1 = 0.012$, $p = 0.911$).

A better way of examining parasite-induced reductions in fecundity is to manipulate parasite burden experimentally. Whilst a large-scale experimental manipulation of female parasite loads has yet to be

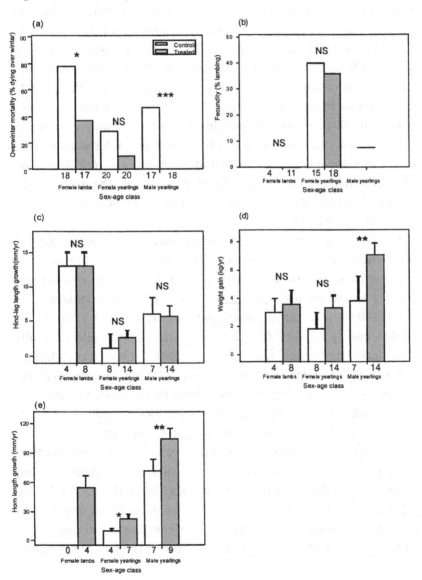

undertaken on St Kilda in a year of low mortality, we can examine the impact of parasite removal on the fecundity of those ewes that survived the population crash of 1991–2 (see above). This indicates that the fecundity of surviving yearlings was marginally enhanced by parasite removal, but not significantly so, and none of the ewes that were bolused as lambs produced offspring that survived to weaning age (Fig. 5.12b). Thus, in this year of high over-winter mortality, there was no evidence that parasites affected fecundity, except via their effects on ewe mortality. These results contrast with those from a recent study of Svalbard reindeer (*Rangifer tarandus platyrhynchus*) infected with gastrointestinal nematodes, in which it was found that the removal of parasites with anthelminthics resulted in a significant increase in fecundity of between 5% and 14% (Albon *et al.* 2002; Stien *et al.* 2002). Associated with this increased fecundity were significant increases in body mass and back-fat depth. Interestingly, however, there was no effect of anthelminthic treatment on the survival of female reindeer. It remains to be determined whether parasite-mediated reductions in fecundity occur on St Kilda in years of low over-winter mortality.

Across all age classes, there were significant negative phenotypic correlations between summer FEC and both body weight (males: $r = -0.15 \pm 0.05$, $p < 0.001$; females: $r = -0.20 \pm 0.04$, $p < 0.001$) and hindleg length (males: $r = -0.16 \pm 0.05$, $p < 0.001$; females: $r = -0.16 \pm 0.04$, $p < 0.001$); phenotypic correlations with autumn FEC (males) and spring FEC (females) were also negative and generally statistically significant (Coltman *et al.* 2001a). Genetic correlations between FEC and body weight/size were also significantly negative (Table 5.5 and above). When broken down by age class, a similar picture emerges, with strong negative correlations between August FEC and both hindleg length and body weight, particularly in young animals (Table 5.6). However, the empirical evidence that parasites affect sheep growth rates is poor, with only the correlation between FEC and male lamb hindleg growth proving statistically significant (Table 5.6).

Experimental evidence for an impact of parasites on growth and development comes from a comparison of the growth patterns of control and treated animals in the year following the bolus experiment

Table 5.6. *Correlations between log faecal egg count and (a) absolute size and (b) growth rate of hindleg length, body mass and horn length in female and male Soay sheep*

Measurement[a] and age-class	Females			Males		
	r	df	p	r	df	p
(a) Absolute size						
Hindleg length						
Lambs	−0.400	85	<0.001	−0.166	182	0.025
Yearlings	−0.217	41	0.161	−0.313	93	0.002
Adults	−0.111	136	0.194	−0.096	94	0.354
Body weight						
Lambs	−0.335	84	0.002	−0.163	178	0.029
Yearlings	−0.214	40	0.173	−0.321	91	0.002
Adults	−0.255	136	0.003	−0.153	94	0.135
Horn length[b]						
Lambs	−0.331	84	0.002	−0.001	178	0.989
Yearlings	0.001	41	0.999	−0.341	93	<0.001
Adults	0.035	135	0.685	−0.019	95	0.850
(b) Growth rate[c]						
Hindleg length						
Lambs	0.128	26	0.516	−0.305	43	0.041
Yearlings	0.235	23	0.258	−0.168	38	0.299
Adults	−0.032	59	0.808	−0.285	21	0.187
Body weight						
Lambs	0.176	26	0.370	−0.203	41	0.190
Yearlings	0.175	22	0.412	0.089	37	0.588
Adults	−0.203	58	0.120	−0.126	21	0.564
Horn length[b]						
Lambs	0.187	26	0.341	−0.242	43	0.108
Yearlings	−0.096	23	0.648	−0.234	38	0.146
Adults	0.133	59	0.308	−0.226	21	0.299

[a] Data cover measurements taken in the years 1985–95. Numbers in bold are significant at the 5% level.

[b] To ensure that correlations are comparable, only normal-horned individuals were included in the analysis.

[c] Growth rates are calculated as the difference between the trait size in the summer when the faecal egg count was taken and the trait size the following summer.

of 1991–2 (see above). This showed that although there was no significant effect of parasite removal on hindleg growth in any of the
sex–age categories (Fig. 5.12c), yearlings given an anthelminthic bolus
gained nearly twice as much body weight over the following year as
animals in the control groups (Fig. 5.12d). For males, this difference
was statistically significant (means ± SE: treated: 7.14 ± 0.65, control:
3.86 ± 0.67; $t = 3.148$, df = 19, one-tailed $p = 0.0026$), but for females
it was not (treated: 3.36 ± 0.84, control: 1.75 ± 0.56; $t = 1.348$, df =
20, one-tailed $p = 0.096$); female lambs that survived the winter had
similar weight gains regardless of treatment. The greater weight gains
of treated yearlings cannot be due to the effects of parasite-induced
host mortality because this is likely to truncate the lower end of the
weight distributions, and hence mask any relative weight gains in the
treated group, rather than accentuate them. The effects of parasites
on the growth of domestic sheep are well documented: reductions in
weight gain of 20–60% have been recorded in domestic sheep (Barker
1973; Sykes and Coop 1976; Sykes et al. 1977), and a reduction of
37% was reported in sheep experimentally infected with just 1500
T. circumcincta per day (Coop et al. 1985). This reduction in weight gain
is due to a combination of anorexia (which can result in a reduction
in food intake of up to 20%) and decreased utilisation of ingested
food.

Horn size is an important determinant of male mating success
in Soay sheep (Chapter 9). In normal-horned animals, the horns are
more than three times longer in adult males (mean ± SD: 392 ±
103 mm) than females (126 ± 62 mm). Both theoretical and empirical studies indicate that the expression of sexually selected traits,
such as horns, may be commonly condition-dependent, such that animals in good condition invest relatively more sexually selected traits
(Anderson 1993). As expected, we found that in all age classes, males
with high faecal egg counts tended to have shorter horns and slower
horn growth rates (Table 5.6), though only the correlation between
horn length and FEC in yearlings was statistically significant. The
1991–2 bolus experiment showed that yearling males in the treated
group grew their horns 25% longer than males in the control group
(means ± SE: treated: 106.0 ± 7.1, control: 78.1 ± 7.7; $t = 2.64$, df = 14,
one-tailed $p = 0.0096$; Fig. 5.12e). Within the treated group, horn

growth rate was significantly negatively correlated with FEC at the time of dosing ($r = -0.708$, df $= 9$, $p = 0.0147$; the equivalent correlations for hindleg growth and weight gain were non-significant, $p > 0.2$). Since parasite establishment rate in Soays is repeatable and males appear to be predisposed to high or low infection rates (K. Wilson, unpublished data), it seems likely that this result is a consequence of variation in amount of time males remained parasite-free following bolusing. An increase in horn growth was also observed in yearling females treated with anthelminthics (treated: 23.0 ± 4.8, control: 8.3 ± 1.5), though statistically this difference was less significant ($t = 2.23$, df $= 9$, one-tailed $p = 0.026$), and there was no correlation between horn growth rate and FEC at dosing ($r = -0.387$, df $= 6$, $p = 0.343$). Thus, parasites appear to restrict the opportunities for horn growth, particularly in males, but the magnitude of their effect is possibly not as great as it is for weight gain (in treated yearling males, weight gain increased by 85% whereas horn growth increased by only 25%). There are no comparable data for domestic sheep.

Another important determinant of male mating success in Soays is their ability to locate and defend oestrous ewes (Chapter 9). These activities are likely to require considerable stamina and a reduction in the amount of time allocated to maintenance activities such as feeding and resting. The effect of parasites on reproductive behaviour was determined by examining the correlation between FEC and male time-budgets during the 1996 rut. Male behaviour was classified as sexual, aggressive, feeding, moving or resting (Chapter 9). In both juveniles and older age classes, heavily parasitised males tended to spend less time engaged in sexual activity, fighting and moving, and spent more time feeding and resting. Although the overall correlations between FEC and the proportions of time engaged in sexual behaviour, feeding and aggression were all statistically significant ($|r| > 0.45$, $p < 0.002$), when time-budgets were analysed separately for each age class (so reducing sample sizes), only the correlations between sexual activity and FEC proved to be statistically significant (Fig. 5.13). Thus, it appears that males with high parasite loads spend less of their time performing sexual behaviours and more of their time feeding and resting, suggesting that parasites constrain opportunities for matings and may limit reproductive success. However, without manipulating

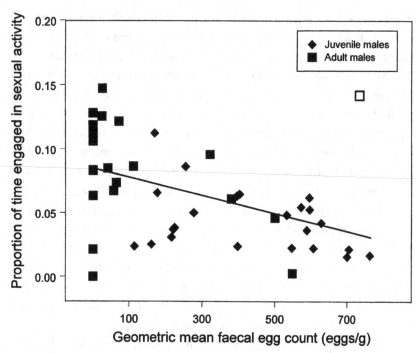

FIG. 5.13. Effects of parasites on male sexual behaviour. Data are for juvenile males (diamonds) and adult males (squares) for the period mid-October to mid-December 1996. The proportion of time engaged in sexual activity is averaged over the eight weeks of the rutting season, based on weekly averages (see Chapter 9). Geometric mean faecal egg counts were calculated over the same period. The overall correlation is statistically significant ($r = -0.458$, df $= 43$, one-tailed $p = 0.0008$), as are the correlations for juveniles alone ($r = -0.363$, df $= 21$, one-tailed $p = 0.044$) and adult males alone ($r = -0.213$, df $= 19$, one-tailed $p = 0.0231$), after excluding the outlier indicated by the open square.

parasite loads directly, the impact of parasites on sexual behaviour must remain speculative.

5.5 Discussion

ST KILDA AS A MODEL SYSTEM FOR PARASITIC INFECTIONS OF WILDLIFE

With one or two notable exceptions (e.g. Elton *et al.* 1931), the study of wildlife diseases was not really taken seriously by ecologists until the

mid to late 1970s, when theoretical and empirical studies demonstrated that parasites were not simply passive passengers in their
hosts, but had the potential to be an important force in their evolution and ecology (Anderson and May 1978, 1979; May and Anderson
1978, 1979). Since then, studies of host–parasite interactions in the
wild have slowly yielded interesting and important new insights into
how animals cope in parasite-rich environments. One of the reasons
why progress in this field has been slow is that the study of wildlife
diseases presents a number of logistical problems that are often difficult or impossible to overcome. Paramount amongst these is simply being able to quantify parasite abundance. There are two main
approaches available: either hosts are destructively sampled and the
parasite burden is counted directly (e.g. Hudson 1986; Stien *et al.* 2002)
or parasite numbers are estimated using indirect measures such as
faecal egg or larval counts (e.g. Wilson *et al.* 2002). The obvious down-
side of destructively sampling hosts is that it is not possible to collect
longitudinal data using this method, and it is precisely this kind of
information that is so useful for quantifying epidemiologically important attributes, such as parasite establishment rates and the strength
of acquired immunity. Indirect measures of parasitism can be used
to collect longitudinal data, but often their relationship to the actual parasite burden is unknown or inconsistent (e.g. lungworm larval
counts in bighorn sheep; Festa-Bianchet 1991b), making their interpretation difficult or impossible.

Fortunately, there is a strong positive relationship between faecal
egg count and worm burden in Soay sheep (Fig. 5.2) and this, combined with the fact that most of the sheep on St Kilda are individually identifiable, has allowed us to track the parasitism trajectories
of known animals virtually from the day they are born until the day
they die. These long time series are virtually unique in wildlife epidemiology and have revealed a number of important new insights. For
example, it is now clear that the characteristic sex difference in parasite burdens seen in adult sheep first appears in lambs when they are
just ten weeks old, and the difference in parasite loads between the
sexes magnifies with age as females, but not males, develop a strong
acquired immune response to their parasites (Figs. 5.6 and 5.10). These

longitudinal data have also allowed us to determine that the characteristic spring rise in faecal egg counts that occurs in pregnant ewes reaches its peak at almost exactly the time of parturition (Fig. 5.9a). Both of these examples illustrate the power of individual-based parasite data to shed light on temporal epidemiological patterns. Similar patterns probably occur in other mammal populations elsewhere, but their detection relies on a level of data resolution that is rarely achievable in other systems because of logistical constraints.

Our studies on St Kilda have revealed that parasites may impact on the weight gain, weapon growth and mortality of their hosts (Fig. 5.12). They may also impact on the stamina of rutting males (Fig. 5.13) and ultimately limit the reproductive success of both sexes. However, the density of Soay sheep on St Kilda is unusually high and their parasite burdens tend to be large. So, are the costs of parasitism exerted on Soay sheep unusually high or are similar costs observed in other vertebrate–macroparasite systems? On the available evidence, it appears that in many instances the costs of parasitism are significant (Grenfell and Gulland 1995; Gulland 1995; see below). However, most of these studies are correlational, showing that individuals with high parasite loads tend to die sooner, produce fewer young and/or grow at a slower rate. Therefore, in general, these types of study may yield biased or incomplete estimates of the true costs of parasitism, because they are potentially confounded by variation in the quality of animals – an association between high parasite loads and low fitness may be due to the detrimental effects of parasites on host fitness, but may also be due to the fact that animals in poor condition not only die sooner and reproduce less but also have increased susceptibility to parasites. Studies on St Kilda, and elsewhere (e.g. Albon et al. 2002; Stien et al. 2002), suggest that reliable estimates of the costs of parasitism may be gained only by experimentally reducing or enhancing parasite loads and observing the subsequent effects on the life-history characters of interest.

Anthelminthic dosing of sheep on St Kilda has also shown that the costs of parasitism may be revealed only under conditions of 'intermediate stress'. For example, the chemical removal of parasites had little impact on overall mortality during the 1988–9 population crash, when over 70% of the sheep died regardless of parasite burden, and it is unlikely that parasite removal would have had much impact on the level of mortality that occurred during a low-density winter like 1989–90, when less than 5% of the sheep died. Only in 1991–2, a year of intermediate mortality (44%), was parasite removal reflected in enhanced survival (Fig. 5.12). The effects of parasites on fitness may also be expressed differently in different sex–age classes. For example, parasite removal improved the survival chances of female lambs, but not female yearlings (Fig. 5.12). Thus, as with studies of the cost of reproduction, it is becoming increasingly clear that the costs of parasitism are not fixed, but are 'context-specific' (Chapter 10). It is likely that similar context-specific costs are prevalent in other host–parasite interactions.

UNDERSTANDING NEMATODE INFECTIONS OF SOAY
AND DOMESTIC SHEEP

A unique attribute of our studies of the epidemiology of nematode infections of Soay sheep is that we are able to draw parallels with comparable studies on their domestic counterparts. The seasonal pattern of parasitism in Soays on St Kilda is remarkably similar to that observed in domestic hill sheep in Scotland (Fig. 5.8). The major difference in their dynamics of infection is in the magnitude of the first spring rise in faecal egg counts, which is substantially larger in Soays than in Scottish hill sheep. However, this difference is probably explicable in terms of a difference in the life-histories of the two breeds of sheep (i.e. the precocial sexual maturity of the Soays). A similarity between the epidemiological patterns of wild and domestic sheep means that reasonable extrapolations can be made from one situation to the other. For instance, it is often extremely difficult to obtain reliable data on worm burdens from wildlife hosts because of restrictions on culling. Because there are fewer such restrictions for domestic animals, it is possible to use data on worm numbers from domestic animals to determine the likely patterns for wildlife hosts. Thus, it has been possible to construct epidemiological models for Soay sheep parameterised using FEC and worm burden data from Scottish hill sheep (Boyd 1999; B. T. Grenfell and K. Wilson, unpublished data). Whilst such models allow us to analyse the early development of parasitism in Soays with a reasonable degree of accuracy, they fail to predict seasonal variation in FEC for animals older than about six to seven months. Understanding the deficiencies of the models is clearly a priority for future research.

Epidemiological studies of domestic animals may also gain valuable insights from studying the infections of wildlife since controlled studies of domestic animals are rarely able to simulate the same degree of stress on their subjects that is often observed in wildlife populations. For example, the spring rise in FEC observed in male Soays in high-density years (Gulland and Fox 1992) strongly indicates that a major factor generating the spring rise in females is (Fig. 5.8) nutritional stress. Most studies of domestic animals tend to concentrate on either young animals (less than one year old) or well-fed adult females (breeding stock). The absence of studies of adult male domestic sheep means that an apparent 'peri-parturient rise' in males has never been recorded in them. This, combined with the understandable reluctance to impose extreme food rationing on pregnant ewes, has meant that the importance of nutrition in generating the spring rise in the FEC of domestic sheep has probably been underestimated.

Concurrence of the epidemiological patterns observed in wild and domestic sheep is not restricted to the parasite stages living within their hosts; they also extend to the free-living stage on the pasture. The Soays provide good evidence for two patterns regularly observed in studies of domestic sheep parasites: first, the biphasic pattern in L3 counts, with one peak in late spring and the other in midsummer (Fig. 5.4a); second, the positive relationship between stocking density and the magnitude of the midsummer peak (Fig. 5.4b). As observed in these other studies, the relationship between stocking density and L3 density becomes less clear at other times of year, probably due to the effects of temperature, rainfall and humidity. On St Kilda, the density of infective larvae on the pasture shows high spatial repeatability both within and across years (Fig. 5.5a). However, the relationship between L3 density and faecal egg count is not consistent (Fig. 5.5b), due to spatial variation in other factors, such as vegetation structure, body condition, genetic make-up, and so on. Untangling these

interacting factors, and their importance in determining spatial variation in sheep mortality and life-history variation, is clearly a challenge
for the future.

SEX BIASES IN PARASITISM AND MORTALITY

One of the most striking patterns to emerge from our studies on
St Kilda is the consistent difference between the sexes in their susceptibility to parasitic infection. Within six weeks of picking up their
first infection, males have significantly higher faecal egg counts than
females, and by their first summer (aged just four months old) their
FEC is more than double that of females. The faecal egg counts of
the two sexes continue to diverge throughout their lives (Fig. 5.10),
except in high-density springs when they temporarily converge (Fig.
5.6). Parasite establishment rate appears to be approximately twice as
great for male lambs as for females (Fig. 5.7) and, whereas females
develop long-lasting acquired immunity in response to nematode infection, males fail to develop any further protective immunity beyond
their second year of life (Fig. 5.10). Thus, males appear to have poorer
acquired immunity than females and higher parasite loads.

This trend, of males being more heavily parasitised than females,
is not unique to the Soay sheep of St Kilda. It is also observed in a
number of other ungulate species, including bighorn sheep infected
with lungworm (Festa-Bianchet 1991b), white-tailed deer infested with
ticks (Kollars *et al.* 1997) and red deer infested with bot flies (Bueno-
de la Fuente *et al.* 1998). Indeed, a number of recent meta-analyses
have found that across a range of host taxa, including mammals,
the two sexes often differ in their susceptibility to parasitism (Poulin
1996; Schalk and Forbes 1997; McCurdy *et al.* 1998; Moore and Wilson
2002). These analyses were prompted by the observation that male
and female sex hormones differ in their effects on the immune system; in females, oestrogens stimulate humoral immunity and inhibit
cell-mediated responses, whereas male androgens (including testosterone) depress both humoral and cell-mediated immune responses
(Grossman 1985; Schuurs and Verheul 1990). As a consequence, parasites may often fare better inside male than female hosts, possibly
generating a sex bias in parasitism rates. The two sexes also differ in

their behaviour and this might lead to variation in their exposure to the parasites' infective stages, as might differences in body mass, since large animals may offer larger targets to the infective stages of parasites or their vectors. All of the comparative analyses conducted thus far indicate that, in mammals, males tend to be more heavily parasitised than females. Moreover, the extent of the male bias in parasitism tends to be greater in polygynous than monogamous species, and is positively correlated, across species, with the degree of sexual size dimorphism (Moore and Wilson 2002). Thus, species in which males are significantly larger than females tend to exhibit male-biased parasitism, whereas the opposite trend is observed in species where females are the larger sex (Fig. 5.14a).

In humans (Owens 2002; Wilson *et al.* 2003), as well as in other mammal species (Promislow and Harvey 1991; Promislow 1992), males tend to suffer not only from higher levels of parasitism than females, but also greater mortality. A recent comparative analysis indicates that these two phenomena may be related since, across mammal species, the extent of male-biased mortality was positively correlated with the

FIG. 5.14. Relationship between sex-biased parasitism and (a) sexual size dimorphism and (b) sex-biased mortality in natural populations of mammals. Each of the closed symbols represents mean values for a mammal species, using data extracted from the literature (see Moore and Wilson (2002) for details). The data for the Soay sheep (open symbols) are based on the sex-specific mean prevalence data displayed in Fig. 5.10, sex-specific body weights from Illius *et al.* (1995), and sex-specific life-expectancy data (unpublished analysis). For each host species, sex-biased parasitism is calculated as the difference between mean prevalence of infection in males and females. Thus, positive values represent species in which males tend to be more heavily parasitised than females. Sexual size dimorphism is calculated as the logarithmically transformed ratio of mean male to mean female body mass, such that positive values represent those species in which males are typically larger than females. Sex-biased mortality is calculated at the log-transformed ratio of female life expectancy to male life expectancy. Thus, positive values represent species for which females typically outlive males. For Soay sheep, sex-biased mortality estimates range between 0.09 and 0.44, for cohorts born between 1980 and 1990. For illustrative purposes, only the minimum value is shown here. The lines are the least-squares regressions to the data published by Moore and Wilson (2002), which exclude the Soay sheep data.

extent of sex-biased parasitism (Moore and Wilson 2002) (Fig. 5.14b), even after the extent of sexual size dimorphism had been controlled for. Thus, it appears that differential susceptibility to parasitism may contribute to the widespread phenomenon of male-biased mortality in mammals (Promislow and Harvey 1991; Promislow 1992). To test

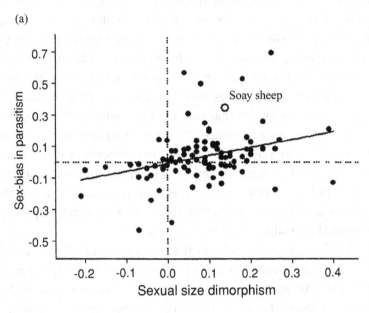

(a)

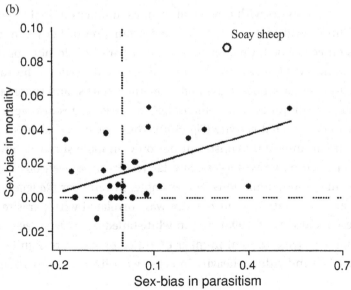

(b)

these ideas further, empirical studies of host–parasite systems are required, and St Kilda provides an ideal model system for doing this. Indeed, to date, Soay sheep on St Kilda provide the best data so far in support of the hypothesis that parasites contribute to sex-biased mortality (Moore and Wilson 2002; Wilson *et al.* 2002). Not only do male Soay sheep have consistently higher parasite loads than females, but they also appear to suffer greater parasite-induced mortality. In all age classes, over-winter mortality is significantly higher in males than females, being up to 100% higher in some years (Clutton-Brock *et al.* 1997a). The observation that this sex-biased mortality was eliminated in a group of yearling males that had their parasites chemically removed (Fig. 5.12) suggests that much of the 'additional' mortality suffered by males is a consequence of their higher parasite loads, at least within the yearling age class. Further experiments will be required before we can determine the extent to which sex-biased susceptibility to parasitism is responsible for generating male-biased mortality in older age classes, and whether this mechanism can help explain male-biased mortality in other host taxa.

Parasites impinge not only on the survival of male Soays, but also on their reproductive success. On St Kilda, where male–male competition is the primary form of sexual selection, mating success is determined by relative stature (as measured by body size and mass), horn size (used in fights with other males) and stamina (required to pursue rival males and oestrous females, and a function of body condition) (Preston *et al.* 2003; Chapter 9). Although parasites do not appear to affect the skeletal growth of either sex, they do reduce the rate at which males gain body mass and grow their horns, and these effects are greater in males than females (Fig. 5.12). Parasites also appear to reduce male sexual activity (Fig. 5.13). Thus, for males, at least, parasites are an important constraint not only on male survival, but also on male reproductive success. Similar results have been observed in other ungulate populations. For example, in reindeer (*Rangifer tarandus*), antler asymmetry but not size was correlated with parasite load (Lagesen and Folstad 1998), and in white-tailed deer (*Odocoileus virginianus*), body size and total number of antler points were significantly reduced in individuals heavily infected with liver fluke, particularly for animals in the youngest age classes (Mulvey and Aho 1993). All of these studies suggest that parasites may impinge on the reproductive success of individual males and that weapons (antlers and horns), in particular, may act as sensitive indicators of a male's ability to resist parasitic infection. As such, it has been suggested that visual attributes of weapons may be used in mate choice to indicate information about parasite burden (Markusson and Folstad 1997). To date, there is little evidence for female mate-choice in Soays (see Chapter 9).

REGULATION OF VERTEBRATE POPULATIONS BY MACROPARASITES

The mathematical models of Anderson and May (e.g. Anderson and May 1978; May and Anderson 1978) clearly demonstrate the potential of parasites to regulate their host population via density-dependent reductions in host fitness. These models show that stable regulation of the host population is particularly likely when a large proportion of the parasite population is aggregated in a small proportion of the hosts (i.e. when k is small; see section 5.3). However, if parasite aggregation is too great, then the host population will escape regulation by the parasite. Since the majority of macroparasites, including *T. circumcincta*, exhibit aggregated distributions (Shaw and Dobson 1995; Shaw *et al.* 1998), it is possible that population regulation by macroparasites is quite common (Tompkins and Begon 1999).

By far the best evidence that parasites may regulate the size of a wild vertebrate population comes from studies conducted in the north of England on the red grouse (*Lagopus lagopus scoticus*) and its parasite *Trichostrongylus tenuis* (Hudson *et al.* 1985, 1992a; Hudson 1986; Hudson and Dobson 1989, 1997; Dobson and Hudson 1992). The red grouse is a medium-sized game bird that inhabits the heather moorlands of Britain, and *T. tenuis* is a nematode parasite that lives in the caecum of its host. Unlike Soay sheep, red grouse do not produce an effective immune response against their main parasite and this appears to have a major impact on the host–parasite dynamics. Populations of red grouse exhibit cycles with periods ranging between four and eight years (Potts *et al.* 1984; Hudson *et al.* 1985; Dobson and Hudson 1992). As with the Soays, high population densities are associated with high parasitism rates and high mortality. Experimental reductions of parasite burdens suggest that *T. tenuis* not only reduces the body weight and survival of red grouse, but also has direct effects on their reproductive success (clutch size, hatching success and chick survival) (Hudson 1986; Hudson and Dobson 1989). Mathematical models indicate that it is the parasite-induced reductions in host reproduction that enhances the ability of *T. tenuis* to destabilise the red grouse population dynamics and to generate population cycles (Dobson and Hudson 1992). These models predicted that if the parasite population could be reduced to a sufficiently low level, then the population cycles would be stopped. The results of a replicated field experiment over ten years clearly showed that when cyclical grouse populations were administered with anthelminthics to chemically remove their parasites, the amplitude of population cycles was severely reduced, whilst control populations continued to cycle as before (Hudson *et al.* 1998). This strongly suggests that parasites are primarily responsible for the instability observed in these populations of red grouse, though the results remain controversial (Hudson *et al.* 1999; Lambin *et al.* 1999; Tompkins and Begon 1999).

So, do parasites have a similar effect on the population dynamics of Soay sheep and other ungulate populations? For the vast majority of ungulate populations, the role of parasites in host dynamics has not been examined. However, a recent study of the Svalbard reindeer living in the high Arctic strongly suggests that the negative impact of nematode parasites on host fecundity may be sufficient to regulate population densities around their observed levels (Albon *et al.* 2002). The evidence that parasites are necessary and sufficient for the regulation of Soay sheep dynamics is less convincing. Experimental manipulations of parasite burdens on St Kilda have clearly demonstrated that trichostrongylids reduce Soay sheep survival during population crashes, particularly in young animals and males of all ages, and there is also some evidence that parasites may reduce the fecundity of females. Thus, it seems likely that the parasites are at least contributing to the depth of the crashes, if not the frequency of their occurrence. However, unlike red grouse, which lack an effective immune response against *Trichostrongylus tenuis*, the Soay sheep acquired immune response is usually extremely effective against *Teladorsagia circumcincta* (in females, at least) and breaks down only in late winter or early spring during years of high population density. Thus, the effects of parasites on growth, survival and reproduction are much more 'focussed' in Soay sheep than in red grouse.

Population crashes on St Kilda are generated by a complex interaction between the sheep, their food supply, their parasites and climate (Grenfell *et al.* 1992, 1998; Coulson *et al.* 2001; Chapter 3). Thus, mortality episodes coincide with periods when the sheep have become malnourished, due to a reduction in: the amount of food available (a function of sheep density and weather), feeding opportunities (limited by weather and parasite-induced anorexia), and gut absorption (reduced by parasite damage). However, the only way to clearly demonstrate the nature of this interaction would be to conduct a factorial experiment in which nutrition and parasite load were independently manipulated and subsequent mortality monitored. Even then, this would not conclusively demonstrate that the parasites are necessary for the generation of population crashes. To do this, we would need to conduct a large-scale experimental reduction of parasite loads similar to that performed by Hudson *et al.* (1998). Although such a manipulation is logistically feasible, the inability to reliably predict crashes in advance (Grenfell *et al.* 1998; Coulson *et al.* 2001; Chapter 3), the lack of suitable control populations, and the inability to replicate the experiment, would severely undermine the usefulness of such a manipulation. Indeed, a large-scale experiment would be inconsistent with the management objectives that The National Trust for Scotland have for St Kilda and its sheep (Johnston 2000). It seems likely, therefore, that progress in determining the importance of parasites in generating population crashes will be made only by combining long-term analyses of the vegetation dynamics (Chapter 4), with small-scale experiments (similar to that conducted by Gulland *et al.* 1993) and mathematical modelling (Grenfell 1988, 1992; Grenfell *et al.* 1995).

6

Mating patterns and male breeding success

J. M. Pemberton *University of Edinburgh*

D. W. Coltman *University of Sheffield*

J. A. Smith *University of Glasgow*

and

D. R. Bancroft *GPC AG Genome Pharmaceutical Corporation, Munich, Germany*

6.1 Introduction

The detailed analysis of breeding success of many individuals over entire lifetimes has proved extremely illuminating for our understanding of natural selection and population dynamics, for example in Soay sheep (Chapter 3), in red deer (Albon *et al.* 2000; Clutton-Brock *et al.* 2002) and in many other species (see studies reported in Clutton-Brock 1988a; Newton 1989). In general, however, this literature is dominated by data on female reproductive success, which, due to the prevalence of maternal care, are relatively easy to collect accurately. In polygynous mating systems, parallel studies of male breeding success are particularly important. First, information on lifetime breeding success tells us how natural and sexual selection shape the strategies males employ to obtain fertilisations. Second, if selection on males leads to heavy energy investment at particular life stages or times of year which make males more likely to die than females, then an understanding of selection contributes directly to our understanding of population dynamics. Finally, if selection on males sets up conflicts of interest between males and females that results in selection on females, knowledge of such conflicts refines our understanding of female reproductive strategies and dynamics.

Prior to 1990, relatively few studies had reported on male lifetime breeding success within polygynous vertebrate breeding systems, and data were based exclusively on observed mating success; examples

166

include red deer (Clutton-Brock *et al.* 1988a), elephant seals (Le Boeuf and Reiter 1988), lions (Packer *et al.* 1988), vervet monkeys (Cheney *et al.* 1988) and savannah baboons (Altmann *et al.* 1988). An important problem was that breeding success estimated from behavioural observations was not securely underpinned with evidence of genetic success. Since 1990, molecular genetic techniques have revealed substantial levels of covert breeding success and sperm competition across a wide range of taxa (Birkhead and Moller 1992; Hughes 1998), and behavioural estimates of male breeding success are now under revision across a broad front. In some cases, such as the red deer on Rum, genetic success is strongly correlated with behavioural success ($r^2 = 0.9$), allowing continued use of behavioural data to estimate male mating success and infer paternity (Pemberton *et al.* 1992; Kruuk *et al.* 2000), while in others substantial revision of previous estimates is required. For example, in the Northern elephant seal Le Boeuf and Reiter (1988) found that in four cohorts of seals followed for entire lifetimes, 4.4% of males obtained 75% of observed matings, giving a standardised variance in male lifetime breeding success (variance divided by the square of the mean) of 21.19, the highest estimated for any organism in a review at the time (Clutton-Brock 1988b). Molecular analysis in the same study population has since revealed that fertilisation success of the alpha males is not as high, and that a maximum of 27.5% of pups are sired as a result of observed matings, the rest being sired by peripheral males (Hoelzel *et al.* 1999). With such problems in mind, most studies of male breeding success in polygynous systems now deploy molecular genetic analysis, and investigations have progressed to investigating how selection acts on male strategies and traits (Hogg and Forbes 1997; Kruuk *et al.* 1999b, 2002; Preston *et al.* 2001; Coltman *et al.* 2002). In this chapter, we describe our efforts to measure male mating success in Soay sheep, first from behavioural observations and second using molecular genetic techniques. We describe the resulting mating system and set the scene for investigations of selection operating through male breeding success which is described further in Chapters 7 and 9.

A natural population experiences continual change in environmental conditions, and varies in size. Where this variation sets up differences between cohorts and differences in the level of male–male competition in different years, it is likely to alter the outcome of male mating competition. Lifetime breeding success may vary more (or less) than would be predicted from estimates drawn from specific cohorts or years, with consequences for the opportunity for selection and the maintenance of genetic variation in a population through random drift. Very few studies have investigated the effects of population change on male breeding success, an exception being Rose *et al.* (1998) and Clutton-Brock *et al.* (1997c). They documented cohort and year effects in the red deer population of Rum over a monotonic increase in population size (and increase in proportion of females), showing that as levels of competition between males decreased, so younger males participated more in the rut and skew in male breeding success increased. With their spectacular repeated fluctuations in population size and coupled changes in sex ratio (see Fig. 3.14), Soay sheep offer one of the best model systems in which to investigate the effects of cohort and level of competition on male lifetime breeding success. In this chapter we investigate a series of interlinked questions about temporal variation in male breeding success in the Soay sheep population. First, how does male breeding success vary with age? Do juvenile males, which participate behaviourally in the rut at seven months of age, actually obtain paternities, and do old males, which have outlived most of their cohort, obtain large numbers of paternities? Second, how does age-specific male success vary across years of different population size? Is the success of young males greater following a population crash that has eliminated many older males? What is the distribution of male lifetime breeding success, and what causes variation in lifetime breeding success? Finally, what implications does the mating system have for the opportunity for selection and the maintenance of genetic variation in the Soay population?

6.2 Measuring male breeding success

Most Soay females spend the central part of oestrus in consorts (see section 2.6). To investigate whether relative male breeding success can be accurately estimated from their involvement in consorts, as it can from harem membership in red deer (Pemberton *et al.* 1992), we compared observations of the frequency with which males consorted with females with paternity success, based on molecular techniques, for the ruts of 1987 to 1989. In these ruts, the Village Bay study area was censused several times per day during daylight hours. During each census, the identity of every consorting male and female was recorded. On average, each oestrous female was seen in consort with 2.03 different males (range 1–8), and each male was with the same oestrous female for an average of 1.5 censuses (range 1–8). In addition, we investigated the extent to which observed consorts could be used to identify the father of the subsequent lamb.

The first indication that observed consorts might not predict paternity was that in 13 of 27 cases (48%) tested by multilocus DNA fingerprinting (Jeffreys *et al.* 1985) all consorting males were eliminated as the genetic father (Bancroft 1993). Later, locus-specific DNA profiling (see next section) was used to infer paternity from amongst all males alive in the 1987–9 ruts. Of 128 cases in which the female was seen in consort and the subsequent lamb was assigned a father, there were only 35 cases (27%) in which the genetic father was seen in consort with the female (Coltman *et al.* 1999a). Clearly, consort censuses failed to record a substantial proportion of pairings, and could not be used to infer the identity of a lamb's father with much confidence. Rank order of males on behavioural and genetic measures of breeding success were correlated in all ruts, though the data are very noisy (Fig. 6.1a–c). For example in 1987, males that were awarded no paternities were seen in consort with up to eight females, while two males that were each awarded two paternities were never seen in a consort observation. Overall, consort censuses provide little predictive power for the measurement of male breeding success (Coltman *et al.* 1999a). The reasons for this include the fact that censuses do not detect nocturnal consorts or brief liaisons, and the fact that covert sexual selection is taking place through sperm competition (see section 9.3). Our analysis of mating patterns on St Kilda has therefore relied on paternities inferred entirely by molecular methods.

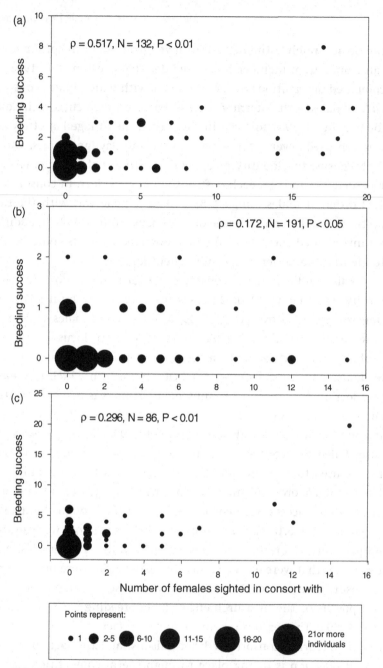

FIG. 6.1. Relationship between number of females sighted in consort and breeding success (number of lambs sired) for all candidate males in the (a) 1987, (b) 1988 and (c) 1989 ruts. The rank order correlation (Spearman's ρ) is positive and significant in each year, but the relationship is not sufficiently strong to accurately estimate male breeding success from observation alone. (From Coltman *et al.* 1999a.)

6.3 Molecular paternity inference

To allow for the promiscuity of both sexes of Soay sheep, we assumed that any tagged male alive during the relevant rut is equally likely to be the father of any subsequent lamb. Since the numbers of individuals compared can be large (e.g. the rut of 1996 involved 294 living males as candidate fathers of 192 subsequent lambs), our analyses used the locus-specific protein and microsatellite polymorphisms described in Chapter 8 (see Table 8.1) and the parentage inference software Cervus 1.0 (Marshall *et al.* 1998). For each lamb, a log-likelihood ratio or logarithm of the odds (LOD) score was calculated for each candidate father, and criteria for assigning paternity to the male with the highest LOD score with 95% and 80% confidence were obtained by simulation. The simulations included realistic features of the data, for example that 20% of candidate fathers were unsampled (i.e. they entered the study area for the rut and left before being caught and sampled), that 10% of genotypes were missing and that there was a 1% genotyping error rate (Pemberton *et al.* 1999).

In total, we analysed the paternity of 1617 lambs born following the ruts of 1986–96 (Pemberton *et al.* 1999). Of these, the father was inferred with 95% confidence for 383 lambs (23.7%) and with 80% confidence for 945 lambs (58.4%), leaving 672 lambs (41.6%) with no inferred father. These figures reflect the problems encountered in paternity analysis when making multiple comparisons using genotype data for loci with a maximum of eleven alleles and often fewer (see Chapter 8). Across years, the proportion of paternity cases resolved at 95% confidence declines as the number of candidate males increases ($r^2 = 0.60$, $p < 0.05$) and there is a similar, but non-significant, trend for cases resolved at 80% confidence ($r^2 = 0.23$, $p = 0.08$) (Pemberton *et al.* 1999). In the remainder of this chapter we describe patterns within the 80% confident paternity data set for the ruts 1986–96 inclusive. This approach maximises the sample size used and, as indicated above, is less affected by variation in the number of rutting males than the 95% confident data set. However, one in five lambs may have been assigned the wrong father, and the data set accounts for only 60% of the lambs born in the study area over the study years. In the rest of this chapter, 'breeding success' is the number of lambs

assigned to a male within a breeding season, while 'lifetime breeding success' is the number of lambs assigned to a male throughout
his lifetime. Because only 60% of lambs born are accounted for, except where stated otherwise, all male breeding success figures quoted
would need to multiplied by approximately 1.7 to give true values,
and our estimates of variance in male breeding success are necessarily provisional.

6.4 Breeding success among age classes and years

Juvenile males participate in the rut, and decrease their over-winter
survival chances by doing so (Stevenson and Bancroft 1995; see also
Chapter 9). To what extent is rutting effort rewarded by paternities?
Answering this question reveals a series of relationships between
the demography of the population, male age and breeding success
(Pemberton *et al.* 1999). We have investigated these relationships using the Village Bay population size as a measure of the ambient level
of competition in any particular rut. Because of the poor survival of
males in population crashes, population size is strongly correlated
with sex ratio (see Fig. 3.14).

For all age classes of males, mean breeding success decreases as
population density increases (Fig. 6.2). Juvenile males average 0.6 of a
paternity in ruts like that of 1986, when the ratio of males : females is
low following a population crash, but only 0.05 of a paternity in high
density, competitive, ruts like that of 1988 (Fig. 6.2a). Adult males (over
two years) average two or more paternities each in low-competition
ruts but only 0.3 of a paternity in competitive ruts (Fig. 6.2c). The relationship is similar but not significant for yearlings, probably because
yearlings were all but wiped out in the 1986 and 1989 crashes and the
remaining three individuals in each cohort give outlying points for
the 1986 and 1989 ruts (Fig. 6.2b). The intercept and slope of the relationship are significantly greater for adults than juveniles (intercept:
$t_{20} = 11.97$; slope: $t_{20} = -17.93$; both $p < 0.001$).

Mean breeding success conceals variation in the distribution of
breeding success among individuals (Pemberton *et al.* 1999). As population density increases, a smaller proportion of males sires at least
one lamb (Fig. 6.3a). Adult males always have a higher probability of

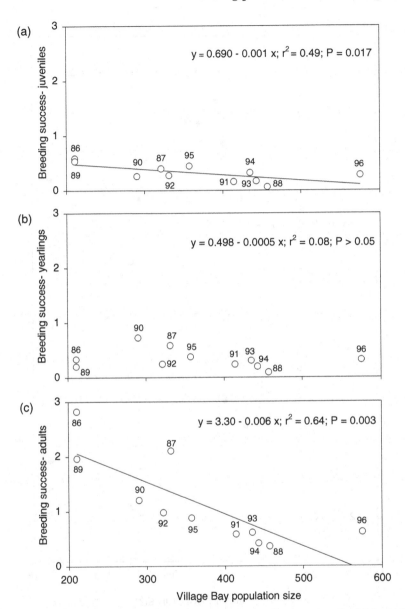

FIG. 6.2. Relationship between mean breeding success for (a) juveniles, (b) yearlings and (c) adult males (over two years) and the level of competition in the rut, measured by the Village Bay population size. Note that in (b) the points for 1986 and 1989 each result from only three yearling males that survived the preceding crash. (From Pemberton *et al.* 1999.)

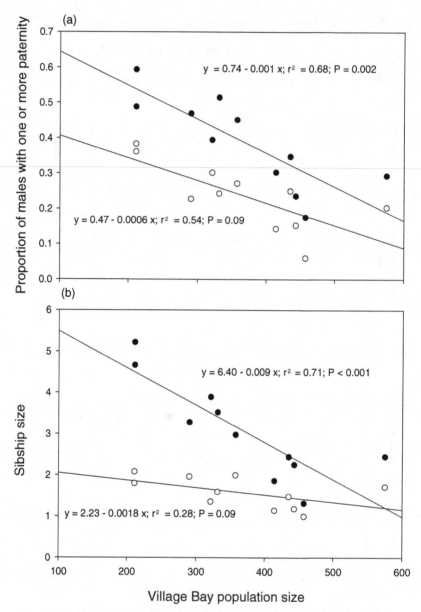

FIG. 6.3. (a) The proportion of young (juvenile and yearling, o) and adult (•) rams that obtained at least one paternity in relation to population size. (b) The mean sibship size sired by young and adult breeders in relation to population size. (From Pemberton *et al.* 1999.)

siring at least one lamb, ranging from 0.2 at high population size to 0.55 at low population size, than juveniles and yearlings (ranging from <0.1 to 0.4). There is no difference in the slope between young males (pooled juveniles and yearlings) and adults, but there was a significant difference in elevation (ANCOVA population size $F_{1,18} = 29.4$, $p < 0.001$; age (young or adult) $F_{1,18} = 5.75$, $p = 0.028$; interaction population size·age $F_{1,18} = 1.19$, $p = 0.291$). Among those males that breed, the mean number of offspring sired, expressed as sibship size, varies with density and age (Fig. 6.3b). Among adult breeders, the mean sibship size sired declines significantly with population size, ranging from five lambs in low-density ruts to around one lamb at high density, while among juveniles and yearlings it is uniformly low at all levels of population size (ANCOVA population size $F_{1,18} = 25.05$, $p < 0.001$; age (young or adult) $F_{1,18} = 25.44$, $p < 0.001$; interaction population size·age $F_{1,18} = 11.22$, $p = 0.004$).

6.5 Lifetime breeding success and cohort effects

As expected for a polygynous species, the distribution of lifetime breeding success (LBS) is highly skewed, with many males dying without being assigned any paternity, and a few males acquiring many paternities over their lifetime (Fig. 6.4). In various analyses, based on different subsets of males, including and excluding animals still alive at the time of analysis, we have estimated that standardised variance on male lifetime breeding success is between 3.4 and 6.6.

Sources of variation in LBS in the eleven cohorts of males born 1986–96 were analysed by Coltman *et al.* (1999c). Large cohort differences in male LBS are to be expected, since some cohorts of males are born into a low-density population, experience a low-competition, successful first rut, have high survivorship in their first winter, and then experience a second successful rut as yearlings. In contrast other cohorts of males are born into a high-density population, experience a highly competitive, relatively unsuccessful first rut, and have low survivorship in their first winter. The mean and standard deviation in LBS in each cohort are shown in Fig. 6.5. As expected, the 1988 and 1991 cohorts, born immediately before a crash, had low LBS, while the 1986 and 1989 cohorts, born into a low-density population, have high LBS. More generally, the population density

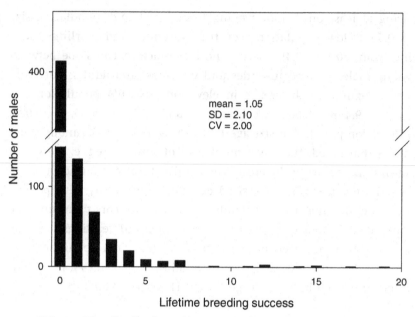

FIG. 6.4. The distribution of male lifetime breeding success for all males born between 1986 and 1996. The data are for 699 paternities inferred at 80% confidence and represent 44% of the sampled lambs born in Village Bay over the study years. Note that this graph includes some males that were still alive at the time of analysis.

into which a cohort is born to some extent predicts the time to the next crash when males are vulnerable. As a result, there is a negative relationship between density in the year of birth and the mean male LBS of a cohort (Fig. 6.6; see also Appendix 3).

Even if a male is lucky enough to be born when density is low, there is still substantial variation in LBS among males born in a particular cohort (Fig. 6.5). For example, though the mean number of paternities in the 1990 cohort is around two, one individual sired nineteen lambs. Some of this variation is explained by lifespan: the longer a male lives, the more offspring he sires (Fig. 6.7) and some by the fact that males become more successful as they age, peaking at about five years of age before declining (Fig. 6.8).

Longevity accounts for much of the variation in male LBS. We used a generalised linear model to identify sources of variation (excluding individual phenotype) in lifetime breeding success. For this model, we restricted the data to individuals born from 1986 to 1991

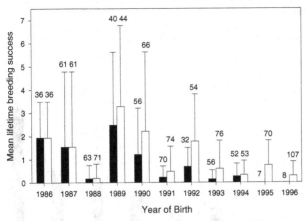

FIG. 6.5. Mean lifetime breeding success (LBS) of cohorts of Soay males born 1986–96, for all sampled males (open columns) and restricted to males that have died (filled columns). Error bars represent one standard deviation. Males born in 1988 and 1991, immediately before a population crash, have lower LBS than adjacent cohorts. Note that the breeding success of recent cohorts was incompletely estimated, since many animals were still alive at the time of analysis. (From Coltman *et al.* 1999c.)

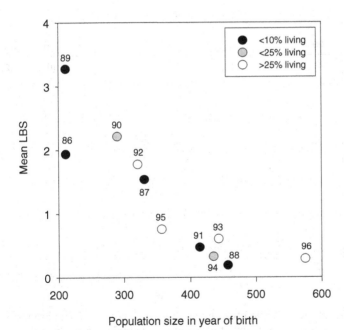

FIG. 6.6. The influence of density in the year of birth on mean lifetime breeding success (LBS) per individual for different cohorts of males on St Kilda. The decline in mean LBS with density in birth year appears robust to variation in the proportion of each cohort still living. (From Coltman *et al.* 1999c.)

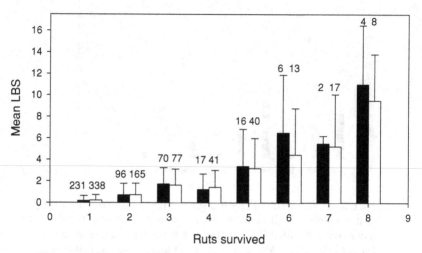

FIG. 6.7. Influence of lifespan on lifetime breeding success (LBS) of males born since 1986. Data are for all sampled males (open columns) and restricted to individuals that have died (filled columns). Bars represent one standard deviation. (From Coltman *et al.* 1999c.)

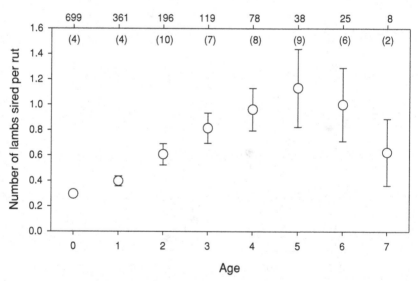

FIG. 6.8. Age-specific breeding success of Soay males. Data are mean (± SD) number of paternities assigned per male per rut by age for all males born since 1986. Numbers in parentheses give the maximum age-specific breeding success, and numbers along the top *x*-axis give sample size represented by that point. (From Coltman *et al.* 1999c.)

Table 6.1. *Generalised linear model[a] of lifetime breeding success in Soay males born between 1986 and 1991*

Term	Coefficient	df	Δdeviance[b]	p
Number of ruts survived	0.429	1	430.57	<0.0001
Juvenile breeding success	0.284	1	45.33	<0.0001
Density in year of birth	−0.0034	1	36.33	<0.0001
Density·juvenile breeding success	0.0023	1	10.51	0.001

[a]This model was fitted assuming a Poisson distribution using a log link function (Crawley 1999). 61.5% of the total variation in lifetime breeding success was explained (null deviance = 1104.1; residual deviance = 425.1; $n = 340$).
[b]Δ deviance is the change in deviance explained by the model when the term in question was fitted last. Note that the total deviance explained by the model is greater than the sum of Δ deviance attributed to each term due to the effects of some terms being aliased.

($n = 340$ males). This limited the data to cohorts for which complete lifetime data were available for over 92% of the individuals considered. Explanatory variables considered included breeding lifespan (number of ruts survived), population density in year of birth (Village Bay population size), and juvenile breeding success (one or more paternities awarded following the rut in the year of birth). This model explains over 60% of the total variation in lifetime breeding success (Table 6.1). The number of ruts survived accounts for most of the explained variation, with increased longevity associated with greater lifetime breeding success. However, the number of ruts survived is highly correlated with density in year of birth ($\rho = -0.65, n = 340, p < 0.001$). Lifetime breeding success decreases with increasing population density in the year of birth, and individuals that are successful in their first rut have greater lifetime breeding success than individuals that are not. The significant interaction between juvenile breeding success and density in year of birth indicates that this difference was greatest for individuals born in years of high population density. This can mostly be attributed to very high over-winter mortality among juveniles following years of high population density, as most individuals born in high-density years fail to survive to the following rut.

6.6 Opportunity for selection and the maintenance of genetic variation

The extent to which individuals vary in breeding success determines the opportunity for selection (Arnold and Wade 1984a, b) or loss of genetic variation by genetic drift (Hill 1972; Nunney 1993). In this section, we investigate the extent to which Soay rams may be under natural and sexual selection and also consider the implications of our findings for genetic drift in the Soay sheep population of Hirta.

We estimated standardised variance in male breeding success, I_{LBS}, over the six cohorts of males born between 1986 and 1991 (i.e. they had completed lifespans), as 3.46, but this overall figure conceals important patterns (Coltman *et al.* 1999c). First, there are systematic differences in I_{LBS} between cohorts born at different population densities (Fig. 6.9a), with variance low in cohorts born at low density and high

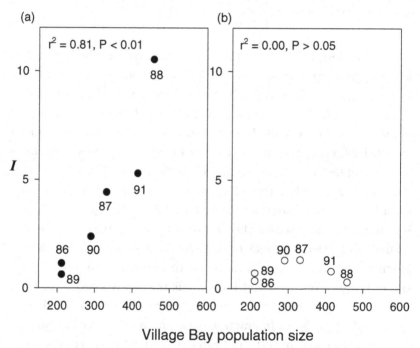

FIG. 6.9. Variation in lifetime breeding success expressed as I, the ratio of the variance to the (mean)^2, (a) for all sampled males within each cohort and (b) restricted to breeders within each cohort. Numbers adjacent to points show year of birth. (From Coltman *et al.* 1999c.)

in cohorts born at high density. Much of the variation at high density comes about because most individuals fail to breed: the first rut that such cohorts experience is at high density and very competitive, and many individuals in the cohort die in the subsequent winter. When I_{LBS} is recalculated omitting males that never bred, values are low for all cohorts and show no obvious pattern with density in the year of birth (Fig. 6.9b). Systematic partitioning of components of I_{LBS} in the 1986–91 cohorts (Brown 1988) (Fig. 6.10) shows that the contributions of different components vary widely, and the greatest opportunity for sexual selection (characterised as variation in fecundity per rut) occurs in cohorts born at low density (Fig. 6.10b). Taken together, these analyses reveal the potential for fluctuation in the selection regime experienced by Soay rams. Males born at low density in 1986 were on average rather successful (Fig. 6.5), had estimated I_{LBS} of only 0.63, of which ~6% was due to variation in longevity of breeders and ~74% was due to variation in fecundity per rut. In contrast the 1988 cohort, born at high density, were on average very unsuccessful (Fig. 6.5), I_{LBS} was high at 10.54, of which ~35% was due to variation in longevity of breeders and only ~6% was due to variation in fecundity (though note also the fluctuations in percentage due to joint variation in the longevity and fecundity of breeders shown in Fig. 6.10d).

The calculations outlined above also define the potential for male breeding success to promote loss of genetic variation by drift. However, it is interesting to consider how these patterns map onto the population dynamics, since crashes are obviously the periods during which the effects of variance in male breeding success will be most focussed. On the whole, our analyses point to mechanisms that ameliorate the loss of variation (Pemberton *et al.* 1996, 1999; Coltman *et al.* 1999c).

The variable breeding success of cohorts of juvenile males probably reduces loss of genetic variation from the population. As we have shown (Figs. 6.2 and 6.3), the extent to which juvenile males are successful in the rut varies inversely with population size. In the lowest-density ruts we have recorded (1986 and 1989), around 40% of juveniles sired an average of two lambs each, accounting for 19% of the subsequent lambs, whereas in high density ruts (e.g. 1988), less

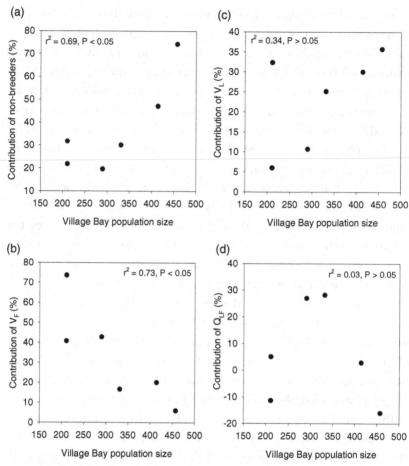

FIG. 6.10. (a) Contributions of non-breeders, (b) variance in the fecundity of breeders, (c) variance in the longevity of breeders and (d) joint variation in fecundity and longevity to the variance in lifetime breeding success in relation to population density in the year of birth for six cohorts of Soay males. (From Coltman *et al.* 1999c.)

that 5% of juveniles sired an average of one lamb each, accounting for 13% of the subsequent lambs. To emphasise the point further, only 31 yearling or adult males were counted on the whole of Hirta following the 1989 crash, about one for every 14 surviving females, but in the 1989 rut at least 19 juvenile males sired one or more lambs in Village Bay alone. As a result of the reduced levels of competition and fewer non-breeders, within-year standardised variance in male

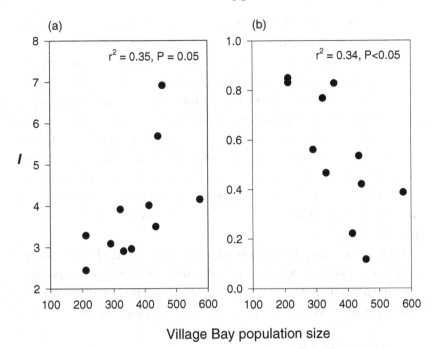

FIG. 6.11. Standardised variance, *I*, in seasonal breeding success in relation to population size during the rut, (a) estimated across all males alive during the relevant rut and (b) estimated across all males that sired at least one lamb in the relevant rut. Note change in vertical scale. ((a) From Coltman *et al.* 1999c.)

breeding success is lower in years of low density than in years of high density (Fig. 6.11a). A crude estimate of the impact of breeding by juveniles and yearlings to standardised variance in lifetime breeding success was made by Pemberton *et al.* (1999). Calculated across all complete lifespans and using all inferred paternities, the standardised variance in male breeding success is 6.6 (mean 0.73); calculated over the complete lifespans but omitting all paternities attributed to juveniles and yearlings, standardised variance in male breeding success is 26.1 (mean 0.31).

Differences in the genetic constitution of male cohorts also seem likely to contribute to the maintenance of genetic variation. Given that they are the most successful in terms of lifetime breeding success, the cohorts of males born immediately after crashes will be influential in terms of determining effective population size. These males are conceived during high-density ruts and *in utero* during the crash, so how representative are they of the pre-crash male population? Variance in male mating success is high during high-density ruts (Fig. 6.11a), but this is mainly due to non-breeders, and a completely opposite pattern emerges when non-breeders are excluded (Fig. 6.11b). In high-density ruts, successful breeders (of all ages) sire few lambs, and variance in breeding success is low. In consequence, most lambs born following a crash are the sole representative of a male and have no paternal sibs in the same cohort (Pemberton *et al.* 1996); they are as representative a sample of the pre-crash male population as it is possible to get.

Finally, the successful cohorts born after crashes are also notable for having the lowest variance in LBS (Fig. 6.9). Their genetic representation in subsequent generations, while large relative to other cohorts, is at least relatively evenly distributed within these cohorts.

6.7 Discussion

Behavioural observation and paternity inference through genetic profiling demonstrate that both sexes of Soay sheep are highly promiscuous. A graphic illustration of this fact is provided by paternity data for twins. Of 80 pairs of twins which both have an assigned father, only 21 (26%) have the same father (Pemberton *et al.* 1999). Ovulation of two eggs, when it occurs, is probably not entirely simultaneous, and the turnover of mating partners is sufficient to ensure that most twins are half sibs. In consequence of their promiscuity, census-based methods for inferring paternity and for estimating the breeding success of individual males in the population are not accurate, and these objectives are best achieved by genetic methods. This is in contrast to red deer, another ruminant for which behavioural and genetic male breeding success have been compared. In red deer, females are not promiscuous, commonly mating once per oestrus, and often stay in the harem of the same male for a number of days around oestrus, so that a single census each day of the rut captures most changes in harem membership (Pemberton *et al.* 1992). Backdating from birth date to conception date and examining harem membership around conception provide an accurate measure of male breeding success, and provides useful information for paternity assignment (Pemberton *et al.* 1992; Slate *et al.* 2000b). These studies illustrate how the details of a polygynous breeding system may affect the appropriate method-ology for measuring relative male breeding success.

The breeding success gained by juvenile and yearling males is a striking and unusual feature of the Soay mating system. The copula-tions obtained during coursing and chasing by these age classes lead to substantial numbers of paternities (section 6.4), which often con-stitute the whole of a male's lifetime breeding success (section 6.5). Coursing is also a well-characterised mating tactic of bighorn sheep, in which genetic profiling has demonstrated that males following this strategy sire up to 44% of subsequent lambs (Hogg 1984, 1988; Hogg and Forbes, 1997). However, in contrast to Soays, coursing behaviour and copulations are never observed in juvenile bighorns and rarely in yearlings, and males aged less than two years do not sire lambs (Hogg and Forbes, 1997). In bighorn sheep, male maturation, in terms of participation in the rut, is much later than in Soays, and coursing is a tactic followed by young or subordinate males until they are cap-able of defending individual females in consorts, which requires the body and horn size only developed at several years of age (Coltman *et al.* 2002). Similarly, although young red deer males occasionally investigate or harass females (Clutton-Brock *et al.* 1982a), there is no evidence that such liaisons ever result in paternity success; rather, all the evidence suggests that those mature males able to maintain ex-clusive harems obtain the greatest number of paternities (Pemberton *et al.* 1992).

We have shown that the distribution of male breeding success in a rut varies systematically with population density and sex ratio. Specif-ically, at high population size when sex ratio is most even and compe-tition is strong, variance in male mating success is high due the fact that a high proportion of males, especially juveniles, sire no offspring, whereas at low population size and low competition levels, variance in male mating success is low due to the fact that a high proportion of males, especially juveniles, sire offspring. Density-related changes in the distribution of male mating success also occur in relation to population size in the red deer on Rum (Clutton-Brock *et al.* 1997c) and provide an interesting comparison with the Soay sheep. In both populations, high density causes male-biased mortality and a female-biased sex ratio in the population. However, the timing of mortality in relation to age at participation in the rut is very different in the two cases and decouples population density from sex ratio in the rut. In red deer, males show strong density-dependent mortality in the first two years of life, so rising population density is associated with a rising proportion of females, and males do not hold harems until they are about five years of age. As density rises, more male deer hold harems and the mean age of harem holders declines. These changes can be taken as an indication that the level of competition for females declines with increasing density (Clutton-Brock *et al.* 1997c), although it is perhaps more accurate to say that in response to reduced competition levels, more males participate. In addition, skew in male breeding success increases, because while many young males hold harems, they do so for short periods and are recorded as having very low breeding success (Clutton-Brock *et al.* 1997c). In contrast, in Soays, the proportion of the population made up by males is highest in high-density years, including during the rut, when juveniles and yearlings attempt to breed before dying (sections 6.5 and 6.6). However, due to the high proportion of males that fail to sire a single lamb, variance in male breeding success is also high in high-density ruts (Fig. 6.11a). Despite their different demographics, both populations show high variance in male success at high densities. This comparison raises a general issue about estimating the distribution of breeding success. Although the patterns reported here seem clear enough, measures of skew and variance are extremely sensitive to the number of participants with zero or low success (see Fig. 6.11a versus Fig. 6.11b, for example) and it is important that the criteria for including individuals in analysis are clearly defined and appropriate. Here, because Soay males start to breed so early, our estimates for Soay males included all males alive in a given rut, and are therefore maximal estimates, even though some poorly grown juveniles may not have competed in high-density ruts (Stevenson and Bancroft 1995). Because male red deer do not start to hold harems until several years of age, estimates only included males that held harems (Clutton-Brock *et al.* 1997c); they are not maximal estimates since in fact many more males are alive and attempt, but fail, to get harems (Clutton-Brock 1982a).

What is the precise mechanism that causes juvenile mating success to vary between years? It may be that all age classes of males put the same effort into breeding each year, and the outcome simply reflects the numbers of individuals in each age class. Alternatively, juveniles may change their level of effort in relation to their own condition or their perception of the level of competition in the rut. Finally, their success may vary with levels of sperm depletion among older males (Preston *et al.* 2001), which begs the question why female Soays are so promiscuous. These issues are addressed in Chapter 9, alongside an appraisal of whether attempting to breed as a juvenile is an optimal strategy.

The lifetime breeding success of a male Soay sheep is profoundly affected by his year of birth (section 6.5). Cohort effects in male lifetime breeding success have been documented in other vertebrates. For example Rose *et al.* (1998) showed that mean LBS varies nearly five-fold between cohorts of male red deer (measured among those that reached two years of age) and that spring rainfall in the year of birth explained some of the variation in male LBS, but not density (although this association was not replicated in a study of individual LBS) (Kruuk *et al.* 1999b). Cohort effects in male Soay LBS are much stronger: there is more than a 15-fold difference in mean lifetime breeding success between cohorts (measured among all males born; e.g. 1988 versus 1989) (Fig. 6.5), and LBS is strongly associated with density in the year of birth (Fig. 6.6). The association with density probably has relatively little to do with food availability during early growth, and much more to do with the fact that density in the year of birth is indicative of the time until the next population crash, and the number of ruts survived is the single largest component of LBS (Table 6.1).

Our analyses of variance in lifetime breeding success yielded a standardised variance for male lifetime breeding success of between 3.4 and 6.6, suggesting the opportunity for selection on males, or the opportunity for genetic drift, are substantial. Comparisons with other studies are hampered by differences in the measurement of breeding success between systems (behavioural, genetics-based) and the differences in the males selected for analysis (see above) (Coltman *et al.* 1999c). However, our values are at the upper end of the distribution reviewed by Clutton-Brock (1988b) and higher than both values estimated for red deer males at that time ($I_{LBS} = 1.43$ including breeders only, and $I_{LBS} = 2.51$ including non-breeders) (Clutton-Brock *et al.* 1988a). Furthermore, through analysing the contribution of components to variance LBS, we have documented at least the potential for the kind of selection to fluctuate between different cohorts.

To what extent does selection actually occur through male breeding success and its components, and to what extent does it fluctuate? As is described in detail in Chapters 7 and 8, survival through population crashes is associated with body size (Milner *et al.* 1999a), horn type and coat colour (Moorcroft *et al.* 1996), inbreeding (Coltman *et al.* 1999b) and other genetic variables (Gulland *et al.* 1993; Paterson *et al.* 1998; Coltman *et al.* 2001b). To the extent that these traits explain variation in longevity, they contribute to variation in male breeding success. Second, as described in detail in Chapter 9, male fecundity is associated with body and horn size and other traits (Coltman *et al.* 1999c; Preston *et al.* 2001). So far, our analyses of body size (measured as hindlimb length) suggest that selection is in the same direction during survival selection and fecundity selection (Coltman *et al.* 1999c; Milner *et al.* 1999a).

Finally, our study throws up a new challenge for those developing methods to estimate effective population size, N_e, the standard metric required for estimating the role of genetic drift in a population. Overlapping generations, fluctuating population size and sex differences in reproductive scheduling have all, to some extent, been incorporated into estimation methods using demographic data (Caballero 1994; Vucetich *et al.* 1997) and Bancroft (1993) estimated N_e for the Hirta population of Soays since introduction to be 250 using the method of Nunney (1991), but a demographic method that has all these features *and* a breeding system that changes systematically with population size has not yet been developed. For a number of reasons, we think it would be valuable to develop such a method. First, a breeding system

that changes with population size does not seem a particularly surprising observation, since the individuals involved are probably simply
maximising their own fitness (see Chapter 9). Second, such systems
may be common – we already know that similar events take place
in red deer (above). Third, in many respects the Soay population is a
model for populations of conservation concern, and it would be de-
sirable to increase the accuracy of effective population size estimates
for such populations. In the future, we hope to estimate the effective
population size of the Hirta Soays more accurately, using a variety of
approaches both demographic and genetic (e.g. Hill 1981), and study
the impact of changes in the breeding system on effective population
size.

7

Selection on phenotype

J. M. Milner *Scottish Agricultural College, Crianlarich, UK*

S. D. Albon *Centre for Ecology and Hydrology, Banchory, UK*

L. E. B. Kruuk *University of Edinburgh*

and

J. M. Pemberton *University of Edinburgh*

7.1 Introduction

The long-term monitoring of selection on phenotypic traits in natural populations inhabiting fluctuating environments can provide valuable insights into interactions between ecological and evolutionary processes (Endler 1986). To date, some of the best-known examples of natural selection in the wild come from investigations in birds, particularly Darwin's finches which inhabit the extreme, and variable, environment of the Galápagos Islands (Grant 1986; Price and Boag 1987; Grant and Grant 2002).

The opportunity for selection is dependent on the extent to which individuals differ in fitness components such as survival and fecundity (Arnold and Wade 1984a,b). Although selection itself is acting on the phenotype of the individual and is measured at that level, for there to be an evolutionary response to selection there has to be genetic variation underlying the phenotype (Endler 1986). The measurement of a genetic response in free-living populations is far more challenging than measuring phenotypic selection alone.

The Soay sheep population is an ideal system in which to study selection because the periodic population crashes provide important opportunities for selection. By following the lives of marked individuals, we now understand some of the ecological factors that contribute to individual differences in both fecundity and survival (Clutton-Brock *et al.* 1991, 1992; see Chapter 3). In addition, we have shown that particular components of fitness are correlated with heritable

190

Soay Sheep: Dynamics and Selection in an Island Population, ed. T. H. Clutton-Brock and J. M. Pemberton.
Published by Cambridge University Press. © T. H. Clutton-Brock and J. M. Pemberton 2003.

morphometric (Illius *et al.* 1995; Coltman *et al.* 1999c; Milner *et al.* 1999a) and polymorphic traits, including coat colour and horn type (Moorcroft *et al.* 1996; Clutton-Brock *et al.* 1997b). To understand the operation of selection and its evolutionary consequences we need to address a number of questions. Does survival or fecundity contribute most to the opportunity for selection? How does the opportunity for selection vary with environmental conditions? Does the intensity of selection on particular traits increase when the opportunity for selection is high? Does the direction or intensity of selection vary between years or between age or sex classes? For instance, is selection stronger in young than old individuals or in males versus females?

In this chapter we first examine the annual opportunity for selection, how variation in survival and fecundity contribute to this, and how it is affected by population density. Second, we examine the evidence for phenotypic selection, mostly through differences in survival, on three measures of body size (body weight, hindleg length and incisor arcade breadth) and on two conspicuous polymorphic traits (coat colour and horn phenotype). We investigate the extent to which selection is density-dependent and influenced by the severity of winter weather, within and between age and sex classes. Then we describe what is known about the genetic response to selection on each trait, which depends on the trait's mode of inheritance. Third and finally, we discuss whether selection can maintain phenotypic variation, and how selection on phenotypes might feed back to influence the population dynamics of Soay sheep.

7.2 The opportunity for selection

The opportunity for selection through annual breeding success (I_{fec}) and annual survival rates (I_{surv}) in a given year can be estimated using the standardised variance of each fitness measure (variance/mean^2) (Arnold and Wade 1984a,b). Mean annual survival rates are calculated assuming non-survivors score 0 and survivors 1, while in females mean annual breeding success is calculated assuming surviving females score 0, 1 or 2 if they produce 0, 1 or 2 lambs respectively. In males, I_{fec} is taken from Coltman *et al.* (1999c), based on all males known to be alive in each year. Only lambs with assigned paternities contribute to these calculations (see Chapter 6).

The annual opportunity for selection due to fecundity I_{fec}, is significantly higher than that due to survival I_{surv} ($t_{28} = 2.989$, $p = 0.006$ in females; $t_{24} = 3.544$, $p = 0.002$ in males) and the average opportunity for selection is several-fold greater in males than females (Fig. 7.1). As population density rises, the annual opportunity for selection increases ($t_{13} = 3.954$, $p = 0.002$), but the increase in opportunity due to fecundity, I_{fec}, is small relative to the highly significant increase due to survival I_{surv} (Fig. 7.1). Consequently, in low-density years (when the Village Bay population size is less than 400 individuals), I_{fec} is significantly higher than I_{surv} in both sexes ($t_{12} = 8.607$, $p < 0.0001$ in females; $t_{11} = 8.948$, $p < 0.001$ in males), while, at high densities, there is no difference in the opportunity for selection due to fecundity versus survival ($t_{14} = -0.549$, $p = 0.591$ in females; $t_{11} = 1.595$, $p = 0.139$ in males).

Overall, the opportunity for selection due to fecundity is greater than that due to survival because of the greater variance in annual breeding success than in survival. Similarly the opportunity for selection in males is greater than in females because of the greater individual variation in male breeding success (Coltman *et al.* 1999c; Chapter 6). However, high density (or low resource availability) generates an increased opportunity for selection, due largely to an increased opportunity for selection through survival. To date, formal analyses of selection have primarily been carried out for survival.

7.3 Selection on body size

EVIDENCE OF PHENOTYPIC SELECTION

There is strong evidence that fitness is related to body size in Soay sheep, with large size being advantageous in terms of both enhanced survival (Fig. 7.2) (Clutton-Brock *et al.* 1992, 1996; Coltman *et al.* 1999c; Milner *et al.* 1999a) and fecundity (Clutton-Brock *et al.* 1996, 1997b; Coltman *et al.* 1999c). Similar results have been found in bighorn sheep (Jorgenson *et al.* 1993; Festa-Bianchet *et al.* 1997, 1998; Coltman

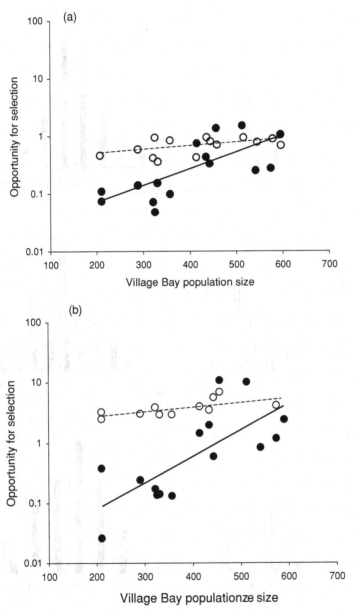

FIG. 7.1. Changes in the opportunity for selection due to survival (solid circles) and fecundity (open circles) in (a) females and (b) males, in relation to the Village Bay population size. Male annual breeding success values from Coltman *et al.* (1999c); note that only lambs with assigned paternities will contribute to these calculations (see Chapter 6).

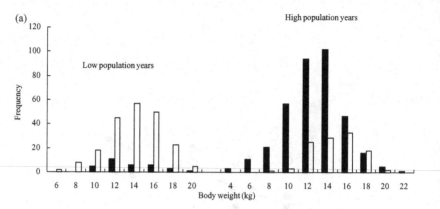

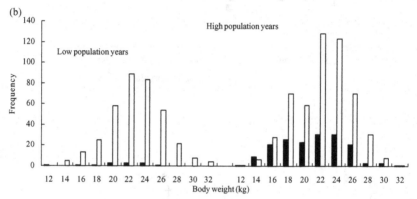

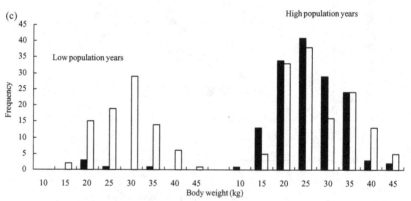

FIG. 7.2. Frequency distribution of body weights, corrected for catch date, of (a) juveniles, (b) yearling and adult females and (c) yearling and adult males during years of low (<400 individuals in the study area) and high (>400 individuals) population size, between 1985 and 1996. Individuals which died (filled bars) are distinguished from individuals that survived (open bars) over-winter. (Reprinted from Milner *et al.* 1999a.)

et al. 2002) and other ungulate species (red deer: Guinness *et al.* 1978c; Albon *et al.* 1986; Kruuk *et al.* 1999b; Loison *et al.* 1999a; mountain goats: Côté and Festa-Bianchet 2001; reindeer: Stien *et al.* 2002).

To compare the intensity of selection on our three correlated body size measures and decide whether body weight, hindleg length or incisor arcade breadth is the target of selection, we calculate standardised directional selection gradients (*S*) (Lande and Arnold 1983). Body weight is the live mass in August, corrected for catch date. Hindleg length is the length of the fibular tarsal bone (long bone of the lower leg) measured in the live animal. Incisor arcade breadth is the distance between the outer left and right edges of the fourth incisor (incisiform canine) on the lower jaw, measured from dental impressions. This size measure has a functional significance for survival through its relationship with bite size and food intake rate (Gordon *et al.* 1996). Following Catchpole *et al.* (2000), the age classes distinguished are juveniles, yearlings, prime adults (two to six years) and older adults (seven years and older) but classes have been merged where differences are not significant. Years are taken as running from spring to spring, so the winter 1985–6, for example, is identified as part of the year 1985.

Body weight is consistently the focus of direct selection among juveniles and adult females (combining yearlings, adults and older adults), and has the largest and most significant selection gradient. Selection on hindleg length and incisor arcade breadth is generally indirect, arising from the fact that these measures are correlated strongly with body weight (Table 7.1) (Milner *et al.* 1999a). Among yearling and adult males combined the pattern is less clear, because sample sizes are smaller and the individual variation in traits is greater, resulting in larger standard errors and few significant results. However, in most years, the trends are similar to those in females. An exception to this occurred in 1991, when selection appears to have focussed on the incisor arcade (Table 7.1).

Across years, there is a general concordance in the direction and strength of selection across sex and age classes, though the intensity of selection is greater in juveniles than in adults. It has long been realised that in winters when the population is high (over 400 sheep in the study area) and mortality is severe, individuals that survive are

Table 7.1. *Standardised directional selection gradients (S'± standard error) resulting from differential over-winter mortality for correlated morphometric characters in juveniles, yearling and adult females and yearling and adult males*

	n	Body weight	Incisor breadth	Hindleg length	r^2
			$S'\pm SE^a$		
Juveniles					
1990	36	$-0.163 \pm 0.071^*$	0.113 ± 0.072		0.09
1991	64	0.653 ± 0.339	0.118 ± 0.282		0.08^*
1992	60	-0.034 ± 0.070	-0.098 ± 0.070		0.03
1993	99	$0.598 \pm 0.152^{***}$	-0.144 ± 0.140		0.17^{***}
1994	65	0.224 ± 0.372	-0.110 ± 0.371		0
1995	46	0.117 ± 0.063	-0.108 ± 0.062		0.05
1996	114	$0.608 \pm 0.176^{***}$	-0.285 ± 0.181		0.08^{**}
Yearling and adult females					
1990	48	0.091 ± 0.047	0.000 ± 0.051	-0.052 ± 0.054	0.02
1991	89	$0.169 \pm 0.069^*$	0.078 ± 0.062	0.017 ± 0.071	0.11^{**}
1992	96	0.029 ± 0.017	-0.001 ± 0.016	0.024 ± 0.019	0.06^*
1993	112	$0.114 \pm 0.037^{**}$	-0.019 ± 0.030	-0.050 ± 0.035	0.06^*
1994	98	0.022 ± 0.051	-0.036 ± 0.044	0.019 ± 0.050	0
1995	82	—	—	—	—
1996	167	$0.075 \pm 0.026^{**}$	0.017 ± 0.020	-0.015 ± 0.025	0.07^{**}
Yearling and adult males					
1990	9	-0.036 ± 0.163	0.070 ± 0.242	-0.189 ± 0.219	0
1991	27	0.214 ± 0.163	$0.392 \pm 0.167^*$	-0.070 ± 0.183	0.18
1992	36	0.021 ± 0.038	0.046 ± 0.041	-0.051 ± 0.061	0
1993	43	0.137 ± 0.133	-0.121 ± 0.106	0.067 ± 0.128	0.01
1994	33	-0.253 ± 0.220	-0.007 ± 0.178	0.244 ± 0.197	0
1995	10	—	—	—	—
1996	55	0.150 ± 0.180	-0.052 ± 0.114	0.022 ± 0.159	0

[a]Values indicate the relative intensity of selection, comparable between traits and between selection events. Among juveniles body weight and hindleg length were too closely correlated for both to be included in the analysis. In 1995 too few adults of either sex died to permit the calculation of selection differentials. All characters have been $\log_e$-transformed. $^*P < 0.05$, $^{**}P < 0.01$, $^{***}P < 0.001$.]

Source: Reprinted from Milner *et al.* (1999a).

larger than those that die in all sex and age classes (Clutton-Brock *et al.* 1996; Coltman *et al.* 1999c; Milner *et al.* 1999a). By contrast, in winters with lower population sizes and higher survival, there is little or no difference in body weight (Fig. 7.2) (Milner *et al.* 1999a) or leg length (Coltman *et al.* 1999c), suggesting that selection is density-dependent. Here, we analyse selection formally, comparing shifts in character means resulting from winter mortality by calculating the selection differential, S', the change in the population mean of a trait before (X_b) and after (X_a) selection, standardised by the variance of the mean before (v_b) for each year (Falconer and Mackay 1996). If annual selection differentials are plotted directly against the population size entering the winter, selection on body weight increases in relation to population size. In general, there appears to be little selection due to differential survival below a threshold population size of about 400 sheep in Village Bay, while at higher population densities the intensity of selection is stronger, but varies between sectors of the population, and between years.

Since survival varies in relation to the prevailing winter weather as well as population density (Milner *et al.* 1999b; see Chapter 3), we might expect the strength of selection to be affected by density-independent factors too. In mild, wet, windy winters, characterised by a high winter North Atlantic Oscillation (NAO) index (see Chapter 2), survival tends to be lower than during colder, drier winters. Although winter weather only has a weak positive effect on the strength of selection, in the case of juvenile females, it interacts with density to explain much of the residual variation in selection in high-density years (Fig. 7.3). Density, NAO and their interaction together explain 78% of the annual variation in selection ($F_{3,7} = 13.21$, $p = 0.003$). A similar, though non-significant, trend is apparent in other sex–age classes, with selection being strongest in years when high NAO coincides with high density. Since both population density and adverse winter weather contribute to over-winter mortality, the proportion of the population surviving the winter provides a useful means of combining these effects. There is a strong negative relationship between the intensity of selection and the proportion of the population surviving for all three morphometric traits (Fig. 7.4). Although the

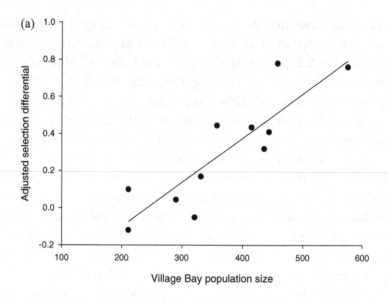

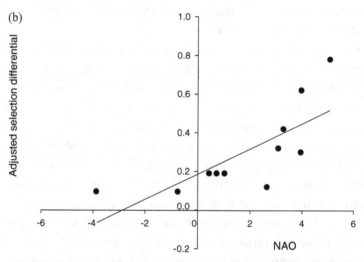

FIG. 7.3. Selection differentials (S') for body weight in juvenile females 1985–1996 (a) in relation to the study area population size when adjusted for the winter values of the North Atlantic Oscillation (NAO) and (b) in relation to the NAO when adjusted for population size. Female juvenile model $S' = 0.001^*p - 0.196^*NAO + 0.001^*p.NAO - 0.229$ where p is the Village Bay population size, '*' denotes multiplication and '.' denotes the interaction term between p and NAO.

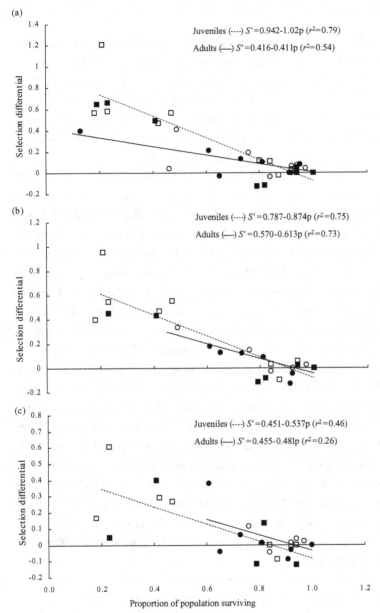

FIG. 7.4. Selection differentials (S') for log-transformed morphometric characters (a) body weight, (b) hindleg length and (c) incisor arcade breadth, in female (open squares) and male juveniles (solid squares), yearling and adult females (open circles) and yearling and adult males (solid circles), in relation to the proportion of each age–sex class that survived over winter. Fitted regression lines are shown for juveniles (dashed lines) and yearlings and adults (solid lines). Differences between the sexes were not significant within age classes. Data for body weight 1985–96, hindleg length 1988–96 and incisor arcade breadth 1990–96. (Reprinted from Milner *et al*. 1999a.)

relationship is linear over the observed range, under very extreme mortality, little or no selection would be expected. We also find that selection on body weight is significantly stronger in juveniles than adults (Fig. 7.4a: $F_{1,39} = 11.0$, $p < 0.01$), but this is not the case for selection on hindleg length or incisor arcade breadth (Fig. 7.4b, c: both $p > 0.2$). The strength of selection does not appear to differ between the sexes within juveniles or adults (all $p > 0.4$).

In addition to repeated selection for increased body size resulting from differential mortality, there is evidence of selection for large body size operating through fecundity. For example, density-dependent selection of hindleg length due to fecundity has been observed in males (Coltman *et al.* 1999c). Longer limbs are associated with increased fecundity and greater lifetime breeding success of males born in low population density years. Over the life cycle, there is a significant positive selection differential of 0.317 standard deviations in leg length at one year of age, following survival to breeding. This is equivalent to a shift of approximately 2.2 mm or 1.2% in the mean leg length of yearling males. By contrast, there is no fecundity advantage of relatively long leg length in cohorts born at high density. Selection on leg length operating through fecundity has not been investigated in females, but there is a strong fecundity advantage of increasing body weight in females (Clutton-Brock *et al.* 1996, 1997b; see Chapter 3). Heavier females have higher conception rates (Clutton-Brock *et al.* 1997b) and a higher probability of twinning, and give birth to heavier offspring at all population densities. These observations suggest that selection for increased body size due to differential mortality is likely to be reinforced by selection due to differential fecundity in both sexes.

GENETIC RESPONSE TO SELECTION ON BODY SIZE

Having demonstrated repeated selection for heavy animals, we might expect a shift in the population mean body weight over a few generations if large body size is heritable. Our approach to estimating heritability of the morphometric traits was to use a pedigree involving over 1800 females (dams) and 400 males (sires, inferred by molecular paternity techniques (see Chapter 6)), combined with maximum likelihood procedures and an 'animal model' (Knott *et al.* 1995; Milner *et al.* 2000). The best fitting model, accounting for maternal effects, shows comparatively low heritability estimates (h^2) for all three morphometric traits (Table 7.2). In males, these range from 0.12 (SE = 0.05) for body weight to 0.29 (SE = 0.10) for incisor arcade breadth and, in females, from 0.17 (SE = 0.06) for incisor arcade breadth to 0.26 (SE = 0.11) for hindleg length (Milner *et al.* 2000). These compare with heritabilities of morphometric traits in red deer ranging from 0.07 for male leg length to 0.60 for jaw length in males (Kruuk *et al.* 2000) and with heritability estimates for body mass ranging from 0 to 0.57 in bighorn sheep, depending on age and season (Réale *et al.* 1999; Réale and Festa-Bianchet 2000). Our estimates are based on pooled data from all age groups and although age is accounted for in the models by fitting growth curves (Milner *et al.* 2000), other studies have shown that the heritability of body mass increases with age (Näsholm and Danell 1996; Réale *et al.* 1999). The heritability of body weight in adult Soay sheep may therefore be higher than the values presented here. We found no significant differences in heritability estimates between the sexes in either hindleg length or incisor arcade breadth. However, in the case of body weight, the total phenotypic variance in male body weight is much greater than that in female body weight (Table 7.2). This, combined with a lower additive genetic variance (V_A) in males, leads to a lower, and less significant, estimate of the heritability of body weight for males (0.12 ± 0.05) than for females (0.24 ± 0.09).

Despite evidence of repeated selection over a twelve-year period, reinforced by directional selection due to differential fecundity, and significant heritability of body size traits, there is no evidence of a systematic change in body weight (Fig. 7.5), after age, sex and population size have been controlled for ($F_{1,2125} = 0.86$, $p > 0.3$), or in hindleg length (Coltman *et al.* 1999c). In practice, the low additive genetic variance compared to the high environmental variance of body weight will always mean that the evolutionary response to selection on body weight will be slow. Indeed, if we assume the mean selection differential due to survival across all sex–age classes in all years is 0.23, the predicted response across the whole population is an increase of only 0.02 kg per year (Milner *et al.* 1999a). This is too small

Table 7.2. Variance components and heritability estimates (h^2) (and their standard errors) made using a maximum likelihood model when maternal effects (M) were taken into consideration. Total phenotypic variation was divided into four variance components, the additive genetic variance (V_A), the permanent environmental variance (V_E), the maternal effects variance (V_M) and the residual variance (V_R). The pedigree contains 2129 individuals and estimates are based on 947 records from 449 individuals and 232 dams for females and 575 records from 375 individuals and 218 dams for males.

Trait	V_A (SE)	V_E (SE)	V_M (SE)	V_R (SE)	h^2 (SE)	M (SE)
Females						
Body weight	1.30 (0.47)	1.48 (0.42)	0.80 (0.25)	1.80 (0.08)	0.24** (0.09)	0.15*** (0.04)
Hindleg length	20.0 (8.66)	32.9 (7.73)	16.2 (5.07)	7.37 (0.34)	0.26* (0.11)	0.21*** (0.06)
Incisor arcade breadth	0.30 (0.10)	0.14 (0.08)	0.21 (0.07)	1.06 (0.05)	0.17** (0.06)	0.12** (0.04)
Males						
Body weight	1.10 (0.45)	2.86 (0.57)	0.45 (0.44)	4.67 (0.25)	0.12* (0.05)	0.05 (0.05)
Hindleg length	15.3 (10.1)	36.1 (8.09)	8.14 (4.89)	18.6 (1.21)	0.20 (0.13)	0.10 (0.06)
Incisor arcade breadth	0.43 (0.09)	0.15 (0.07)	0.00 (0.02)	0.89 (0.05)	0.29*** (0.10)	0.00 (0.01)

*, $p < 0.05$, **, $p < 0.010$, ***, $p < 0.001$.
Source: Reprinted from Milner et al. (2000).

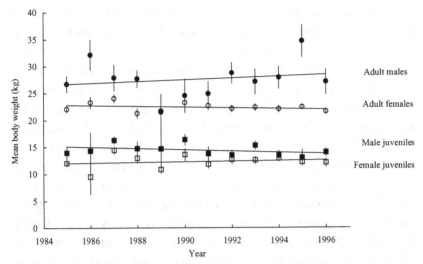

FIG. 7.5. Changes in mean body weight (corrected for catch date), in female (open squares) and male juveniles (solid squares), in yearling and adult females (open circles) and yearling and adult males (solid circles), during the period 1985 to 1996. Year-to-year variability in adult weights is largely due to variation in the age structure. (Reprinted from Milner *et al.* 1999a.)

to detect given measurement errors arising from factors such as gut fill and wetness of the fleece, and we should not be surprised that no response was observed.

7.4 Selection on horn phenotype

EVIDENCE OF PHENOTYPIC SELECTION

There is also evidence of phenotypic selection on horn phenotype. Soay sheep have an inherited polymorphism for horn development (Chapter 2): in females, horns may be normal (goat-like), scurred (vestigial and deformed) or polled (hornless), while in males, only normal (larger, heavier and spiralled) and scurred phenotypes occur. Female horned and polled phenotypes share similar fitness parameters (Clutton-Brock *et al.* 1997b) so have been combined in our analyses. The observed frequencies of these phenotypes, their proposed genotypes and the proposed inheritance model is given in Appendix 2.

The strength of selection on horn phenotype due to differential over-winter survival varies between years in both sexes (Moorcroft *et al.* 1996). Selection is strongest in winters of high mortality when scurred phenotypes are favoured in both sexes (Fig. 7.6) (Moorcroft *et al.* 1996). Across the first eight years of the study, a strong density-dependent relationship was apparent, with survival of scurred phenotypes being significantly higher than that of non-scurred in both sexes when density was high, while there were no differences when population density was low (Moorcroft *et al.* 1996). However, the absence of population crashes in more recent high-density years, means that over a twelve-year period from 1985 to 1996, we only find weak evidence for density-dependent selection in females ($F_{1,10} = 4.36$, $p = 0.063$). In addition there is a weak, but not significantly independent, relationship with variation in winter weather, indexed by the NAO, again only in females ($F_{1,10} = 4.16$, $p = 0.072$). Nonetheless, as with the morphometric traits, there is a strong negative relationship between the strength of selection and the proportion of the population surviving in both sexes (Fig. 7.6b).

The pattern of selection in males and females is generally concordant across years (no significant sex × year interaction) during the twelve-year study period. Selection significantly favoured scurred horns in both sexes in 1985 (males: $\chi^2 = 5.55$, $p = 0.019$; females: $\chi^2 = 8.53$ $p = 0.004$) and 1988 (males: $\chi^2 = 9.41$, $p = 0.002$; females: $\chi^2 = 9.74$, $p = 0.002$), in males only in 1991 ($\chi^2 = 8.23$, $p = 0.004$) and in females only in 1994 ($\chi^2 = 4.52$, $p = 0.034$). Within the sexes, there is also a general concordance in the direction and strength of selection between age classes in males (no significant horn × age class interaction), while in females there is a tendency for the strength of selection to differ between age classes (horn × age class interaction $F_{3,2774} = 3.39$, $p = 0.017$). Within females, selection for scurredness is strongest in juveniles, while non-scurred individuals are favoured among older adults. Although selection is more variable in males than females, the strength of selection is similar in the two sexes (no significant difference in elevation of regression lines, $t_{1,21} = 1.36$, $p > 0.1$) (Fig. 7.6b).

(a)

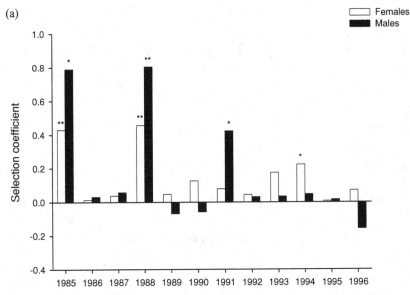

(b)

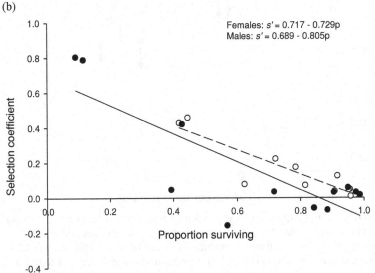

FIG. 7.6. (a) Selection coefficients (s') for horn phenotypes in males (filled bars) and females (open bars), with positive values indicating selection for scurred phenotypes. Significance based on χ^2 values, *$p < 0.050$, **$p < 0.010$. (b) The relationship between selection coefficient and the proportion of the population surviving in males (solid circles) and females (open circles). Fitted regression lines are shown for males (solid line) and females (dashed line).

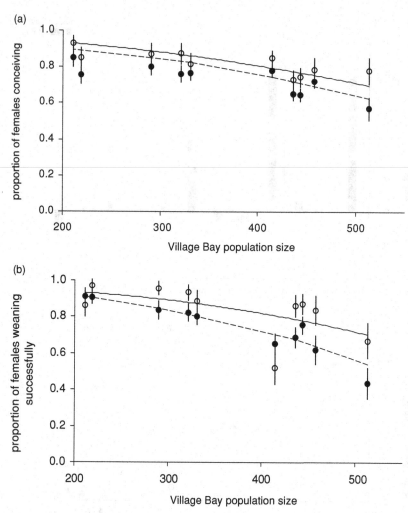

FIG. 7.7. The proportion of scurred and non-scurred females of all ages (a) conceiving and (b) having produced living lambs, successfully weaning them, at different population sizes between 1985 and 1992. Population size was estimated the previous autumn. Scurred females are shown by open circles and solid lines, non-scurred are shown by solid circles and dashed lines. (Redrawn from Clutton-Brock 1997b.)

Selection operating through fecundity also appears to favour scurred phenotypes in females. Although formal analysis of selection has not yet been carried out, scurred adult females have consistently higher breeding success than non-scurred females, due both to higher conception rates and weaning rates in their offspring (Fig. 7.7) (Clutton-Brock *et al.* 1997b). The apparently higher fecundity of scurred phenotypes in females may be related to the fact that they are significantly heavier at all population densities (Clutton-Brock *et al.* 1997b). Differences in fecundity between horn types disappear when body weight the previous August is included in models (Clutton-Brock *et al.* 1997b). In addition, increased body weight may account for the higher survival of their offspring since mother's weight affects offspring birth weight and offspring survival (Chapter 2). There is also a tendency for the birth weight of offspring born to scurred females to be less affected by high population density than the birth weight of offspring born to non-scurred females, which could account for the improved survival to weaning among offspring born to scurred females at high density. In general, these results suggest that selection for scurred horn types operating through fecundity will reinforce selection due to differential survival in adverse years.

While selection appears to favour scurredness in females, there is evidence that this may be opposed by sexual selection for normal horns through male mating success (Clutton-Brock *et al.* 1997b). Scurred males cannot fight as successfully as normal horned males during the rut, rarely have exclusive access to oestrous females and have significantly lower annual breeding success (see Chapter 9). The relative benefits of different horn types in males are examined in Chapter 9.

GENETIC RESPONSE TO SELECTION ON HORN TYPE

In the earlier studies, researchers combined polled and scurred horns to look at the inheritance of horn phenotypes (Doney *et al.* 1974). They classified 47% of females as normal horned and 53% as polled or scurred, and assumed that the inheritance pattern was a simple dominance of normal horned over the others. However, this early work only considered females, and did not take into account that alleles encoding normal horns are more dominant in males than females. A more plausible model is one in which the presence or absence of normal horns in a sheep is due to three alleles with different expression in the sexes. Further details, including observed and predicted frequencies

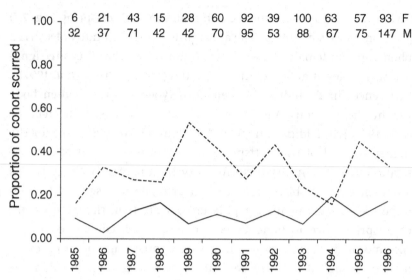

FIG. 7.8. Year-to-year changes in the proportion of scurred individuals born within a cohort of males (solid line) and females (dashed line). Sample sizes are given for each sex in each year. Females horn types can generally only be classified in individuals over one year old.

of the different phenotypes under this model of inheritance, are given in Appendix 2.

There is considerable variation in the proportions of the different horn morphs born each year (Fig. 7.8). Weak evidence suggests there may be an increase in the proportion of scurred males born since 1985 (logistic regression of proportions $\chi^2 = 4.15$, $p = 0.042$) but there is no significant long-term change in phenotype frequencies in females (Fig. 7.8). The mechanisms by which horn phenotype frequencies are maintained are not yet fully understood but probably involve opposing natural and sexual selection in males (see Chapter 9), as well as inertia arising from the complex inheritance system for scurredness, especially in females.

7.5 Selection on coat colour

EVIDENCE OF PHENOTYPIC SELECTION

As described in Chapter 2, coat colour can be either light or dark, with dark coats (wild-type and self-coloured combined) being approximately three times as common as light-coloured coats (Boyd *et al.* 1964). The observed phenotype frequencies and their proposed genotypes are given in Appendix 2. The overall percentage of self-coloured individuals is so low that in the following analyses we compare selection for light- and dark-coloured coats, combining wild-type and self-coloured individuals. An early suggestion that light self-coloured sheep were less fit than other morphs (Ryder *et al.* 1974) was based on very small samples of captive animals and is not statistically proven.

There is temporal variation in selection on coat colour due to differential over-winter survival (Fig. 7.9a) (Moorcroft *et al.* 1996). In winters when selection is strongest, dark-coloured coats are favoured, while in other years, either there is no selective advantage, or weaker selection favours light-coloured phenotypes (Moorcroft *et al.* 1996). Over the period 1985–92, significant density-dependent selection occurred in females, favouring dark phenotypes at high density. In males, both coat colours had equally low survival probabilities at high density, while at low density, light-coloured males have higher survival than dark ones, but survival of both phenotypes is generally high and selection is not significant (Moorcroft *et al.* 1996). Over the time period 1985–96, the occurrence of high density, non-crash years means annual differences in selection for coat colour are not directly associated with population density or winter weather in either sex. Nevertheless, it is clear that there is a strong negative linear relationship between the strength of selection for dark coats and the proportion of the population surviving the winter in females (Fig. 7.9b). Amongst males, the relationship is linear over the range of survival observed in females, but there are two years (1985, 1988) when survival was so low (approximately 10%) that differences between the phenotypes are lost, giving a non-linear relationship overall (Fig. 7.9b).

In contrast to selection on horn type, selection on coat colour is not concordant between the sexes across years (sex × year interaction $F_{11,3958} = 3.84$, $p < 0.001$) and there is some evidence of opposing selection. Indeed, in 1993, significant selection occurred in opposite directions in each sex. In 1991, there was significant selection for dark coats in females ($\chi^2 = 6.60$, $p = 0.010$) while in 1993 and

(a)

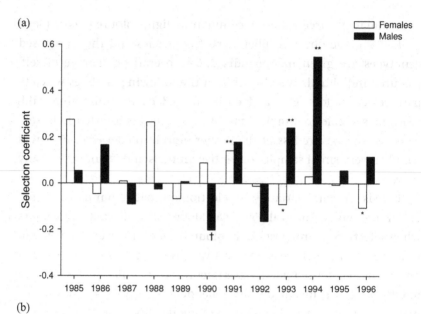

(b)

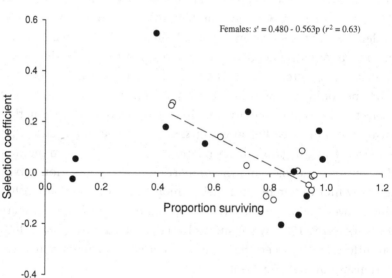

FIG. 7.9. (a) Selection coefficients (s') for coat colour phenotypes in males (filled bars) and females (open bars), with positive values indicating selection for dark coats. Significance based on χ^2 values, $^\dagger p < 0.10$, $^* p < 0.050$, $^{**} p < 0.010$. (b) The relationship between selection coefficient and the proportion of the population surviving in males (solid circles) and females (open circles). A fitted regression line is shown for females (dashed line) but the relationship is not significant in males.

1996 there was significant selection for light coats (1993: $\chi^2 = 4.34$, $p = 0.037$; 1996: $\chi^2 = 4.30$, $p = 0.038$). There is also significant variation in selection between age classes in females ($F_{3,2726} = 2.737$, $p < 0.05$), with selection tending to favour light coats in juvenile females more often than in other age classes. Within males, dark coats were favoured in 1993 and 1994 (1993: $\chi^2 = 8.37$, $p = 0.004$; 1994: $\chi^2 = 8.23$, $p = 0.004$), while light coats were only weakly favoured in 1990, with selection just not significant at the 5% level ($\chi^2 = 3.78$, $p = 0.052$). There is no evidence of different patterns of selection between age classes in males. Overall, there are three cases of highly significant selection for dark coats (two in males and one in females) and two instances of less significant selection for light coats, both in females.

Further research is required to determine the biological basis of reversals in the direction of mortality selection on coat colour. Dark coated phenotypes are significantly heavier than light-coloured phenotypes in both sexes (Clutton-Brock *et al.* 1997b). It seems likely that this explains the higher survival of dark-coloured animals in some years since the effects of coat colour disappear when body weight is fitted in models of individual survival (Milner *et al.* 1999a). To date we have no similar explanation for the reversal of selection favouring light-coloured females. As dark-coated individuals are heavier than light-coloured ones, this might also be expected to confer fecundity advantages (see above). However, the evidence suggests that female conception rate, lambing date and offspring birth weight do not vary with coat colour (Clutton-Brock *et al.* 1997b). The relationship between male fecundity and selection on coat colour is not yet known.

GENETIC RESPONSE TO SELECTION ON COAT COLOUR

The inheritance of coat colour appears to involve a single autosomal locus with two alleles, of which the dark allele is dominant to the light (Doney *et al.* 1974; see Appendix 2). The direction and strength of selection on coat colour are variable between years and sexes, which may partly explain why the phenotypic frequencies appear essentially unchanged over three decades. The proportion of dark-coated

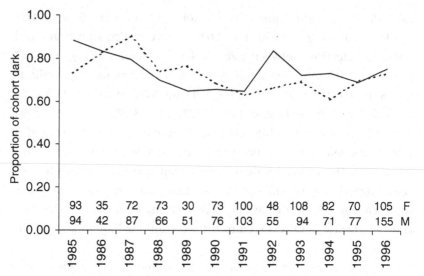

FIG. 7.10. Year-to-year changes in the proportion of dark-coated individuals born within a cohort of males (solid line) and females (dashed line). Sample sizes are given for each sex in each year.

individuals born has varied between cohorts from 65% to 88% in males and 61% to 90% in females between 1985 and 1996, but there has been no consistent long-term change in the proportions (Fig. 7.10). Although the selection coefficients favouring dark coats are always larger than those favouring light coats, without taking the size of sex and age classes into account it is unclear whether or not the reversals in mortality selection we have measured would be sufficient to maintain observed morph frequencies in the long term. Furthermore, because the light allele is carried as a recessive allele in heterozygous dark individuals, the light allele is sheltered from selection in these animals, and may be retained in the population for longer than if it was always expressed. A more thorough analysis of selection through different fitness components, including breeding success, and modelling are needed to determine the expected evolutionary outcomes.

7.6 Discussion

A primary aim of the Soay sheep study is to document the extent of natural selection experienced both during population crashes and through breeding success (section 1.3). Our work shows that the annual opportunity for selection is much greater in males than females and increases with population density, largely due to a rise in the opportunity for selection due to survival. However, it is inevitable that there will be a greater opportunity for selection through fecundity in non-crash years. In future, comparisons of the opportunities for selection between the sexes and between cohorts born in low- and high-density years would therefore most usefully be carried out using lifetime breeding success data, as initiated for males by Coltman *et al.* (1999c) (see Chapter 6).

It is clear that there are considerable temporal fluctuations in the intensity of phenotypic selection on specific traits within the Soay sheep population. As in Darwin's finches in the Galápagos Islands (Price *et al.* 1984), selection is most intense when mortality is highest, both between years and between sectors of the population. Consequently, selection is strongly affected by population density and factors, such as winter weather, that influence survival. Selection operating through fecundity appears to reinforce selection through survival but would benefit from further research.

In contrast, we have found little temporal fluctuation in the direction of selection. Instead we show evidence of repeated directional selection and the consistent favouring of certain phenotypes in successive selection events. In the case of morphometric body size traits, large individuals are consistently favoured, both through differential survival and fecundity. Oscillations in the direction of selection under varying environmental conditions tend to be weak on St Kilda because survival across phenotypes is generally high except at high population density. By comparison, in the Galápagos, selection fluctuates in response to environmental extremes such as drought and El Niño events, both of which can cause considerable mortality but favour different phenotypes (Price *et al.* 1984; Gibbs and Grant 1987; Grant and Grant 1989, 2002). In terms of the polymorphic traits we have examined, scurred individuals repeatedly show better survival than normal horned or polled individuals and also show enhanced female fecundity, while dark-coated individuals are generally more strongly favoured than light-coloured ones through differential survival. We have found no evidence of countervailing selection within body size traits or horn type, and in coat colour, opposing selection may not to be strong enough to maintain the observed phenotypic variation.

Finally, we have found that, in some traits, the strength of selection varies between sex and age classes. Selection on body weight is stronger in males than females and, within sexes, is stronger in juveniles than adults. By contrast, selection on horn phenotypes is similar between the sexes but varies between age classes in females while selection on coat colour appears to show no consistent pattern across age and sex classes. In addition, sexual selection for normal horns in males may oppose natural selection for scurredness because of the superior breeding success of normal-horned males. However, it is not yet clear how the lifetime breeding success of scurred males (which tend to be longer-lived) compares with that of normal-horned males (Chapter 9). These observations have important implications for both genetic and ecological studies, in particular, the maintenance of genetic variation (Clarke and Beaumont 1992; Ellner and Hairston 1994) and the feedback into population dynamics (Begon 1992; Endler 1992), and are discussed in detail below.

GENETIC RESPONSE AND THE MAINTENANCE
OF PHENOTYPIC VARIATION

Repeated directional selection towards large body size has been found in all sectors of the population. Yet, no genetic response is detectable despite the enhanced fitness of larger individuals, the lack of fluctuation in the direction of selection and a significant genetic component of variation. A number of other studies have demonstrated strong directional selection on heritable traits but found no evidence of evolutionary change (reviewed by Merilä *et al.* 2001c). It has long been realised that trade-offs between traits, resulting in negative genetic correlations between them, may reduce the response to selection (Roff 1992). In addition, long-term studies of collared flycatchers (*Ficedula albicollis*) and red deer (*Cervus elaphus*) suggest that other processes including environmental change and selection on the environmental component of trait variation can also decouple predicted and observed responses (Kruuk *et al.* 2001, 2002b; Merilä *et al.* 2001a,c). It is not yet clear which, if any, of these processes may account for the lack of evolutionary change in the Soay sheep. In future, it would be ideal to analyse the genetic response within a quantitative genetic framework by estimating breeding values of individuals and to track changes in this measure.

A complementary approach for inferring selection is to work at the level of one or a few individual polymorphic loci, and ask whether individuals with different alleles or genotypes differ in fitness (see Chapter 8). Using this approach, genetic responses can be measured if allele frequencies systematically change in the predicted direction. Some of the best case studies of evolution in action have been documented in species with inherited phenotypic polymorphisms controlled by one or a few loci, for example wing colour in the peppered moth (Kettlewell 1973), and the number and size of coloured spots in male guppies (Endler 1980). No strong systematic change in the phenotypic frequency of the polymorphic coat or horn traits has been detected within the Soay sheep population over three decades. Presumably, as in the case of morphometric traits, this is partly because changes in selection pressures vary with year-to-year changes in population density or winter weather (Milner *et al.* 1999b) and relative fitness changes with environmental conditions (see above). In addition, countervailing selection in other components of fitness (Rose 1982; Pemberton *et al.* 1991) such as a fecundity advantage to normal-horned males due to sexual selection, or heterozygous advantage (Lewontin *et al.* 1978), could also contribute to maintaining phenotypic variation.

Studies of phenotypic polymorphisms are, in general, limited by their availability, and it was not until the development of techniques for detecting genetic variation at the molecular level that a locus-specific approach to studying selection became widely available. A further advantage of many molecular polymorphisms is that they are co-dominant and heterozygous genotypes can be detected. Molecular genetic variation and selection on genotype are discussed in detail in Chapter 8.

SELECTION AND THE FEEDBACK TO POPULATION DYNAMICS

The distribution of heritable variation in fitness may also have important consequences for population dynamics (Begon 1992; Clarke and Beaumont 1992; Endler 1992). A striking feature of the patterns of phenotypic selection, particularly within females (and males in the case of body size traits), is that the magnitude of selection varies across age classes. It has already been shown that fluctuations in the age and sex structure affect the population's response to density and weather conditions (Coulson *et al.* 2001). Understanding the feedback of interactions between differential fitness of phenotypes, age–sex classes and environmental change may also be important if we hope to predict future fluctuations in population size (Chapter 3) (see also Moorcroft *et al.* 1996).

8

Molecular genetic variation and selection on genotype

J. M. Pemberton *University of Edinburgh*

D. W. Coltman *University of Sheffield*

D. R. Bancroft *GPC AG Genome Pharmaceutical Corporation, Munich, Germany*

J. A. Smith *University of Glasgow*

and

S. Paterson *University of Liverpool*

8.1 Introduction

The application of molecular techniques to population biology can be traced back to the origins of allozyme gel electrophoresis (Smithies 1955) and the invention of histochemical stains targeting specific enzymes (Hunter and Markert 1957). Initial surprise at the high level of variation revealed was rationalised by the understanding that much molecular variation must be selectively neutral (Kimura 1968; King and Jukes 1969), and making this assumption, molecular markers were applied to samples of anonymous individuals to quantify population-level processes such as inbreeding, differentiation and phylogeography (Avise 1994).

Alongside these population genetics studies, there was a series of detailed studies in which phenotypic and spatial data about the individuals sampled was also collected, and the enzyme kinetic properties of alleles were quantified, allowing tests of selective neutrality at specific allozyme loci. Examples of allozyme polymorphisms that appear to be under selection include phosphoglucomutase in *Colias* butterflies (Watt *et al.* 1983), leucine aminopeptidase in the mussel *Mytilus edulis* (Koehn and Hillbish 1987) and lactate dehydrogenase-B in the marine killifish *Fundulus heteroclitus* (Powers *et al.* 1983). Similarly, several studies demonstrated correlations between average allozyme heterozygosity and phenotypic measures, which might be due

217

Soay Sheep: Dynamics and Selection in an Island Population, ed. T. H. Clutton-Brock and J. M. Pemberton.
Published by Cambridge University Press. © T. H. Clutton-Brock and J. M. Pemberton 2003.

to overdominance (heterozygote advantage) at specific loci or inbreeding depression (Allendorf and Leary 1986; Ledig 1986; Mitton 1997).

Molecular markers were not applied within long-term studies of individually marked animals until the mid 1980s. The need for a reliable method for assigning parentage and relatedness was widely recognised, but the enormous investment required to develop the large number of markers necessary (due to their low individual resolution) could not be justified for natural populations. Where allozymes were applied to paternity assignment in natural populations, their poor resolution could easily be seen: in a ringed population of indigo buntings (*Passerina cyanea*), a minimum of 14.4% extra-pair offspring were identified in the nest, but the true value could have been as high as 27% or 42% in different years, due to the potential for putative parent offspring pairs to match by chance (Westneat 1987). The situation was transformed by the discovery of the hypervariable minisatellite loci that underpin DNA fingerprinting (Jeffreys *et al.* 1985) and of the highly polymorphic microsatellite loci that underlie modern DNA profiling (Litt and Luty 1989; Tautz 1989; Weber and May 1989). For example, in indigo buntings, DNA fingerprinting unambiguously revealed that the 35% of chicks were the result of extra-pair matings (Westneat 1990). As discussed in Chapter 6, the application of these techniques to identify parentage within long-term studies has had a profound effect on our understanding of some breeding systems, including that of the Soay sheep. Furthermore, the pedigrees that can be built from known parent–offspring links allow us to investigate new topics, including the genetic architecture of quantitative traits (see Chapters 5 and 7) and the rate of inbreeding between close relatives (Marshall *et al.* 2002).

As well as enhancing the investigation of selection on quantitative traits, the application of both old and new molecular tools within studies of known individuals can contribute much to the study of selection at the level of the genotype. The combination of direct measures of individual fitness and precise genetic differences between individuals offers a potentially powerful way to investigate natural selection in action. For example, we have shown differences in juvenile survival between different allozyme and microsatellite genotypes of red deer calves on Rum (Pemberton *et al.* 1988; Coulson *et al.* 1998a), and several studies of birds and mammals have used microsatellite-based estimates of individual inbreeding to document inbreeding depression (Coltman *et al.* 1998; Coulson *et al.* 1998b, 1999b; Slate *et al.* 2000a; Amos *et al.* 2001; Hansson *et al.* 2001; Rossiter *et al.* 2001; Hoglund *et al.* 2002; Slate and Pemberton 2002).

In this chapter we describe allozyme and microsatellite genetic variation in the Soay sheep population and evidence for selection operating on this variation. It would not be surprising if Soays had low levels of molecular variation, since the history of the population includes a number of founder events and potential genetic bottlenecks which might have caused loss of genetic variation through the sampling effect. For example, the first introduction to the islands, unrecorded by history, and the 1932 introduction to Hirta, were founder events, while the repeated crashes of unmanaged populations on Soay and Hirta throughout their history may have been bottlenecks, particularly among males (though see Chapter 6). However, there are also reasons for expecting appreciable diversity in the Soays. First, up to and including 1932, the population could have received occasional influxes of genes from more modern breeds, for example the Scottish blackface that was farmed on the islands. Second, variation could be maintained by various kinds of selection.

In this chapter we investigate selection on molecular genetic variation at two levels. First, we ask whether there is evidence for selection against inbred individuals. Inbreeding depression is the reduction in fitness of individuals due to identity of alleles by descent, and probably occurs due to homozygosity of deleterious recessive alleles (i.e. dominance) or heterozygote advantage (overdominance) (Charlesworth and Charlesworth 1987). Inbreeding depression is a widespread phenomenon, and is highly relevant to several aspects of biology including the evolution of life-histories, agriculture and conservation. Although we now know that inbreeding depression is often severe in free-living populations, there remain many important questions about it, in particular its relationship with environmental heterogeneity: does inbreeding depression increase when conditions get worse, and what is the shape of the relationship (Frankham 1995; Crnokrak and Roff 1999; Keller and Waller 2002)? Inbreeding depression is not usually thought of in the context of maintaining genetic variation in populations, but if the most homozygous individuals in the population are least fit, it may slow the rate of loss of genetic variation from the population.

A typical approach to measuring inbreeding depression in free-living populations is to use pedigrees to estimate inbreeding coefficients and compare these with trait values (e.g. Keller 1998; Kruuk *et al.* 2002a). Alternatively, since there is a negative correlation between inbreeding coefficient and heterozygosity, individual heterozygosity measured across several loci can be used as a metric of inbreeding. In section 8.4 we use the latter approach to investigate evidence for inbreeding depression in Soay sheep.

Second, in section 8.5 we investigate selection acting in relation to individual loci. Here, our studies have concentrated on expressed loci or loci linked to them, and our investigations have become more sophisticated over time. Ascribing selection to variation at a specific polymorphic locus is actually extremely difficult. Any polymorphic locus is linked to others, and due to the reduced recombination between closely linked markers, alleles at neighbouring loci do not segregate independently of each other (they are said to be in linkage disequilibrium). This means that any apparent selection at a polymorphic locus could be due selection at a linked locus, though obviously it will be strongest at the locus which is the target of selection. Thanks to the development of a high-density linkage map for sheep (Crawford *et al.* 1995; de Gortari *et al.* 1998; Maddox *et al.* 2001), we have been able to choose groups of linked markers in order to pinpoint the target of selection more accurately with time.

As described in Chapter 5, there is strong selection for resistance to gastrointestinal helminths (measured as faecal egg count, FEC) in the Soay sheep population, and heritable variation for this trait. Coevolution between hosts and parasites has attracted a lot of interest from evolutionary biologists, since frequency-dependent selection for resistant alleles in the host and infective alleles in the parasite can, in principle, maintain genetic variation in both (Anderson and May 1982b). In consequence, our studies of selection, both on heterozygosity and on individual loci, concentrate on the evidence that the genetic variation studied is associated with parasite resistance, and on whether selection occurs as a result of this association.

8.2 Loci screened

Prior to the studies described in this chapter, there was relatively little evidence for molecular polymorphism in the Hirta Soay population. Specifically, Soays were known to be polymorphic at three biochemical polymorphisms, haemoglobin (Hb), transferrin (Tf), and red cell potassium (Hall 1974).

All Soays captured from 1985 onwards have been tissue sampled for genetic studies and screened at a range of loci. At the start of the study, the main technique available for population genetic studies was protein or allozyme electrophoresis, in which the products of specific functional loci are examined. Soays were surveyed for polymorphism at a standard range of allozyme loci and polymorphic loci were screened in all individuals sampled (Bancroft *et al.* 1995b).

In the 1980s, hypervariable DNA loci were discovered in most eukaryotes, in the form of minisatellites (Jeffreys *et al.* 1985) and microsatellites (Litt and Luty 1989; Tautz 1989; Weber and May 1989). After a brief phase of multilocus DNA fingerprinting using minisatellites, we concentrated on microsatellite DNA profiling, because this approach is more reliable and because it generates single locus genotypes. Microsatellites consist of tandemly repeated arrays of simple DNA sequences, for example $(CA)_{18}$, with extensive variation generated by a high mutation rate to mutations that add or subtract repeat units (creating, for example, alleles of length $(CA)_{17}$ and $(CA)_{19}$). Located between genes or in non-transcribed introns within genes, microsatellites have no obvious function and are generally considered to be neutral to selection. Microsatellite markers are the tool of choice in genetic mapping studies, and we have taken advantage of large-scale genome mapping programmes in domestic ruminants to find microsatellites with many alleles or in specific locations in the genome. For example, when seeking more markers for paternity analysis, we screened 174 mapped bovine microsatellite loci in Soay sheep (Slate *et al.* 1998).

Information about the loci that have been screened in Soay sheep, including the cohorts in which they have been screened, is summarised in Table 8.1. The loci chosen can be divided into four categories: allozymes, randomly located microsatellites, and microsatellites chosen for their location in or near known functional genes,

Table 8.1. *Loci screened in a large sample of Hirta Soay sheep*

Locus	Cohorts screened	Chromosome	Number of alleles	Observed heterozygosity	Reference to source/ method
Allozymes[a]					
Transferrin (Tf)[b]	85–94	1	7	0.782	Bancroft et al. (1995a)
Haemoglobin (β-Hb)[b]	85–94	15	2	0.497	Bancroft et al. (1995a)
Glutamate oxaloacetate transaminase (Got-2)[b]	85–94	unknown	2	0.357	Bancroft et al. (1995a)
Isocitrate dehyrogenase (Idh-1)[b]	85–94	2	3	0.488	Bancroft et al. (1995a)
Adenosine deaminase (Ada)[b]	85–94	unknown	2	0.365	Bancroft et al. (1995a)
Plasminogen (Plg)[c]	85–94	unknown	2	0.567	Tate et al. (1992)
Randomly located microsatellites					
MAF18[d]	85–94	13	3	0.562	Crawford et al. (1990)
MAF35[d]	85–98	23	4	0.569	Swarbrick et al. (1991)
MAF45[d]	85–98	X	6	0.759	Swarbrick et al. (1992)
MAF65[d]	85–94	15	4	0.517	Buchanan et al. (1992)
Ovine pituitary adenylate cyclase activating peptide (OPACAP)[d]	85–94	23?	3	0.387	Bancroft (1995)
Retinol binding protein 3, interstitial (RBP3)[d]	85–94	25	3	0.643	Fries et al. (1993)
OarFCB304[e]	85–98	19	4	0.588	Buchanan and Crawford (1993)
OarVH34[e, f]	85–98	3	5	0.559	Pierson et al. (1993)
OarRM106[e]	85–94	16	4	0.454	Kossarek et al. (1993)
OarCP26[e]	85–98	4	5	0.711	Ede et al. (1995)

BM1314[g]	95–98	22	8	0.819	Bishop et al. (1994)
BM203[g]	95–98	26	11	0.749	Bishop et al. (1994)
INRA5[g]	95–98	10	9	0.731	Vaiman et al. (1992)
TGLA13[g]	95–98	2	6	0.737	Georges and Massey (1992)
TGLA263[g]	95–98	1	7	0.793	Georges and Massey (1992)
MHC microsatellites					
OLADRB[h]	85–98	20	7	0.777	Creighton et al. (1992)
OLADRBps[h]	85–94	20	6	0.788	Crawford et al. (1995)
OMHC16[h]	85–94	20	5	0.581	Crawford et al. (1995)
BM1815[h]	85–94	20	3	0.506	Bishop et al. (1994)
BM1818[h]	85–94	20	7	0.655	Bishop et al. (1994)
Gamma interferon microsatellites					
BL4[i]	88–94	3	5	0.603	Smith et al. (1997)
o(IFN)-γ	88–94	3	2	0.489	Schmidt et al. (1996)

[a] In the primary allozyme screen another locus, Peptidase-C (Pep-C) was found to be polymorphic and was included in estimates of polymorphism and heterozygosity. However, it proved unreliable to score and was dropped from the main screening programme.

[b] Allozymes screened in primary survey (Bancroft et al. 1995a).

[c] Protein marker screened in secondary survey (Smith 1996).

[d] Microsatellite markers screened by Bancroft et al. (1995).

[e] Microsatellite markers screened by Smith (1996) to enhance paternity analysis.

[f] OarVH34, listed under 'Random microsatellites', is actually in the gamma interferon region (see text).

[g] Microsatellite markers screened by Coltman et al. (1999b) to enhance paternity analysis.

[h] Microsatellite markers within or adjacent to the Major Histocompatability Complex screened by Paterson (1998).

[i] Microsatellite markers within or adjacent to the gamma interferon gene screened by Coltman et al. (2001b).

the ovine major histocompatability complex (the MHC) and the ovine gamma interferon locus (o(IFN)-γ). Since one of the main motivations for molecular genetic studies of Soays is paternity analysis (see Chapter 6), among the randomly located microsatellites we have moved, over time, towards screening more informative loci, i.e. with more alleles and more even allele frequencies. Remaining parts of this chapter present investigations based on the loci listed in Table 8.1.

When all individuals typed are included in the analysis, nearly all loci listed in Table 8.1 are in Hardy–Weinberg equilibrium and there is little overall evidence for temporal change in allele frequencies over time (Smith 1996; Coltman *et al.* 1999a). The only exception is the allozyme locus plasminogen, which has a significant excess of heterozygotes (Smith 1996). Only two randomly located loci, OPACAP and MAF35, show evidence of linkage (Smith 1996), suggesting that OPACAP may map to chromosome 23, although it has not been formally assigned to date (Table 8.1). OPACAP has been dropped from recent analyses to avoid problems of non-independence from MAF35.

8.3 Amount and distribution of molecular genetic variation

Compared with other mammals, Soay sheep are relatively polymorphic at protein loci (Bancroft *et al.* 1995a). Six of 42 (14.2%) allozyme loci originally screened were polymorphic compared with a mean of 4.1% across mammals (Nevo *et al.* 1984). The number of alleles and heterozygosity of individual loci is shown in Table 8.1. Average protein heterozygosity for Hirta Soays, including monomorphic loci, is 0.078 ± 0.033 (SE) compared with a mean of 0.041 ± 0.025 across mammals (Nevo *et al.* 1984). This places Soays in the top 17% of the mammal distribution (Fig. 8.1) (Bancroft *et al.* 1995a). Furthermore, Soays have relatively high interlocus variance in heterozygosity relative to average heterozygosity compared with the predictions of a mutation–drift model, possibly indicating that selection plays a role in maintaining variation (Bancroft *et al.* 1995a).

In contrast, Soays appear to have relatively low variation at randomly located microsatellite loci. Across-species comparisons of levels of microsatellite variation are far less valid than for allozymes, since

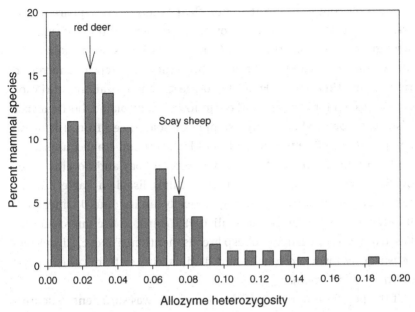

FIG. 8.1. The distribution of average allozyme heterozygosity for 184 mammal species (Nevo *et al.* 1984). Arrows indicate the average heterozygosity found in a survey of 22 populations of the abundant, widespread European red deer (Gyllensten *et al.* 1983) and for the Hirta population of Soay sheep (this study). (From Bancroft *et al.* 1995a; Nevo *et al.* 1984.)

so many more loci are available and different authors use different criteria for selecting loci for screening. One direct comparison is that in our screen of 173 bovine microsatellite loci, Soays had less variation than the abundant, outbred, Scottish red deer, and similar levels to the introduced, bottlenecked UK Japanese sika deer (Slate *et al.* 1998). In another study comparing Soays with nine other European sheep breeds across 20 microsatellite loci (some of which are the same as those shown in Table 8.1), Soays had the second lowest mean number of alleles per locus (4.5; range across breeds 4.0–8.25) and the lowest observed mean heterozygosity (0.45; range across breeds 0.45–0.75) (Byrne *et al.* in press). Compared with many other species of mammal, we have to test more loci to find microsatellites suitable for paternity analysis in Soay sheep.

Relative levels of genetic variation measured by allozymes and microsatellites (or other DNA-level polymorphisms) have not been extensively compared, so it is not clear whether the apparent contrast seen in Soays is unusual. However, a number of lines of evidence strongly suggest that the Hirta Soay population is as isolated as its known history suggests and has not been subject to repeated introductions from other breeds. First, the low level of microsatellite variation is as expected for neutral markers following an isolated, bottlenecked history. Second, when a large sample of Soay sheep from Hirta was compared with 50 modern Scottish black-face and small numbers of post-mortem specimens from Soay sheep on Soay and Boreray sheep on Boreray at several of the microsatellites listed in Table 8.1, the two Soay populations appeared genetically similar and distinct from Boreray or Scottish black-face (Ball 1998). Third, in the microsatellite-based analysis of European sheep breeds mentioned above, Hirta Soays consistently form a coherent and distinct cluster from other breeds (Byrne *et al.* in press).

If the population history of the Soay sheep was sufficient to depress variation at neutral microsatellite loci, then a similar effect would be predicted at the allozyme loci, if they were selectively neutral. The fact that relatively so much allozyme variation is observed suggests a selective mechanism for the maintenance of variation at these loci.

8.4 Selection against inbred individuals

We examined the influence of inbreeding on three aspects of fitness in Soay sheep: parasite resistance (measured as faecal egg count, FEC; see Chapter 5), over-winter survival and female breeding success.

Individual inbreeding can be estimated as inbreeding coefficients measured from pedigrees obtained, in the case of Soays, from observation of mother–offspring relationships and paternity inference using molecular markers (see Chapter 6). Using this approach, the proportion of inbred individuals we are able to detect in Soays is very low: just ten individuals of 898 investigated (5.4%) have inbreeding coefficients of 0.125 and above (Marshall *et al.* 2002). However, this analysis only considered a number of simple pedigree routes by which inbreeding occurs, and if more complete and deeper pedigrees were available we would probably reveal additional inbred individuals, including many with inbreeding coefficients below 0.125.

Instead of inbreeding coefficient, we therefore used individual heterozygosity across up to 14 loci to obtain a measure that is inversely correlated with inbreeding coefficient. To avoid loci that could be the direct target of selection, we measured heterozygosity from randomly located microsatellite loci only (see Table 8.1). To take account of the changes in the panel of markers typed in different cohorts of sheep, individual heterozygosity was standardised by the mean heterozygosity of the loci typed (Coltman *et al.* 1999b).

We analysed August FEC using generalised linear models. FEC has been measured repeatedly on some individuals in the population, so we randomly selected one observation for analysis to ensure statistical independence of data. Several variables are associated with parasite resistance and were fitted in our models (see Chapter 5) and due to differences in the variables involved, different models were constructed for lambs (aged four months) and yearlings and adults. There is no association between heterozygosity and August FEC among lambs, but among yearlings and adults August FEC decreases with increasing heterozygosity ($p = 0.0021$), and there is also an interaction between year, heterozygosity and FEC ($p = 0.021$) (Coltman *et al.* 1999b). The interaction with year is due to population density: faecal egg counts are higher in years of high population density (Chapter 5), and the effect of heterozygosity is relatively stronger in high-density years. In low-density years individuals in the top heterozygosity quartile have approximately 25% lower FEC than individuals in the bottom quartile, while in high-density years the top quartile has less than half the FEC of the bottom quartile (Fig. 8.2) (Coltman *et al.* 1999b).

Since parasites adversely affect survival (Chapter 5), inbred sheep may be less likely to survive the subsequent winter due to their greater susceptibility to parasites. Our analyses of over-winter survival in relation to inbreeding support this hypothesis (Coltman *et al.* 1999b). We analysed survival of sheep experiencing their first high-mortality winter, as juveniles, yearlings and two-year-olds, in the three years 1989, 1992 and 1995, using logistic regression analysis. Probability of survival is associated with sex, age, year and (at $p < 0.009$) heterozygosity; more inbred sheep are less likely to survive (Fig. 8.3). Although juveniles are less likely to survive than older age classes, there was

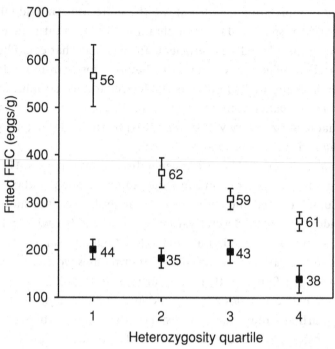

FIG. 8.2. Fitted faecal egg count (FEC) of adult Soay sheep in relation to population density (filled squares represent low-density years (<400 sheep in the Village Bay) and open squares represent high-density years) and individual standardised heterozygosity. Fitted values are those predicted by a generalised linear model with negative binomial error structure of observed FEC measured during August from 1988 to 1997. Mean data are shown blocked by heterozygosity quartiles. Numbers indicate sample size and bars represent one standard error of the mean. (From Coltman *et al.* 1999b.)

no statistical interaction between age class, heterozygosity and the probability of survival (Fig. 8.3).

Taken together, the above analyses suggest that selection against inbred sheep during population crashes may operate through parasite resistance. However they provide only correlative evidence for this connection, and it could be the case that heterozygosity affects parasite resistance and survival independently. Anthelminthic experiments performed in 1989, 1992 and 1995 provide an opportunity to test the causal relationship suggested between inbreeding, parasite

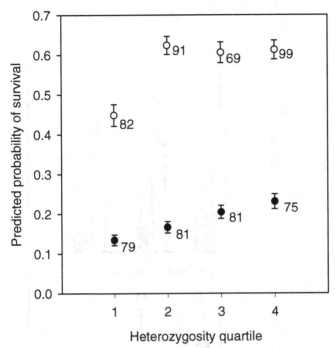

FIG. 8.3. Over-winter survival probabilities predicted by a generalised linear model of juvenile (filled circles) and adult Soay sheep (open circles) that were not treated with anthelminthic in the previous summer, in relation to individual standardised heterozygosity. The data are for the three high-mortality winters 1989, 1992 and 1995. Means are shown blocked by heterozygosity quartiles. Numbers indicate sample size represented by that point and error bars represent one standard error of the mean. (From Coltman *et al.* 1999b.)

resistance and survival (Coltman *et al.* 1999b). In each of these years, some sheep were given anthelminthic treatment that temporarily cleared their parasite burden and generally improved their probability of survival over the subsequent winter (see Chapter 5). If parasite resistance were involved in selection against inbred sheep, we would expect anthelminthic treatment to reduce the difference in survival between inbred and outbred sheep. This prediction is unequivocally confirmed: in treatment years, among untreated sheep, survival is associated with heterozygosity (see above and Fig. 8.4a), while among

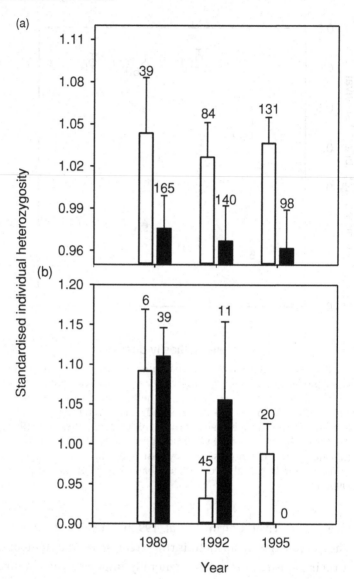

FIG. 8.4. Mean standardised heterozygosity of (a) untreated and (b) anthelminthic-treated Soay sheep that either died (filled columns) or survived (open columns) in years of high over-winter mortality. Bars represent one standard error of the mean. Numbers above columns indicate sample size. (From Coltman *et al.* 1999b.)

treated sheep there is no such relationship (Fig. 8.4b). The experiments show that selection against inbred sheep during over-winter mortality is due to the their lower parasite resistance.

We have also investigated the influence of individual heterozygosity on lifetime breeding success (LBS) of female Soay sheep (Turpie 1999). In preliminary analyses using generalised linear models, LBS appeared to be influenced by heterozygosity. However, when longevity is included in the model of LBS, heterozygosity no longer explains variation in LBS, suggesting that the influence of heterozygosity on female LBS is a consequence of its influence on survival through high-mortality winters, as described above.

If, as we have shown, inbred sheep are at a disadvantage, we might expect the Soays to have evolved a mechanism to avoid mating with relatives. We investigated this question among 806 lambs with known mothers and inferred fathers (see Chapter 6); we estimated relatedness between the parents using microsatellite genotypes (Queller and Goodnight 1989). Mean pairwise relatedness was normally distributed and not significantly different from zero ($r = -0.01 \pm 0.19$ SD), suggesting mating is random with respect to relatedness. A similar conclusion was reached in a study of 152 mother–father–offspring trios screened at additional loci to those shown in Table 8.1 (Ockendon 1999).

8.5 Selection at specific loci

ALLOZYMES AND RANDOMLY LOCATED MICROSATELLITES
As reported in section 8.3 above, allozyme electrophoresis revealed a surprisingly large number of relatively polymorphic loci in Soay sheep (Table 8.1). In studies of the population crashes of 1986, 1989 and 1992, genotypes at allozyme loci and microsatellites screened early on in the project, were fitted as potential explanatory variables in logistic regression models of survival. Similarly, genotypes were tested in some models of faecal egg count, August weight and female fecundity.

Four of the six allozyme loci shown in Table 8.1, and two of the six microsatellites tested, were implicated in models of survival at least once. At transferrin (Tf), male homozygotes survive better than heterozygotes, an observation that was traced to better survival of the two commonest homozygotes, *mm* and *gg* (Bancroft *et al.* 1995b). Glutamate oxaloacetate transaminase (Got-2) genotype was implicated in male survival in a study of all three early crashes (Bancroft *et al.* 1995b) and again in a analysis of the 1992 crash, but this time as an interaction with haemoglobin genotype (Hb) (Illius *et al.* 1995). Genotype variation at the microsatellite loci MAF18 and OPACAP was also associated with survival (Bancroft *et al.* 1995b).

In analyses of parasite resistance, variation at Tf and the microsatellite MAF45 were associated with FEC (Smith 1996). Interestingly, at Tf, the *mm* homozygotes appeared to have lowest FEC, although this result was found for females in spring (Smith 1996), rather than among the males that were previously shown to survive better (Bancroft *et al.* 1995b).

By far the most consistent evidence for selection in relation to an allozyme locus is for adenosine deaminase (Ada). Ada is an enzyme involved in purine metabolism and immune function: severe combined immune deficiency (SCID) children are homozygous for non-functional Ada alleles (Stryer 1988). In Soays, there are two detected alleles, *f* (frequency 0.249) and *s* (0.751), and the three Ada genotypes appear to be associated with variation in several traits. For survival, the overall pattern revealed is that heterozygotes survive over-winter mortality best and that *ff* homozygotes survive worst (Gulland *et al.* 1993; Bancroft *et al.* 1995b; Moorcroft *et al.* 1996). Consistent with this, Ada *ff* individuals appear to have the lowest parasite resistance, having higher FEC (Gulland *et al.* 1993; Smith 1996). Ada genotype is also associated with variation in female August weight and conception rate: amongst adults, *ff* females tend to be heavier and conceive at a higher rate (Clutton-Brock *et al.* 1997b), though this may be dependent on population size (Pemberton *et al.* 1996).

The results summarised above are of considerable interest, since they indicate that at least some of the allozyme loci are associated with fitness variation between individual Soays. Also, they suggest three mechanisms by which genetic variation could be maintained in the population: through antagonistic effects in different fitness components, through interactions with other variables such as density (Pemberton *et al.* 1996) and, in the case of Ada, through heterozygote advantage (overdominance) (Gulland *et al.* 1993). However, the evidence should be treated with some caution: some of the results are based on small sample sizes, were only marginally significant, or should have been treated more stringently due to the multiple tests involved. Ideally, the results would be replicated in larger, independent data sets. Finally, notice that we cannot exclude the possibility that the results reported are due to associations at linked loci. As described below, in recent studies we have explicitly addressed this problem.

MAJOR HISTOCOMPATIBILITY COMPLEX
In vertebrates, the Major Histocompatibility Complex (MHC) comprises a group of closely linked expressed genes that play a central role in immune surveillance and response (Klein 1986). Very high levels of variation have been documented at the MHC loci in a wide range of species (Klein 1986) and several lines of evidence suggest that selection maintains this variation, including relatively even allele frequencies which cannot be explained under neutral evolution (Hedrick and Thomson 1983), high rates of non-synonymous substitution at the antigen-binding sites (APS) (Hughes and Nei 1988, 1989) and the conservation of alleles over large spans of evolutionary time, including through speciation events (Klein *et al.* 1993). The mechanism maintaining this variation has been the subject of intense debate. Some authors have suggested it is a recognition system enabling disassortative, non-inbred matings, and found evidence for this hypothesis (Potts and Wakeland 1990; Potts *et al.* 1991). A more common view is that the variation is maintained through overdominance or negative frequency dependence, resulting from interactions with an evolving parasite and pathogen community (Ebert and Hamilton 1996). A number of studies have reported associations between MHC variation and resistance to pathogens and parasites (Briles *et al.* 1977; Keymer *et al.* 1990; Hill *et al.* 1991; Schwaiger *et al.* 1995), although the precise mechanism of maintenance has not been confirmed.

We genotyped three microsatellite loci located within the MHC (OMHC, OLADRBps and OLADRB) and two microsatellite loci to either side of the MHC region (BM1818 and BM1815) in a large sample of Soay sheep (Table 8.1) (Paterson 1998). The loci within the MHC show linkage disequilibrium ($p < 0.01$) amongst each other but not with the flanking markers. In ruminants the locus OLADRB is of particular interest since it is located immediately adjacent to the expressed Ovar-DRB exon 2, and alleles at this microsatellite are indicative of particular sequence variants at the exon (Schwaiger *et al.* 1994). In Soays, we found five expressed Ovar-DRB exon 2 sequence variants and they were each associated with specific alleles at the DRB microsatellite (Paterson 1998). We have used a range of analyses to investigate selection acting on MHC variation in Soays, including fitting genotypes at each microsatellite locus to generalised linear models of survival and parasite resistance (FEC) after fitting other variables known to be associated with these traits (Paterson and Pemberton 1997; Paterson 1998; Paterson *et al.* 1998).

The Soay MHC region has two main signatures of natural selection (Paterson 1998). Allele frequencies at OLADRBps and OLADRB are more evenly distributed than expected under neutral evolution (Watterson test, $p < 0.01$ (Watterson 1978)), and the Ovar-DRB exon 2 sequence variants show an excess of non-synonymous substitutions in the antigen-binding sites ($p < 0.05$) (Paterson 1998).

Soay sheep do not mate assortatively with respect to variation at the MHC (Paterson and Pemberton 1997). MHC microsatellite genotypes of mated pairs (identified by paternity analysis; see Chapter 6) were compared, and there was no evidence for deviation from mating at random with respect to genotype at any of the loci. This result does not exclude the possibility that sperm or zygote selection could be occurring as has been demonstrated in mice (Wedekind 1994).

Over-winter survival of juvenile Soay sheep is related to genotype at all three microsatellite loci within the MHC (p between 0.024 and 0.041), but not to genotype at the two flanking markers (Paterson *et al.* 1998). Furthermore, enhanced or reduced survival is related to the possession of one or two copies of particular alleles at these loci (i.e. an additive model), rather than to being heterozygous. These associations are not main effects in the models, they emerge as interactions with August weight. In yearlings, the three MHC loci, but not the flanking markers, again show additive effects on survival, this time as main effects (p between 0.016 and 0.044). In yearling survival there are also marginally significant associations ($p = 0.049$ and 0.044) at OMHC1 and one of the flanking markers, BM1815 under a heterozygous-advantage model.

The associations with survival are strongest for the locus OLADRB, and since associations at the MHC loci are not independent, due to linkage disequilibrium (see above), we investigated allele effects in survival and parasite resistance further at this locus. At OLADRB, the 257-bp allele, which is associated with reduced juvenile over-winter survival, is also associated with higher August FEC (Table 8.2a). Among yearlings, two other alleles appear to confer differences in over-winter survival and August FEC, the most consistent observation being that the 263-bp allele is associated with enhanced over-winter yearling survival and lower August FEC (Table 8.2b). Finally, the OLADRB associations found are consistent with the expectations of negative frequency-dependence in that the 257-bp allele, associated with reduced parasite resistance and reduced juvenile over-winter survival, is the commonest allele at OLADRB (frequency 0.236), and the 263-bp allele, associated with enhanced parasite resistance and over-winter survival in yearlings, is one of the rarer alleles for which we have any statistical power (frequency 0.133).

The evidence from studies of MHC variation in Soays is consistent with frequency-dependent selection by parasites, since particular alleles, rather than heterozygosity itself, are related to survival and parasite resistance (Paterson *et al.* 1998). However, we cannot exclude the possibility that selection at different host life-history stages or by different parasites may produce overall heterozygote advantage, particularly since the alleles under selection do appear to differ between juveniles and yearlings (Table 8.2).

GAMMA INTERFERON (O(IFN)-γ)

Gamma interferon is a cytokine involved in switching between the Th$_1$ and Th$_2$ cell responses. The Th$_1$ response, when up-regulated, protects against intracellular parasites but provides relatively beneficial

Table 8.2. *Associations between alleles at the MHC microsatellite OLADRB and over-winter survival and faecal egg count for (a) juvenile and (b) yearling Soay sheep on Hirta*

OLADRB allele (bp)[a]	Cofficient of survival (±SE)	Coefficient of FEC (±SE)
(a) *Juveniles*		
205	0.026 (±0.055)	−0.119 (±0.089)
213	0.005 (±0.050)	−0.070 (±0.106)
257	**−0.181 (±0.061) ($p < 0.01$)**	**0.252 (±0.077) ($p < 0.01$)**
263	0.085 (±0.053)	−0.036 (±0.099)
267	0.091 (±0.080)	−0.009 (±0.091)
276	−0.029 (±0.057)	−0.179 (±0.101)
(b) *Yearlings*		
205	**−0.898 (±0.310) ($p < 0.01$)**	0.115 (±0.117)
213	−0.004 (±0.438)	−0.121 (±0.177)
257	0.311 (±0.293)	−0.159 (±0.117)
263	**0.726 (±0.359) ($p < 0.05$)**	**−0.330 (±0.145) ($p < 0.05$)**
267	−0.423 (±0.400)	**0.340 (±0.139) ($p < 0.05$)**
276	0.384 (±0.382)	0.134 (±0.150)

[a] Six alleles at OLADRB were fitted under an additive model of survival and FEC; the remaining two alleles present at this locus are too rare for analysis. Coefficients shown are main effects except for juvenile survival, which are for weight interactions with OLADRB alleles. The coefficients are derived from generalised linear models assuming a binomial error distribution for survival and a negative binomial distribution for FEC. Significant associations are shown in bold. In juveniles, the 257-bp allele is associated with both lower survival and increased FEC. In yearlings, the 263-bp allele is associated with both increased survival and reduced FEC; the 205-bp allele is also associated with reduced survival and the 267-bp allele with increased FEC.
Source: From Paterson *et al.* (1998).

conditions for extracellular parasites such as nematodes, while the reverse is true for the Th$_2$ response (Else *et al.* 1994; Wakelin 1996; Grencis 1997). In independent searches for quantitative trait loci (QTL) conferring resistance to parasitic helminths in sheep, two studies found a QTL mapping to the region of sheep chromosome 3 where this candidate gene is located (Beh *et al.* 1998; Crawford and McEwan 1998). At a diallelic microsatellite locus located in the first

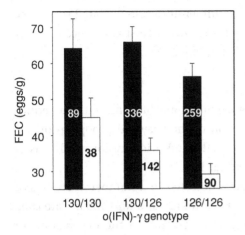

FIG. 8.5. Mean FEC observed in lambs and yearlings (solid and open bars respectively) plotted by gamma interferon microsatellite genotype. Numbers within the columns give sample size and bars indicate standard error. (From Coltman *et al.* 2001b.)

intron of gamma interferon, which had been previously identified in sheep (Schmidt *et al.* 1996), sheep lines selected for high and low resistance have different allele frequencies (Crawford and McEwan 1998).

In Soays, the gamma interferon microsatellite also has two alleles, at 126 bp and 130 bp, and we investigated whether genotypes at this marker or at flanking microsatellites BL4 and VH34 (Table 8.1) conferred resistance to nematode parasites on St Kilda in individuals born between 1988 and 1998 (Coltman *et al.* 2001b). In both lambs and yearlings, FEC decreases with the number of 126-bp alleles carried by the host (lambs, $p = 0.047$; yearlings, $p = 0.0042$) (Fig. 8.5). The effect is stronger in yearlings than lambs: in lambs 126/126 individuals have on average 9.4% lower FEC than 130/130 individuals, whereas in yearlings the difference is 31.2%. These differences are accompanied by differences in the level of immunoglobulin A (IgA) circulating in the blood, which is positively correlated with the number of 126-bp alleles in lambs ($p = 0.011$). No such associations are apparent at the flanking markers. The increasing effect with age is consistent with the development of immune function, though clearly this interpretation does not apply to the MHC results reported above. In contrast to the other associations with parasite resistance described above, there is no evidence of a correlation between gamma interferon genotype and the probability of over-winter survival (Coltman *et al.* 2001b). Given the function of gamma interferon, this result raises the interesting possibility that the gamma-interferon-mediated resistance to gastrointestinal helminths is traded off against resistance to other kinds of parasites with intracellular stages.

8.6 Discussion

We have described evidence for a general mechanism that may reduce the rate of loss of genetic variation in the Soay sheep population, selection favouring outbred individuals, and evidence for specific association with variation at individual loci.

Inbreeding depression has been detected in a number of studies of free-living organisms (Keller 1998; Crnokrak and Roff 1999; Slate *et al.* 2000a; Keller *et al.* 2002; Kruuk *et al.* 2002a), so the observation of inbreeding depression in Soay sheep is not especially unusual. So far, however, the impact of the inbreeding depression detected is not very high – heterozygosity and its interaction with year explain around 5% of the total variance in yearling and adult FEC and less than 1% of the total deviance in over-winter survival. It is possible that more inbreeding depression will be detected when the analyses can be repeated using inbreeding coefficients based on a pedigree with more generations than we have currently monitored.

Our analyses of inbreeding depression in parasite resistance confirm the emerging consensus that inbreeding depression is more severe under adverse environmental conditions (Dudash 1990; Meagher *et al.* 2000; Keller *et al.* 2002). The difference between the most inbred and most outbred quartiles of the population increased with rising density (Fig. 8.2). We cannot make the same assertion about survival, since there is so little mortality between crashes that we only analysed crash survival.

Inbreeding depression in parasite resistance (here measured as FEC) has not been widely reported. Further research is needed to understand the mechanism behind this observation. Is it the result of heterozygote advantage at loci involved in defence against specific parasites, or does an overall heterozygote advantage arise as a consequence of superiority of defence against multiple parasite strains or species as suggested by recent experimental work in mice (Penn *et al.* 2002)? Or perhaps FEC is acting as an indicator of an individual's overall condition, and the loci underlying the observed relationships have nothing to do with specific parasite de-fence mechanisms.

Whatever the underlying mechanism for the inbreeding depression detected, it has two interesting consequences at the population level. First, population crashes are associated with *increases* in mean population heterozygosity, and then, as the population reproduces after a crash, homozygous genotypes are restored and heterozygosity *decreases* (Fig. 8.6). This is entirely consistent with the observations described above, but it is not the response normally expected by population geneticists for heterozygosity at neutral loci, which, if anything, are expected to stay the same or lose variation during population declines. It seems possible that the observed effect will reduce the rate of loss of rare alleles from the population, though this requires further investigation.

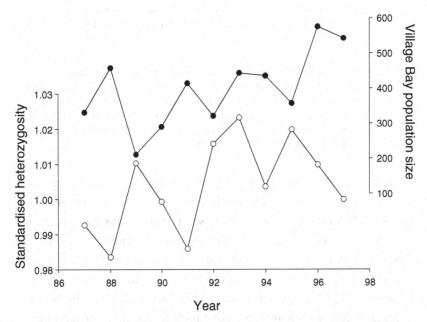

FIG. 8.6. Mean standardised heterozygosity of the population alive in each year (open circles) varies inversely with population size (closed circles). Population size is the number of individuals of all ages known to be alive on 1 October and sighted during at least one census of the study area in that year. The number of sampled individuals represented by heterozygosity each point ranges from 154 (1989) to 608 (1997).

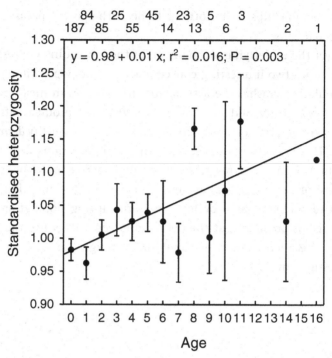

FIG. 8.7. Relationship between mean standardised individual heterozygosity of marked Soay sheep known to be alive in August 1997 and age. Bars represent standard error and numbers along the top axis represent sample size represented by that point. Data are restricted to individuals that had never been treated with anthelminthics during their life. (From Coltman *et al.* 1999b.)

Second, among the sheep alive at any one time in the population, there is a positive correlation between age and heterozygosity (Fig. 8.7) (Coltman *et al.* 1999b). Again, this observation is consistent with the results obtained for selection on individuals reported above, and along with the selection reported in Chapter 7, indicates that the animals that survive a crash are genetically different from those that did not. The age of and reproductive differences between males and females clearly explain much about their differences in crash survival and longevity (see Chapters 2, 3 and 9), but a future avenue for investigation is whether the survival rate of prime-aged females, which influences the dynamics of the population (sections 3.5 and 3.8) can be partly attributed to their being genetically superior as a result of selection in earlier crashes experienced in their lifetime.

Our studies of associations between genotype at individual loci and fitness traits have grown more sophisticated with time, and only for the MHC and gamma interferon can we claim to have narrowed the associations found to a particular genome region. In future studies, we need to examine markers surrounding the allozyme loci, a project currently hampered by the fact that three of the six allozymes genotyped have not been mapped. By studying adjacent markers and a greatly enlarged data set, we should be able to determine whether the effects noted for allozymes (and indeed random microsatellites) are genuine.

At individual loci, we have concentrated on loci with a known function in immune defence (i.e. Ada, MHC and o(IFN)-γ). It is exciting to have found so many associations involving FEC, but again it should be borne in mind that FEC is a relatively crude index of parasite resistance, and it is possible that it reflects the general condition of the host better than it reflects a specific resistance to nematode infection.

For two loci (Ada and MHC) we have also found associations with survival that are consistent with the associations with parasite resistance, in that less resistant genotypes or carriers of less resistant alleles are less likely to survive. In consequence, one might predict a change in allele frequencies over time. We have not observed such changes yet, which suggests one of two possibilities. First, the time elapsed may have been too short to observe such changes. Second, there may be other selective forces that maintain polymorphisms at the observed allele frequencies. There is some evidence in favour of the latter. At Ada, the reduced survival of the *ff* genotype may be compensated for by enhanced fecundity of *ff* females. At OLADRB in the MHC, different alleles appear to confer parasite resistance and survival differences at different host ages (juveniles versus yearlings), and experimental work in mice suggests a role for parasite species diversity in maintaining MHC diversity (Penn *et al.* 2002). At o(IFN)-γ we have not detected differences with survival; since o(IFN)-γ is involved in switching between Th$_1$ and Th$_2$ cellular responses, one possibility is that the 130-bp allele confers protection to an intracellular parasitic threat which we have not yet identified (Coltman *et al.* 1999b).

Despite the many further questions raised by our studies of molecular genetic variation in Soay sheep (above), we have revealed enough associations to suggest that selection plays a role in explaining why Soays have rather higher levels of variability at some loci than would be expected from their population history.

9

Adaptive reproductive strategies

I. R. Stevenson *University of Stirling*

P. Marrow *BTexact Technologies, Ipswich, UK*

B. T. Preston *University of Stirling*

J. M. Pemberton *University of Edinburgh*

and

K. Wilson *University of Stirling*

9.1 Introduction

When in its life should an individual first attempt to reproduce? How often should it breed thereafter? How much effort should it invest in each attempt? And does this vary between individuals in the same population? These questions are central to the many studies investigating adaptive life-history strategies, across the taxonomic spectrum, yet detailed answers are provided by few, particularly in large, free-ranging species. This is because comprehensive data on the costs and benefits of reproduction throughout life are essential for the task, but are difficult to collect in the wild, especially for males in polygynous species. With Soay sheep, however, we have the detailed information with which to investigate the reproductive benefits and costs for both sexes, and the way these vary with the environment and individual phenotype.

In Soays, the costs and benefits of reproduction take on even greater significance given the domestic roots of the population. A brief examination of the Soay life-history reveals unusual patterns that have led many to question whether these are simply the maladaptive legacy of past domestication. Soays certainly bear the hallmarks of artificial selection for high productivity (Chapter 1). Both sexes, but particularly males, mature early (Fig. 9.1) despite incurring high survival costs; and females continue to display fatally high fecundity at peak population densities (Chapter 2). These costs suggest that Soays are too

243

Soay Sheep: Dynamics and Selection in an Island Population, ed. T. H. Clutton-Brock and J. M. Pemberton.
Published by Cambridge University Press. © T. H. Clutton-Brock and J. M. Pemberton 2003.

fecund for their own good. Yet it remains possible that these are not the result of domestication, but are adaptive responses to their environment. Life-history theory provides a framework within which we can use the observed costs and benefits of reproduction to predict optimal fecundity schedules and to compare these with the patterns we observe.

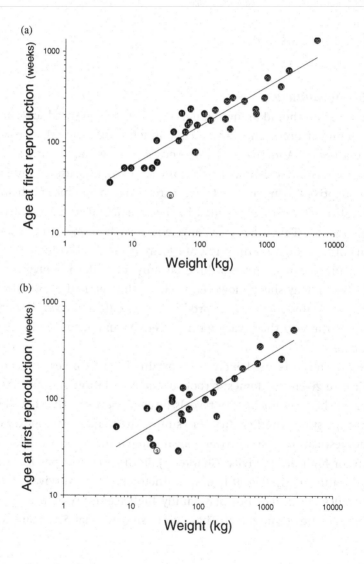

FIG. 9.1. Allometric plots of age at first mating/conception for (a) males and (b) females in the Orders Perissodactyla, Artiodactyla and Proboscidea. Age at maturity is taken as the age at which males are first noted to be actively involved in the breeding season, as distinct from the age at physiological maturity; for females it is taken as age at conception. The equations of the lines are (a) log(age) = 0.46.log(weight) + 1.24, $r^2 = 0.82$ (b) log(age) = 0.42.log(weight) + 1.17, $r^2 = 0.76$. The species represented are as follows. Males: 1 *Madoqua kirkii* (Kellas 1954); 2 *Neotragus moschatus* (Kingdon 1979); 3 *Hyemoschus aquaticus* (Kingdon 1979); 4 *Muntiacus reevesi* (Corbett and Harris 1991); 5 *Tayassu tajacu* (Sowls 1966); 6 *Capreolus capreolus* (Corbett and Harris 1991); 7 *Gazella thomsonii* (Robinette and Archer 1971); 8 Soay sheep; 9 *Ovis musimon* (Schaller 1977); 10 *Antilocapra americana* (Burt and Grossenheider 1976); 11 *Tragelaphus scriptus* (Kingdon 1979); 12 *Aepyceros melampus* (Jarman 1979); 13 *Gazella granti* (Walther *et al.* 1983); 14 *Oreamnos americanus* (Geist 1964; *et al* Houston *et al.* 1989); 15 *Ovis canadensis* (Blood *et al.* 1970; Hogg 1987); 16 *Phacochoerus acthiopicus* (Kingdon 1979); 17 *Kobus kob* (Buechner *et al.* 1966); 18 *Rangifer tarandus* (Leader-Williams 1988); 19 *Damaliscus lunatus* (Estes 1992); 20 *Cervus elaphus* (Clutton-Brock *et al.* 1982a); 21 *Kobus defassa* (Spinage 1982); 22 *Equus burchelli* (Kingdon 1979); 23 *Hippotragus equinus* (Kingdon 1979); 24 *Camelus dromedarius* (Kingdon 1979); 25 *Alces alces* (Burt and Grossenheider 1976); 26 *Syncerus caffer* (Sinclair 1974); 27 *Taurotragus oryx* (Kingdon 1979); 28 *Bison bison* (Burt and Grossenheider 1976); 29 *Diceros bicornis* (Schenkel and Schenkel-Hulliger 1969); 30 *Hippopotamus amphibius* (Laws and Clough 1966); 31 *Ceratotherium simum* (Owen-Smith 1974); 32 *Loxodonta africana* (Perry 1953; Laws 1966). Females: 1 *Rhynchotragus kirkii* (Kellas 1954); 2 *Sylvicapra grimmia* (Haltenorth and Diller 1977); 3 *Soay sheep*; 4 *Gazella thomsonii* (Robinette and Archer 1971); 5 *Tayassu tajacu* (Sowls 1955; MacDonald 1984); 6 *Ovis musimon* (Bon *et al.* 1993); 7 *Gazella granti* (Walther *et al.* 1983); 8 *Aepyceros melampus* (Jarman 1979); 9 *Odocoileus hemionus* (Leberg and Smith 1993); 10 *Antilocapra americana* (MacDonald 1984); 11 *Kobus kob* (Buechner *et al.* 1966); 12 *Cervus elaphus* (Clutton-Brock *et al.* 1982a); 13 *Kobus leche* (Sayer and van Lavieren 1975); 14 *Ovis canadensis* (Schaller 1977); 15 *Alcelaphus buselaphus* (Kingdon 1979); 16 *Kobus defassa* (Haltenorth and Diller 1977; Spinage 1982); 17 *Conochaetes taurinus* (Millar and Zammuto 1983); 18 *Equus burchelli* (Kingdon 1979); 19 *Taurotragus oryx* (Kingdon 1979); 20 *Syncerus caffer* (Sinclair 1974); 21 *Giraffa camelopardalis* (MacDonald 1984); 22 *Diceros bicornis* (Schenkel and Schenkel-Hulliger 1969); 23 *Hippopotamus amphibius* (Laws and Clough 1966); 24 *Ceratotherium simum* (Owen-Smith 1974); 25 *Loxodonta africana* (Perry 1953; Laws 1966). ((a) from Stevenson and Bancroft (1995); (b) from Stevenson (1994)).

Optimal life-history patterns are determined by the fitness costs and benefits resulting from different reproductive strategies. These costs and benefits underlie the trade-offs between different components of fitness that result if resources used for reproduction are diverted from somatic maintenance or future reproduction. For example, high fecundity may allow individuals to produce a large number of offspring, but, if this greatly reduces their future survival, lifetime reproductive output may be reduced. In this case, reduced early fecundity could evolve as an adaptive strategy; but we must be certain that it does not result, instead, from non-adaptive constraints, such as the routing of resources towards growth rather than reproduction (see section 1.4). To understand whether the life-histories of Soay sheep are adaptive we must understand how costs and benefits change with age at first reproduction and investment at each reproductive opportunity.

The benefits of reproduction for young animals are often low, and the costs high, since maturation is an expensive process (Bernardo 1993) and must compete with growth for allocation of resources (Gadgil and Bossert 1970; Stearns 1989). Fecundity is therefore often low for young individuals (Stearns 1992), and any offspring that do result may be smaller or unviable because of poor nutrition before or after birth, or because of parental inexperience in species that care for young (e.g. Røskaft *et al.* 1983; Bell 1984; Festa-Bianchet *et al.* 1995). Young males also tend to be relatively unsuccessful breeders, since access to females is often strongly dependent on body size (studies in Clutton-Brock 1988a; Roff 1992). Moreover, females may actively avoid the attentions of young, inexperienced and unproven males (Cox and Le Boeuf 1977; Owen-Smith 1993; Sabat 1994). In addition to inferior performance, young breeders may experience additional costs such as reduced growth and lower future fecundity (Bercovitch and Berard 1993; Festa-Bianchet *et al.* 1995), greater parasitism (Festa-Bianchet 1989a), or lower survival (Harvey and Zammuto 1985; Reiter and Le Boeuf 1991).

While the 'decision' of when to mature is taken only once, individuals must then choose how much effort to invest at each subsequent breeding attempt. Again, there are costs and benefits to be considered, both for parent and offspring, since there are physical and physiological limits to how hard an individual can work (Drent and Daan 1980). When these bounds are reached, the per capita investment in each offspring will decline and, with it, the viability of the offspring (Reid 1987). For example, egg quality often declines as clutch size increases in birds (Monaghan *et al.* 1995), while in many taxa the long-term consequences of high fecundity can include reduced future success (Luckinbull and Clare 1985), lower parasite resistance (Ots and Hõrak 1996) and poorer survival (Bryant 1979; Seigel *et al.* 1987; Westendorp and Kirkwood 1998). For males, too, reproductive costs can be substantial, particularly in polygynous species (Promislow 1992). Perhaps the most extreme example of costly mating activity is observed in the Australian marsupial mouse *Antechinus* spp., in which every male dies following the first breeding season (Woolley 1966; Lee and McDonald 1985).

Reproductive costs vary widely with both environmental conditions and individual circumstance (section 1.4). Early maturity may be favoured where offspring can make a large contribution to population growth (Fisher 1930; Stearns 1992), or where variable life expectancy reduces the marginal survival costs of heavy reproductive investment. In the latter case, high levels of non-reproductive mortality may mean that the reproductive costs of early breeding are never experienced. When infected by castrating nematodes, for example, water snails massively increase investment in current reproduction at the expense of future survival (Minchella and Loverde 1981), and similarly, parasitic wasps increase reproductive investment when faced with adverse weather likely to increase their mortality (Roitberg *et al.* 1993). In addition, the balance between costs and benefits is strongly influenced by variation in individual condition and quality. Red deer hinds in good condition suffer few costs and give birth to offspring with improved survival chances compared to those of poor-condition hinds, which in turn have higher mortality rates and are also less likely to breed the following year (Mitchell and Lincoln 1973; Clutton-Brock *et al.* 1982a, 1983; Albon *et al.* 1986). Again parasitism can alter the balance between costs and benefits of reproduction as in great tits where a trade-off between offspring quality and number was found only in the presence of ectoparasites (Richner *et al.* 1993) or bighorn sheep where young ewes only showed a mortality cost of reproduction if also infected by pneumonia (Festa-Bianchet 1989a).

COSTS AND BENEFITS OF REPRODUCTION FOR SOAYS

The costs of early reproduction for Soay sheep on St Kilda are clearly high in some years (Chapter 2). For example, when mortality is high, conception reduces a juvenile female's chance of survival further (section 2.9; Fig. 2.14), and any offspring produced are low quality, around one-third lighter than those of adult females and with lower survival (Clutton-Brock *et al.* 1996). Twinning also has high costs for Soay sheep, resulting in reduced survival rates for both offspring and mother (section 2.7–2.9). High fecundity is not always costly, however, and in non-crash years neither juvenile reproduction nor twinning by adults affects mortality risk.

Early maturity of Soay rams also appears to have high costs, but few benefits. Mortality of juvenile males is higher than that of females, following the rut, despite their greater body size (section 2.8). Even though rutting behaviours are costly, juvenile Soays are rarely successful in holding consorts with females and indeed are often actively avoided by females at the peak of oestrus (Grubb and Jewell 1973; I. R. Stevenson, pers. obs.)

Although the costs of reproduction are high for both sexes, they vary widely both with environmental conditions and phenotypic quality. For females, body weight strongly affects reproductive performance: as in many species, heavier individuals have a greater chance of breeding successfully at all ages, and produce heavier offspring, which in turn have superior survival (sections 2.7 and 2.8) (Saether and Heim 1993; Sand and Cederlund 1996; Festa-Bianchet 1998). For males, the benefits of reproductive activity are closely linked to changes in the social environment, which is highly variable due to the frequent periods of high male-biased mortality (Chapter 3) (Clutton-Brock *et al.* 1991; Stevenson 1994). Following population crashes, young males experience a release from competition and may gain considerable reproductive success (Chapter 6) (Bancroft 1993; Stevenson and Bancroft 1995; Pemberton *et al.* 1996). Thus, assessment of the optimal life-history for Soay sheep is not a simple task, and must address variation in individual quality and the environment. This can only be addressed by modelling the system mathematically.

9.2 Optimal life-histories for female Soay sheep

LIFE-HISTORY MODELLING

Using optimality modelling (Parker and Maynard Smith 1990) it is possible to calculate the reproductive strategies that would maximise individual fitness and then compare these with observed behaviour. In particular, we investigate how the optimal strategy varies with an animal's 'state' (a combination of phenotype and environment) by using state-dependent life-history modelling (McNamara 1991; McNamara and Houston 1992, 1996). This technique uses stochastic dynamic programming (Bellman 1957) to predict the long-term fitness consequences of life-history decisions. The key state variables incorporated in these optimality models are body weight, age and population density. Using empirical data, we parameterised the model with the survival costs and reproductive benefits for females making different reproductive decisions (skip reproduction, produce one offspring or produce two offspring) (Clutton-Brock *et al.* 1996; Marrow *et al.* 1996).

We initially modelled population density by assuming that population crashes occurred on a three-year cycle, and that stages of the cycle succeeded each other regularly. We termed this the 'perfect information model' since, in effect, females could predict population fluctuations. Subsequently, we modified the assumption of a regular cycle to produce a 'random years' model. In this, different levels of population density succeeded each other at random and so females had no information about impending population fluctuations. The random years model is more realistic since we have no evidence of adaptive mechanisms that allow females to make predictions about future population fluctuations. Since the peak of mortality occurs several months after the breeding season (Chapter 3), females make their reproductive decisions when forage is still plentiful and before they have to suffer the consequences of their actions. In addition, this model reflects more accurately the unpredictable population dynamics of recent years rather than the apparently regular 'cycles' that were observed in the 1980s and early 1990s (Chapter 3).

PREDICTED OPTIMAL STRATEGIES

If females cannot predict population fluctuations (the random years model), then they cannot fine-tune their behaviour but, instead, must follow a strategy that is optimised for average conditions across years. In this case, fecundity cannot respond to population density, but can still be modified according to an individual's weight and age (Fig. 9.2) (Perry 1953; Laws 1966).

The model predicts that the optimal strategy for adult females is to breed every year, with the heaviest individuals giving birth to twins. Juvenile females are predicted to conceive single offspring, unless they are relatively heavy, in which case they are expected to avoid reproduction. These predictions do not differ significantly from the

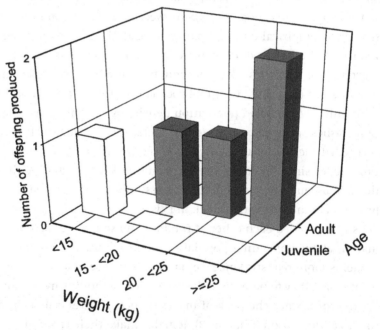

FIG. 9.2. Predicted optimal strategies for females of different ages and weights (random years model). The predicted behaviour is the same across all stages of the population cycle. (From Clutton-Brock *et al.* 1996.)

reproductive patterns observed in the field across the range of population density (G-test $= 1.8$, 4 df, $p > 0.5$) (Marrow *et al.* 1996). The agreement is particularly good for adult ewes, suggesting that their reproductive behaviour is close to optimal (Marrow *et al.* 1996). The fit is less good for juveniles, however, with the model over-estimating the fecundity of low-weight individuals and under-estimating that of heavy juveniles. This is likely to reflect its simple nature and, as we show below, these predictions for juveniles are very sensitive to small changes in conditions (see 'Early reproduction and twinning as adaptive strategies for females').

A further test of our model can be gained by comparing the theoretical population dynamics, calculated from optimal strategies, with the dynamics displayed by the real population. The predicted dynamics show a regular cycle (Fig. 9.3), but no long-term changes in population

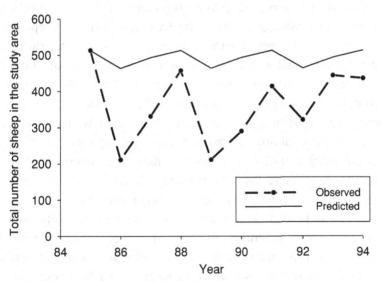

FIG. 9.3. Predicted and observed population dynamics. This shows the predicted population dynamics that result when all females are assumed to follow the random years optimal strategy. The model was started with one animal in each state, and the total population density evaluated over 100 annual decisions. The last nine years are shown scaled so that the population size in 1985 is the same as that observed in the real population. The 'observed' plot is the total number of sheep recorded in the study area between 1985 and 1994. (After Marrow *et al.* 1996.)

density. This is similar to the dynamics observed, though the amplitude of fluctuations is less. The fact that the simulation is more regular than observed in the field is likely to be due to the action of the weather (Chapter 3) (Grenfell *et al.* 1998; Coulson *et al.* 2001), which is excluded from our model. Combined with the similarities between the optimal strategy and observed levels of fecundity, and between the levels of fecundity predicted across population fluctuations, this information suggests that female Soays follow a close-to-optimal reproductive strategy based upon their weight and age and limited cues.

EARLY REPRODUCTION AND TWINNING AS ADAPTIVE
STRATEGIES FOR FEMALES

We can identify the selective forces that restrict the reproductive tactics of females by examining the sensitivity of the optimality results to changes in the reproductive costs (Clutton-Brock *et al.* 1996; Marrow *et al.* 1996). By changing the mortality costs of different reproductive actions, we can determine the threshold required to alter the optimal strategy for females in a given state. This analysis demonstrates that, although the costs are often high, early breeding probably represents an optimal strategy for female Soays (Fig. 9.4). The mortality risk of breeding for low-weight juveniles must increase by about 12% before delayed reproduction is favoured (Fig. 9.4a). By contrast, a *reduction* in mortality risk of 4% is needed to make reproduction optimal for heavy juveniles (Fig. 9.4b). Such a small change in mortality costs may explain the discrepancy between the predicted strategy ('don't breed': Fig. 9.2 and Perry 1953; Laws 1966) and observed fecundity (where up to 72% of heavy juveniles are observed to breed: Marrow *et al.* 1996), since environmental variation not included in our model may overcome the mortality costs which make giving birth sub-optimal.

FIG. 9.4. Results of sensitivity analysis of mortality costs of reproduction. Fitness is in arbitrary units. 'Mortality increase' shows the extent to which the mortality costs of reproduction need to change from values predicted from field data, in order to produce the changes in fitness shown. The graphs are drawn for: (a) juveniles of <15 kg; (b) juveniles of 15–20 kg; (c) adult females of 15–20 kg; (d) adult females of 20–25 kg; (e) adult females of >25 kg. (From Clutton-Brock *et al.* 1996.)

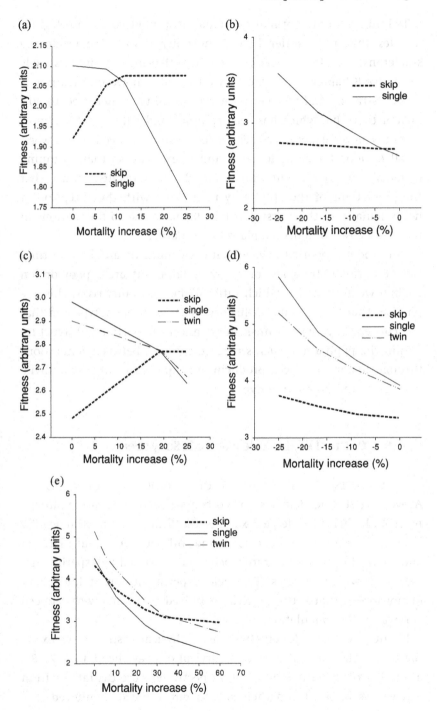

Twinning is only optimal for females in good condition. Lighter females that twin suffer higher mortality than those producing singletons, and also produce poor-quality offspring that are less likely to survive (Chapter 2). For the heaviest females, however, twinning is a good strategy and will persist until mortality is increased to 40% greater than that which has been recorded and a level that is almost certainly never experienced (Fig. 9.4e). Failing to breed is the worst of all options for adult females, and is even worse than twinning by females in poor condition (Fig. 9.4c, d). These results confirm that the predictions of the optimality model are, with the exception of heavy females in their first year (Fig. 9.4b), robust to fluctuations in mortality costs, of a degree plausible in nature.

Our model supports the view that early maturity and high fertility can be optimal strategies for heavy females, but are a poor option for lightweight females, which suffer higher mortality risks. This emphasises that the optimal life-history varies with phenotype and that any concept of a 'species life-history' ignores important substructure within the population. Such structure can pass between generations through maternal effects, producing unexpected results for individual fitness and population dynamics.

9.3 Optimal life-histories for male Soay sheep

ESTIMATING THE SURVIVAL COST OF REPRODUCTION FOR MALES

As we have seen for females, the costs associated with early maturity must be known in order to examine the fitness of the strategy. For males, however, the existence of a trade-off between age at maturity and survival of males has rarely been demonstrated by experimental means in wild ungulates. Two field experiments on St Kilda have demonstrated that rutting activity is indeed associated with reduced survival in this population.

In the late 1970s, Jewell (1997) used physical castration to examine the marked sex difference in survival of Soay sheep. In 1978–80, a number of males ($n = 14$, 8, 50 respectively) were castrated within a few days of birth and their subsequent survival monitored. The

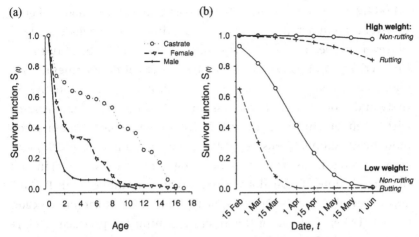

FIG. 9.5. (a) Survival plots for castrates, control males and females from the same cohorts. (Adapted from Jewell 1997.) (b) Survival probabilities for males excluded from rutting by progestogen treatment and control males. Survival plots are shown for 'heavy' and 'light' juveniles (weights in August preceding the rut = 12 kg and 20 kg respectively; mean weight of treated and control males was 15 kg).

results were spectacular, with the survival of castrates outstripping not only their male peers, but also the females of those cohorts and, indeed, several of the following cohorts (Fig. 9.5a). A stark illustration of the cost of being an intact male was observed in March 1989 when, following a particularly severe crash (Chapter 3), there were more castrates alive (the youngest being nine years of age) than there were intact males, of any age, on the entire island (38 castrates, 31 intact males, excluding males *in utero*). Behavioural observations conducted in November 1983 (Jewell 1986) and 1989 (Stevenson 1994) confirmed that the castrates took no part in the rut and fed for around 70% of daylight hours, similar to the feeding level of females, and considerably above the intact males at 30–40%. The last castrate, almost seventeen years of age, died in March 1997; this is the oldest sheep on Hirta since records began. The marked longevity of castrates is in accordance with Fisher (1948) who recorded castrates from the original 1932 introduction (Boyd 1953), which were therefore sixteen years old when spotted in 1948.

Jewell's experiment (1986, 1997) showed that castrates had far better survival than intact males suggesting that it was androgen-dependent. However, it does not specifically connect this male mortality to rutting activity. Androgens control many aspects of development, physiology and behaviour and so a causal link between reproductive activity and mortality cannot be assumed. This link was demonstrated with a refinement of the castration technique, using temporary hormonal castration (Stevenson 1994; Stevenson and Bancroft 1995). Progesterone and its associated steroids (together, 'progestogens') inhibit the production of gonadotrophin-releasing hormone (GnRH) (Knol and Egberink-Alink 1989) which is the start of a 'hormone cascade' (Karsch 1984), stimulating the anterior pituitary to produce luteinising hormone (LH) and follicle-stimulating hormone (FSH) (Knol and Egberink-Alink 1989). Blocking these hormones removes libido, sexual behaviour and aggression (Diamond 1966; Zumpe and Michael 1988; Knol and Egberink-Alink 1989; Zumpe *et al.* 1991), much as physical castration does but with the advantage that the changes are fully reversible (Ericsson and Dutt 1965; Jochle and Schilling 1965). The advantage of this approach is that sexual behaviour can be blocked for the duration of the rut, with minimum effect on 'maleness' either before or after this period. Using this method, in 1991, a group of 18 juvenile males was excluded from the rut in their first year and survival compared with that of controls over the following winter of high population density.

Mortality was some 46% higher amongst the controls than the treated males, though by this stage, plasma testosterone and testicular measurements had returned to normal and were indistinguishable between the groups (Stevenson and Bancroft 1995). Body weight the previous summer strongly influenced mortality and amplified the treatment effect for low-weight males (Fig. 9.5b). Rutting activity was thus demonstrated to reduce over-winter survival probability for young males.

As yet, however, there is no examination of how this cost varies with age. Juveniles, together with very old individuals, are the most likely to die during population declines, and reproductive behaviour may therefore be more risky for these groups than for intermediate age classes.

The cost of reproduction demonstrated by this experiment was an immediate one, but future costs may also exist. If, for example, growth is affected by rutting activity in the first year, males may be unable to attain such a large size and so may be less competitive in future; those males that skip reproduction may grow faster, as well as having higher survival, and so may gain greater reproductive success in subsequent years (unfortunately, too few experimental males were recaught the year after the experiment for this to be examined here). The relative importance of long-term growth costs compared to the immediate over-winter survival costs will vary between years, achieving greater significance at low population density when the survival cost is low. However, alterations of future reproductive success are likely to be relatively unimportant because any benefits of delayed maturity have to be devalued by the risk of not surviving to realise them.

LIFE-HISTORY MODELLING

We used our knowledge of the costs and benefits of reproduction, and their variation with condition, age and density, to predict optimal age at maturity for males. Rather than use stochastic dynamic programming, however, we employed a deterministic, age-structured simulation model (Keen and Spain 1992). The model incorporated a fixed, three-year population cycle (cf. Fig. 9.3), with age- and density-dependent mortality and fecundity rates parameterised from field data. The assumption of a fixed three-year cycle is the same as that in the 'perfect information' model for females, but, as we will see later, the assumptions of the perfect information model are more appropriate for juvenile males. The details of the model are presented in Stevenson and Bancroft (1995), and only a brief summary will be given here.

In order to model the fitness of strategies where first reproduction was delayed, we assumed that the over-winter mortality rate of non-breeders was 0.64 times as great as that of breeders of the same age. This reduction represents the cost of reproduction calculated from the rut exclusion experiment (see above). We varied age at maturity from 0.5 to 5.5 years, according to the strategy adopted. Following maturity, the number of male offspring produced by a sheep of a given age was estimated from the density and age-dependent reproductive success estimates calculated by genetic paternity analysis (Chapter 6). The fitness of an average male, following a given maturation strategy, is estimated by the cumulative·number of male descendants that survive to the strategy-dependent age at maturity, produced over 15 years. This measure incorporates the variable 'discounting' (Charlesworth 1980) that must be applied to the fitness of successive offspring in growing and declining populations. The overall fitness of a strategy was estimated by the geometric mean of the fitnesses calculated for individuals born at each of the three phases of the population cycle. This model allowed us to examine the basic patterns of fitness for individuals following different maturation strategies, but, in future, it would be interesting to examine the effects of stochasticity, as was done for females.

EARLY REPRODUCTION AS AN ADAPTIVE STRATEGY

In years of low and intermediate population density, early maturity is favoured. In such years, mortality is very low and mainly confined to the very young or very old and so there can be little or no *short-term* cost to reproductive activity for juveniles. Moreover, the benefits gained by young males can be appreciable (Fig. 6.2a; Chapter 6). Juveniles should always take part in these ruts, even though reproductive success declines quickly with population density: in the absence of costs, any success is worthwhile, no matter how small.

At high population density, by contrast, juvenile males should skip reproduction. Mortality for juvenile males is always high in these years (even though adult mortality may be low and the population does not decline) (Chapter 3), reaching 99% in extreme years such as 1986, 1989 and 1999 (Chapter 3). Almost half of this mortality risk is due to reproductive activity (see above) and so those males that skip reproduction can greatly increase their probability of survival and do not trade the higher reproductive success as an adult for the low return of a juvenile at high density. Thus, males should display phenotypic plasticity in their juvenile rutting behaviour, responding to population density as a reliable indicator of mortality risk in the coming winter. Observational evidence supports the model predictions, since

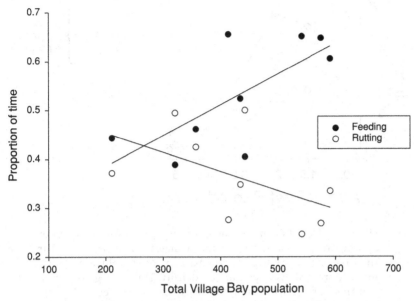

FIG. 9.6. Proportion of time spent feeding and rutting by juvenile males 1989–98 (excluding 1990 when no rut censuses were conducted).

males do indeed display rutting activity that is negatively correlated with population density (Fig. 9.6).

While the average male should follow this strictly density-dependent strategy, variation may be introduced by differences in condition. When the survival cost of reproduction is reduced for particular categories of males (for instance early-born, heavy males (Chapter 2 and Fig. 9.7) with low parasite burdens (Chapter 5)), early maturity may be favoured, even at high population density; but for those in poor condition, delayed reproduction may increase fitness (Fig. 9.7). Thus, we should not expect to observe a single maturation strategy, but individual, state-dependent variation. This prediction mirrors the findings for females (Marrow *et al.* 1996), and is supported by ongoing work comparing the activity budgets of juvenile males of differing quality (see next section).

In summary, there is no single optimal age at first reproduction for male Soays on St Kilda. Rather, the strategy that maximises individual contribution to future generations is influenced by external

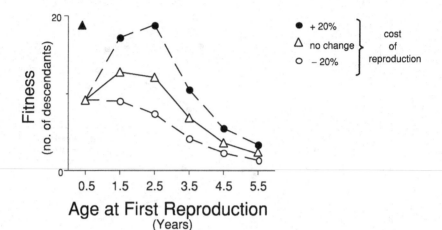

FIG. 9.7. Effect of changing the mortality cost of reproduction on optimal maturation strategy for Soay males. The fitness of maturing in the first year is predicted for the basic model in which males rut as juveniles regardless of population density. This strategy is conservative, and the more realistic plastic strategy that skips precocial reproduction in high-density years has the highest fitness (this is shown by the filled triangle). (From Stevenson and Bancroft 1995.)

factors such as population density, and by individual circumstances like parasite burden, body size, energy stores and date of birth. In the next section, we examine in more detail how variation in phenotype influences the reproductive strategies of males.

PHENOTYPE AND RUTTING BEHAVIOUR

Two broad classes of rutting behaviour can be identified in Soay males. Males either exhibit a 'dominant' or 'subordinate' strategy, determined largely by age, size and horn phenotype, although demography may also play a role. Large, normal-horned adults follow the typical dominant strategy in which the male 'consorts' with individual oestrous females (Grubb 1974b). Small, young and scurred males more often follow an opportunistic 'coursing' strategy, also found in other sheep species (Hogg 1984; Hogg and Forbes 1997). Coursing males rely on hurried matings with undefended females, rather than on direct defence. The subordinate strategy is perhaps better viewed as several convergent strategies rather than a single suite of behaviours. This is because it encompasses young males, which may in future go on to becomes dominants, and also scurred adults, which are 'locked' into the subordinate strategy by virtue of their horn type. Many differences between the dominant and subordinate strategies can be readily identified from activity budgets (Fig. 9.8).

Body size and horn size are central to an individual's success at employing the 'dominant' strategy, not only between horn types and age classes, but also within them. For normal-horned males, for example, horn size, skeletal size and body condition are all positively, and independently, associated with the frequency and duration of consorts (Preston *et al.* 2003). This greater time spent in consorts in turn translates into increased success in gaining paternities (Preston *et al.* 2003).

The importance of horn size is even more apparent between the two male horn morphs, which differ hugely in annual reproductive success, but do not differ in body size (I. R. Stevenson, unpublished data). The small horns of scurred males makes them of little use in the defence of oestrous females, and, as a result, they only invest about 5% of their time in consorts at the peak of the rut, compared to about 50% for normal-horned adults (Mann–Whitney U-test $z = 3.18, n = 33, m = 19, p < 0.001$). Instead, they attempt to locate females coming into oestrus before dominant males can sequester them. They then attempt to gain matings, with the minimum of courtship, before continuing their search for the next female. As a result, scurred males are extremely active, especially when the peak number of females are in oestrus, when they spend about 40% of their time moving about the island, and travel twice the distance covered by normal-horned males (695 m/h versus 324 m/h, $t_{14} = 4.9\ p = 0.0002$). In the absence of unguarded females, scurred males join juvenile males harassing normal-horned adult males and their consorts. If they are successful in separating the female from her consort, a chase frequently results, as the female attempts to escape the attentions of the subordinates. If the defending male cannot fend off the harassing males and regain control, forced copulations are likely to result, often as the female is still running and, when a ewe stops, a succession of males often copulate.

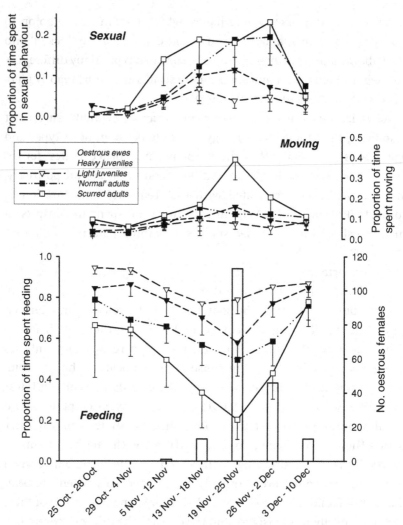

FIG. 9.8. Activity budgets during the rut for scurred and normal-horned adult males and heavy and light juvenile males (>1 standard deviation above or below mean August weight). The data are from 1995 and show the proportion of time spent in sexual behaviour, feeding (including rumination) and moving around the island. The bars indicate the number of females in oestrus each week, and the error bars show the 95% confidence intervals. Note that scurred males display peak levels of sexual activity for a greater duration than normal-horned males and are extremely mobile, searching for unguarded females at the peak of the rut. Scurred males consequently feed significantly less than all other groups. Amongst juveniles, rutting activity is positively related to body weight.

Young males show an opportunistic strategy similar to that of scurred adults, attempting to find unguarded females and harassing consorting adults. They invest less time in rutting behaviour, however, are less mobile, and maintain a much higher level of feeding activity than do adults (Fig. 9.8). All forms of rutting activity are depressed in young males in poor condition: throughout the rut, juveniles of below-average weight are involved in fewer agonistic interactions, devote less time to sexual activity and have a higher feeding rate than heavy males (Fig. 9.8). Lightweight males also begin to revert to pre-rut levels of feeding before the peak of ewes in oestrus. This appears to be a facultative difference in behaviour, rather than one imposed by delayed physical maturity. All but one of twenty-nine juvenile males examined in 1991 were physically capable of mating (as judged by the scoring method of Wiggins and Terrill 1953), and all subsequently exhibited a variable level of sexual activity during the rut (I. R. Stevenson, unpublished data). This is confirmed by investigations of captive Soays, which found that hormonal and testicular responses in juvenile males during the rut were reduced by poor nutrition, but that their expression was not delayed (Adam and Findlay 1997).

PHENOTYPE, REPRODUCTIVE SUCCESS AND ADAPTIVE LIFE-HISTORY
Normal-horned males are virtually always attributed more paternities than juvenile or scurred males in a given year (Chapter 6 and Fig. 9.9b). In the case of juveniles, this is not surprising, but for scurred adults it raises important questions. Scurring is heritable (Appendix 2) (Dolling 1960 a,b; Montgomery *et al.* 1996), yet appears to be highly disadvantageous in males and is uncommon or absent in wild sheep (Geist 1971; Schaller 1977). Having been introduced to Soays by past domestication, why is it still prevalent, at a level apparently unchanged over many years (Chapter 7)? Does it confer benefits, or is it in the process of being lost, slowly, from the population?

One possibility is that scurred males may display an alternative mating strategy, with equal fitness to the normal-horned strategy. Current evidence suggests that the fitness of normal-horned and scurred males may be frequency-dependent (Fig. 9.9a). Although scurred males may gain fewer paternities per breeding season, they may make up for

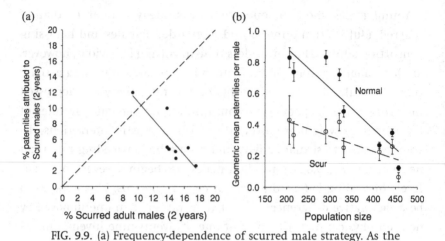

FIG. 9.9. (a) Frequency-dependence of scurred male strategy. As the proportion of scurred males in the population increases, the proportion of paternities gained by them decreases.
(b) Density-dependence of normal/scurred reproductive success. As population density rises, the per capita success of the two strategies converges.

this by greater longevity and so have the opportunity to take part in a greater number of ruts (Chapter 7) (Moorcroft *et al.* 1996). The greater longevity of scurs is paradoxical given their greater mobility, which might be assumed to be costly. We are currently investigating this, but direct measurement of rutting energetics (using the doubly labelled water technique) suggests that appearances may be deceptive and that the expenditure of the two horn types is the same (normal 5544 kJ/day, scurred 5417 kJ/day, $F_{1,37} = 1.08$, ns) (I. R. Stevenson, K. Wilson and D. M. Bryant, unpublished data).

An alternative explanation for the maintenance of scurring in the population is that scurred males may indeed be disadvantaged throughout their life, but this may be opposed by cross-gender balancing selection favouring higher-fecundity scurred females (Chapter 7). At present, it is not possible to say whether male scurring is adaptive, but this is the aim of current research.

SPERM COMPETITION IN SOAY SHEEP

Juvenile and scurred males are heavily dependent on sperm competition for any success they achieve. Since females have prolonged oestrus and, in contrast to many deer species, will mate repeatedly with many males, subordinates with larger ejaculates will be at an advantage if this allows them to swamp those of other males and fertilise the ovum (Gomendio *et al.* 1998). Dominant males, in turn, might be expected to increase ejaculate investment in each female to minimise the loss of paternities to later-mating males (Hogg 1988). All sheep species display prolonged oestrus, and studies of bighorn sheep suggest that, as in Soays, sperm competition is probably very important. Hogg and Forbes (1997), for example, found that coursing males achieved 40–50% of paternities in two separate populations, a success that must be due largely to sperm competition.

Paradoxically, it appears to be the very success of large males in gaining matings that allows smaller, subordinate males to achieve success through sperm competition. Although high weight and large horns both independently increase the success of a male in gaining copulations and, across the whole rut, increase the number of paternities gained (Preston *et al.* 2001), the relative advantage of large males over subordinates decreases later in the rut (Fig. 9.10b). This appears to be due to sperm depletion of the dominants and a consequent increased success of subordinates in sperm competition (Hogg and Forbes 1997; Preston *et al.* 2001). This effect occurs even though the decline in overt reproductive behaviour over this period is greater for subordinate than dominants (Fig. 9.10). Contrary to the normal pattern, sperm count is negatively correlated with testes size (and body size) late in the rut (Fig. 9.11) and the proportion of abnormal sperm increases (Preston *et al.* 2001).

The importance of sperm competition in Soays is backed by anatomy. Species in which sperm competition is important typically have large testes relative to body mass (Harcourt *et al.* 1981), and Soays have testes that are larger than a wide range of other ungulates (Fig. 9.12) (Lincoln 1989). Though even sheep with large testes suffer depletion late in the rut, testes size is positively related to paternity success across the breeding season as a whole (B. T. Preston, unpublished data). Moreover as the number of oestrous ewes increases, the importance of large testes relative to large horns increases since the

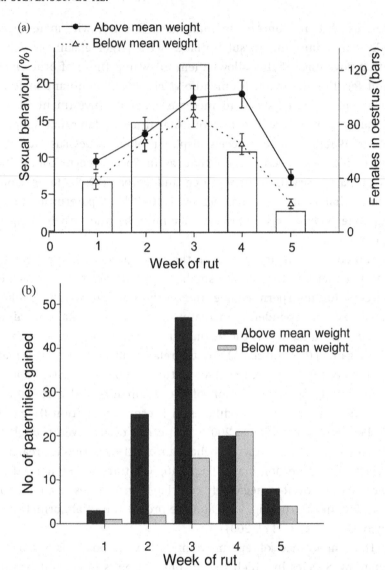

FIG. 9.10. Change in (a) overt sexual activity and (b) resulting reproductive success, over the course of the rut for normal-horned males of above-and below-average weight. Data are from (a) 1996–8 (b) 1985–96; error bars show standard errors. Note that heavier males continue at peak rutting activity for longer than lighter (a), despite the relative increase in success of lighter males later in the rut (b).

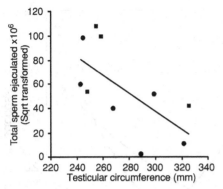

FIG. 9.11. Decline in ejaculated sperm number with scrotal circumference measured at the maximum point. Counts were made in weeks 4 and 5 of 1999 (see Fig. 9.10). (From Preston *et al.* 2001.)

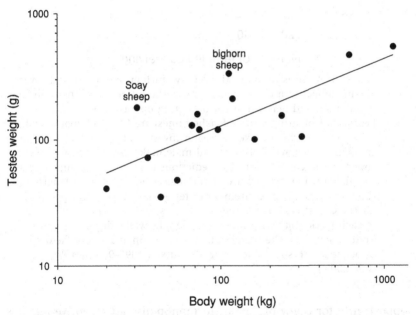

FIG. 9.12. The allometric relationship between body weight and combined testicular weight over fifteen genera of ruminants. Adapted from Hogg (1984). Data for Soays are taken from Lincoln (1989). The equation of the line is $y = 0.5593 + 0.7535x$. (From Stevenson 1994.)

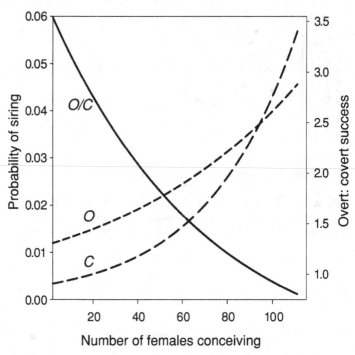

FIG. 9.13. An illustration of the relative influence of overt and covert sperm competition on siring success as the number of females in the population, and hence number of mating opportunities, increases. Predictions of siring success (dashed lines) are for two hypothetical males, one of which makes a greater investment in overt contest competition (O), whilst the second male makes a larger investment in covert sperm competition (C). Predictions for the Overt competitor are calculated using the upper quartile measurement of horn length and the lower quartile measurement of testes size in the adult population; for the Covert competitor the situation is reversed. The relative success of each tactic (O/C) with increasing female availability is also plotted (solid black line). The predictions are based on analysis of factors associated with siring success in the ruts of 1997–9. (From Preston *et al.* 2003.)

opportunity for a few big males to monopolise all receptive females declines (Fig. 9.13) (Preston *et al.* 2003).

The degree of sperm competition will also vary as population density changes. At high density, females frequently mate with many males: in 1996, for example, a single female was observed to be mounted 160 times by seven different males in the course of five-and-a-half hours (though it is highly unlikely that intromission was successful in every case; Lindsay (1966) estimated that the success rate per mating varied between 10% and 20%). Consorting males suffer extremely high levels of harassment from subordinates when population density is high and, under these conditions, dominant males are less able to defend females and maintain exclusive mating rights. This is mirrored by the fact that the reproductive success of normal-horned and scurred males converges as density increases (Fig. 9.9b). At high population density, the dominant strategy tends to break down and dominants pursue a more opportunistic strategy, similar to subordinates, relying largely on sperm competition. This is similar to bighorn sheep, where dominant males are more likely to follow the opportunistic coursing strategy in populations where the sex ratio is less female-biased and so competition for mating opportunities is higher (Hogg and Forbes 1997).

FEMALE PREFERENCE

The reproductive success of different male phenotypes is also affected by intersexual selection (Komers *et al.* 1999). Females rarely stand still to allow mating by scurred or young males, except late in oestrus, and often actively avoid them (Grubb and Jewell 1973) as is found in other species (Cox and Le Boeuf 1977; Hogg 1984; Byers *et al.* 1994; Sabat 1994). Females benefit from being in consort with dominant males since they are largely protected from harassment by subordinate males, and oestrus chases are less likely to occur. These chases are both energetically expensive and carry the risk of injury: cases of forced copulation have occasionally resulted in death following accidental rectal rupture. In fallow deer, females may delay oestrus when only young males are present and incur considerable energetic costs avoiding their attentions (Komers *et al.* 1999). Perhaps the most extreme example of the costs of harassment comes from the introduced mouflon population of the Kerguelen archipelago (Réale *et al.* 1996). Unusually for an ungulate, the operational sex ratio is highly male-biased and rams seem unable to form stable hierarchies. It appears that females in oestrus are harassed by many males and in many cases suffer fatal injuries as a result (Réale *et al.* 1996).

If females would prefer to avoid such attention, why do they display prolonged oestrus and accept multiple matings? For elephant seals, bighorn sheep and pronghorn it has been suggested that females may actually be stimulating male intrasexual competition by drawing the attention of dominant males while escaping the attention of subordinates (Cox and Le Boeuf 1977; Byers *et al.* 1994). Certainly, oestrus chases in Soays attract males from the surrounding area and provoke competition. However, dominant males commonly have less stamina and are more likely to give up than the many juveniles that are frequently involved in a chase. Such chases also occur in feral goats, and again dominant males are less agile and often abandon the chase (Pickering 1983). Rather than stimulating competition to select the fittest male, oestrus chases often result in a female being mated by many juvenile, unproven males who may not survive long enough to see a second breeding season. Perhaps a more likely explanation of prolonged oestrus is that female reproductive behaviour reflects phylogenetic history. Wild sheep do not form large family groups, and together with harassment by coursing subordinates, this may render dominant males unable to defend harems. Under such circumstances, it may be difficult for a female to guarantee that she will be first-mated by a high-quality ram; prolonged oestrus and multiple matings may increase the chance that females will be mated by males able to produce large numbers of high-quality sperm.

9.4 Discussion

COMPARISON OF MALE AND FEMALE STRATEGIES IN A FLUCTUATING ENVIRONMENT

Fluctuating population density results in highly variable costs of reproduction for both sexes. At peak density, high fecundity for females and significant rutting activity for males reduces the probability of over-winter survival for an individual (Chapter 2) (Stevenson and Bancroft 1995; Clutton-Brock *et al.* 1996). Conversely, at low and intermediate densities, there appears to be little cost to current reproductive success. There is, however, a fundamental difference between the sexes in the relative timing of realised reproductive success and the proximate and ultimate costs that are incurred. For females, there appears to be little cost involved in the rut itself. Even in years when the sex ratio is relatively equal, and harassment is common, a female is likely to be the subject of males' attentions for no more than two to four days. This level of disruption seems unlikely to have a major effect on winter survival, since feeding and ruminating still accounts for around 79% of daylight hours during oestrus, compared to around 87% for non-oestrous females (K. Wilson, I.R. Stevenson and B.T. Preston, unpublished data). As a result, females only begin to suffer reproductive costs towards the end of gestation in late winter/early spring (Clutton-Brock *et al.* 1989, 1996; Coulson *et al.* 2001) when their reproductive success is about to be realised. Males, by contrast, contribute no parental care and suffer the costs of reproduction immediately after they have completed their reproductive activities resulting in high mortality earlier in the winter than experienced by females (section 3.5) (Stevenson 1994).

The absence of any form of parental investment by males may, in contrast to females, favour increased investment in the rut at the expense of over-winter survival. Since male mortality rates can be high even when reproductive costs are removed (reaching 44% in the non-rutting juveniles, as described in section 9.3), the low probably of surviving through multiple breeding seasons may tend to favour a semelparous life-history (Promislow and Harvey 1991). In this case, juveniles should invest everything possible in the rut since surviving the winter will not benefit their offspring. By contrast, the minimum 'useful' lifespan for females is almost a year longer since they must survive long enough to rear their offspring to independence. In this respect, Soays bear a certain similarity to *Antechinus* (section 9.1) in which females often have more than one breeding attempt, but males never survive their first year.

Given that females must survive the winter in order to breed successfully, it is strange that they appear unable to respond to population density, while, to some extent, males do (Fig. 9.6). This may be because each sex uses different cues for assessing density, or indeed a reflection of the fact that females require active assessment of the environment, while males may have an 'assessment' made for them by the simple method of physical exclusion from oestrous females. If males cannot gain access to females then, rather than waste time trying, they may just give up. If the ability to reduce rutting activity (whether by response to density *per se* or through the proxy of competitive exclusion) has a genetic component then this can give rise to the evolution of phenotypic plasticity (Komers *et al.* 1994b, 1997; Komers 1997).

Why do females not tailor their reproductive effort to the costs associated with high population density? Although there is a reduction in the extremes of fecundity (twinning and juvenile conception), there is little response in the proportion of adult females conceiving (Chapter 3, section 3.3). It seems that the main cues that sheep can use to adjust their fecundity – body weight and nutrition – are not good indicators of future conditions on St Kilda. In no year does vegetation availability greatly limit the growth of sheep during the summer (Clutton-Brock *et al.* 1997a) and so body weight is largely independent of population density (Clutton-Brock *et al.* 1996). As a result, body condition in the autumn, which strongly influences fecundity (Gunn *et al.* 1986), is also independent of population density. Furthermore, the increased attention from males at high population density might even increase fecundity through the 'ram effect' (the induction of female puberty, oestrus and ovulation by the presence of males) (Pearce and Oldham 1984), though increased harassment can lower ovulation rates (Raynaud 1972, cited in Thibault 1973).

If the mechanisms that control conception cannot respond to population density, then we might expect the female to display post-conception adjustment of fecundity. Some evidence for the ability of female mammals to perform fine-scale adjustment of embryo sex and number is found in red deer and coypus (Clutton-Brock *et al.* 1986; Gosling 1986; Kruuk *et al.* 1999a). In Soay sheep, however, there is no evidence of a reduced number of foetuses being carried late in gestation, suggesting that there is no incidence of either embryo abortion or resorption. This is backed by research in domestic sheep that finds little link between nutrition and embryo wastage after the first month of gestation (Gunn *et al.* 1979; Kelly 1984). Indeed in ungulates in general, there is little information suggesting that abortion (other than through disease) is common. It seems that although females may have more to gain from a density-dependent fecundity response, they cannot express it, perhaps because of phylogenetic constraints or a relatively recent incidence of large population fluctuations.

To return to the original question, 'Are Soays too fecund for their own good?", the answer appears to be no. Given the constraints of limited information about conditions during the forthcoming winter, and an inability to carry out post-conception adjustment of reproductive investment, our modelling indicates that female Soays exhibit close to optimal behaviour. Males, too, appear to be close to optimal, maturing early when conditions are favourable but tending to delay when the mortality risks are highest. So if Soays are more than just the maladapted flotsam of domestication, what are the broader implications of our study?

SOAYS AS A MODEL SPECIES

It is often tempting to concentrate on the ways in which Soays and their environment are atypical: feral sheep living on a small island that supports only a depauperate range of plant communities, lacking competing grazers and predators, and with no opportunity for immigration or emigration. Yet in many ways Soays are a good model. Their reproductive biology is similar to many wild ungulates, and increasing numbers of these 'wild' populations now live in predator-free environments. In this context a detailed knowledge of reproductive behaviour and the factors affecting fecundity is of growing importance in the management of these populations. Our understanding of these topics is rudimentary, however, and the 'simple' system presented by Soays offers a starting-point for testing many of these ideas.

A good example of the use of Soays as a model is to test the widespread assertion that the rutting behaviour contributes to the male-biased mortality observed in many temperate-zone ungulates (Geist 1971; Clutton-Brock *et al.* 1982a; Pickering 1983; Leader-Williams 1988; Festa-Bianchet 1989b; Miquelle 1990). The suggestion of a link between rutting and differential mortality implies that widely differing reproductive trade-offs are experienced by each sex, yet no other study has demonstrated this conclusively, let alone examined how the trade-offs vary between individuals. With Soay sheep we can conduct experiments that confirm this link (section 9.3) and can use the wealth of individual-based data to begin an examination of how optimal life-history differs between individuals both within and between the sexes (sections 9.2 and 9.3).

This investigation of individual variation in optimal life-histories is perhaps the most important aspect of our study of Soay reproductive behaviour. It underlines the fact that talk of a 'species life-history' refers to that rarest of beasts: the 'Average Individual'. For real animals, the optimal strategy is a moving target that varies constantly with the state of the animal and its environment (Stearns 1989). Ignoring this 'substructure' within the population may, for example, result in the wrong prediction being made about how the dynamics of a population will respond, say, to management or other perturbations (Moorcroft *et al.* 1996; Coulson *et al.* 1999a).

Our studies also demonstrate that the pattern of natural selection is rarely simple, and that even when certain phenotypes appear to be without doubt the fittest, 'inferior' individuals may be more successful than we at first imagine. For example, our behavioural observations of rutting males support the typical polygynous model of 'biggest is best' (section 9.3), but this does not translate fully into fitness since young or apparently inferior phenotypes may be more successful than simple behavioural data suggest (section 9.3; Chapter 6). This is particularly important in a fluctuating environment, where the periodic relaxation of competition, due to male-biased mortality (Chapter 3), allows these other males to express full reproductive behaviour (Komers *et al.* 1994a) which in turn may bring considerable success (Chapter 6). Even at high population density, when *overt* competition is at its strongest, *covert* sperm competition may be busy undermining the behavioural advantage of the 'classically successful' large, prime-aged males (section 9.3) (Preston *et al.* 2001). This is an area that deserves further investigation, for although there are many detailed studies of male reproductive behaviour and tactics, information about the ultimate fitness of these is much less common (e.g. Weatherhead *et al.* 1995; though see Hogg 1984, 1987).

In conclusion, Soay sheep present a model system that is more tractable than many others, but it remains a single example of an unusually simple system. Ultimately, there is a need for similar long-term studies on a wide range of species that can confirm or refute the generality of our findings.

10

The causes and consequences of instability

T. H. Clutton-Brock *University of Cambridge*

10.1 Introduction

This final chapter reviews our understanding of the causes and consequences of fluctuations in sheep numbers on St Kilda and sets our results in the context of other studies. It is structured around the three main questions that we have tried to answer (see section 1.1): how are sheep numbers regulated and what physical and biological processes are responsible for changes in population size? How do changes in population density affect selection? And how do they affect the optimal reproductive strategies of individuals? The first section (10.2), examines the causes of death and the reasons why, unlike many other ungulate populations, the population of Soay sheep on Hirta shows such large fluctuations in size. In addition, it reviews the demographic consequences of changes in population size, including the effects of variation in density on development. Section 10.3 examines the impact of fluctuations in sheep numbers on the opportunity for selection and the intensity of selection on particular traits and speculates on the role these changes may play in maintaining genetic and phenotypic diversity. Fluctuations in sheep numbers also affect the costs and benefits of reproduction: section 10.4 reviews evidence of these changes and examines their consequences for the optimal breeding strategies of the sheep. Finally, section 10.5 examines the relevance of work on St Kilda to understanding the dynamics of other resource-limited populations of mammals, arguing that the

276

Soay Sheep: Dynamics and Selection in an Island Population, ed. T. H. Clutton-Brock and J. M. Pemberton.
Published by Cambridge University Press. © T. H. Clutton-Brock and J. M. Pemberton 2003.

demographic processes operating in the sheep are probably found in many other resource-limited populations.

10.2 Comparative demography

THE PATTERNS OF FLUCTUATION

Between 1985 and 2000, the Soay sheep population of Hirta continued to show periods of rapid growth interspersed with population crashes, as it did between 1960 and 1968 (Grubb 1974a, b, c, d), and similar fluctuations also occurred in numbers of black-faced sheep on Boreray (Grenfell *et al.* 1998). These fluctuations in population size were caused by variation in winter mortality and were unusually large for a large herbivore. For example, between 1985 and 2000, sheep numbers on Hirta increased twice by more than 60% in the course of a single year and declined by more than 50% in three separate winters (Fig. 1.5). Over much the same period, the naturally regulated red deer population of the North Block of Rum never increased by more than 15% in the course of a year and never declined by more than 17% (Clutton-Brock *et al.* 1997a).

Years of high sheep mortality do not occur more than once every three or four years (Fig. 1.5). This is presumably because it normally takes at least two breeding seasons for a sheep population that has been reduced to 600–900 animals to regain the size at which another crash is likely. During the period of less intensive monitoring from 1969 to 1984, the incidence of population crashes appears to have been lower and years of rapid increase were also less common. It is not easy to understand why this should have been the case and, in particular, why numbers failed to rise after years of low density. These counts were often carried out by a single observer and our experience of counting sheep on Hirta is that this is not easy even with three groups of observers. Consequently, we believe that these figures may not provide a reliable estimate of changes in sheep numbers and are probably best omitted from analyses of the dynamics of the population.

CORRELATES OF MORTALITY

Three main factors appear to be responsible for variation in the number of sheep dying between years. Much of the variation is accounted for by differences in population size: when the total number of sheep on Hirta exceeds approximately 1150, a high proportion of animals may die – in some cases over two-thirds of the population (Fig. 3.1). However, crashes do not always occur when density is high and their magnitude is not closely correlated with autumn density. For example, although sheep numbers were unusually high in 1997, there was little mortality and the population continued to increase (Fig. 1.5). In contrast, numbers were not especially high in 1988, but more than half the population died (Fig. 1.5). Much of this variation is accounted for by differences in the winter weather. As in many other northern ungulates, mortality is relatively high when high population density coincides with wet, windy conditions in winter (see section 3.5) (Albon et al. 1991, 2000; Forchhammer et al. 1998a, b; Post and Stenseth 1999; Coulson et al. 2001). Variation in winter weather also generates synchronous fluctuations between sheep populations on the different islands of St Kilda, as well as between other populations of northern ungulates (Grenfell et al. 1998; Coulson et al. 1999a; Post and Forchhammer 2002).

The magnitude of crashes is also related to variation in the age and sex structure of the population (Fig. 3.17). When adult mortality has been low for several years, older animals (which are particularly susceptible to starvation) constitute a large proportion of the population so that a minor deterioration in environmental conditions may generate heavy mortality. Conversely, when adult mortality has been high in the preceding year, there are few older animals left in the population and adverse environmental conditions have less effect on overall mortality rates. Similar age-related differences in susceptibility to environmental conditions occur in many other mammals (Festa-Bianchet et al. 1998; Gaillard et al. 1998). Comparisons with red deer suggest that population structure may be unusually variable in the sheep as a result of the alternation of years when few animals die with years when mortality is high and almost all of the older animals are eliminated (Clutton-Brock and Coulson 2002) (see Chapter 3).

CAUSES OF DEATH

The effects of high sheep density on food availability and quality in winter appear to be the principal cause of the relationship between population density and winter mortality. The immediate cause of death in late winter is almost always starvation (Gulland 1991): in crash years, the weight of mature females falls from 30 kg in autumn to less than 18 kg in March (Clutton-Brock *et al.* 1991). As expected, heavier individuals are more likely to survive, and energetic calculations successfully predict variation in mortality among sex and age categories (Clutton-Brock *et al.* 1997a; Milner *et al.* 1999a).

Though heavy grazing pressure may have long-term consequences for the diversity of plant species and for sward structure, the immediate relationship between sheep numbers and winter food availability appears to be a relatively simple one (see Chapter 4). Variation in sheep density has little or no effect on primary production in spring and summer and high sheep density probably influences winter mortality mainly through its effects on food availability in later winter. As the number of sheep present in August increases, food availability in March declines, the sheep feed to a greater extent on dead grass and their fat reserves are gradually used up.

Wet windy weather in winter probably affects the sheep in several different ways. It may lead to an early cessation of plant growth so that food availability is depressed. In addition, it may affect condition and mortality through its direct effects on heat loss, energy requirements and habitat use. Studies of other northern ungulates show that wind chill can lead to substantial increases in heat loss and energy expenditure and may restrict grazing time and reduce use of unsheltered habitats (Staines 1976; Grace and Easterbee 1979; Clutton-Brock and Albon 1989; Stevenson 1994).

Fluctuations in parasite populations also contribute to density-dependent changes in mortality (see Chapter 5). When population density is high and the proportion of lambs is relatively large, both the density of infective larvae in the sward and the number of adult worms carried by female sheep increase. Experimental reduction of parasites increases the survival of sheep in years of moderate mortality but has no effect on the number of sheep dying when mortality is high, though animals whose parasite loads have been reduced die later in the winter (Gulland *et al.* 1993). This suggests that, while fluctuations in parasite load may contribute to the magnitude of crashes, they are not the principal cause of their occurrence (Chapter 5). In this respect, fluctuations in sheep numbers appear to differ from the cycles shown by some red grouse populations, which appear to be driven directly by interactions with their parasites (Hudson 1986; Hudson *et al.* 1992a, b).

RELATIVE MORTALITY

When overall mortality is high, some classes of sheep are more likely to die than others (see above). In many mammals, the survival of juveniles is more strongly affected by adverse conditions than that of adults (Eberhardt 1977, 2000; Saether 1997; Gaillard *et al.* 1998). Soay sheep are no exception: the survival of juveniles is lower and more variable than that of adults and is more strongly affected by high population density and unfavourable winter weather (Fig. 10.1a, c). Similar differences occur in red deer (Fig. 10.1b, d). In both species, older females are also more likely to succumb to adverse conditions than animals in their prime, as in several other ungulates (Gaillard *et al.* 2000). Since rates of mortality are probably affected by a wide range of environmental variables (see Chapter 3), a more accurate impression of the relative effects of adverse conditions on different categories of animals can be gained by plotting increases in the mortality of different age classes against proportion of prime-aged females that died each year (Fig. 10.2a, c). These comparisons show the relatively high mortality of juveniles and old females, and also reveal that younger females of two to three years show disproportionately small increases in mortality compared to four- to six-year-olds in the sheep.

As in many other polygynous ungulates (Gaillard *et al.* 2000), mortality rates throughout much of the lifespan are also higher in males than females in the sheep and male mortality increases disproportionately in adverse years (Fig. 10.2b). Though males are no more likely to die during the neonatal period than females (see section 2.7), they are more likely than females to die during their first winter of life,

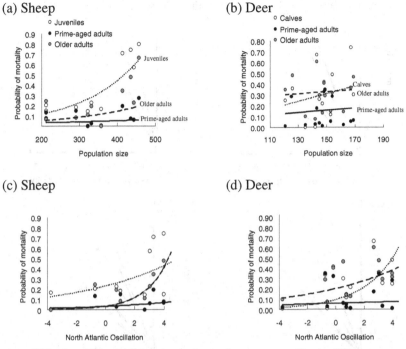

FIG. 10.1. Effects of density and weather on mortality in different age classes of females. (a) Effects of density on sheep, (b) effects of density on deer, (c) Correlations between mortality and variation in the North Atlantic Oscillation (NAO) in sheep, (d) similar correlations in the deer. Winters with high NAO values are relatively wet and windy, those with low NAO values are drier and colder. Open circles and dotted lines represent juveniles, black circles and solid lines represent prime aged adults and grey circles and dashed lines represent older adults Older adult sheep are over six years; deer over ten years. (From Clutton-Brock and Coulson 2002.)

especially in years when population density is high and weather conditions are adverse (Clutton-Brock *et al.* 1997a). These sex differences in first-year survival parallel those in red deer (Fig. 10.2d) and several other ungulates (Clutton-Brock *et al.* 1985b) but differ from the situation in bighorn sheep (Bérubé 1997) and mountain goats (Côté and Festa-Bianchet 2001) possibly because, in wild sheep, juvenile males do not expend as much energy in the annual rut as in Soays. Sex differences in mortality throughout the rest of the lifespan are also unusually large in Soays (Fig. 10.2b) and, after population crashes,

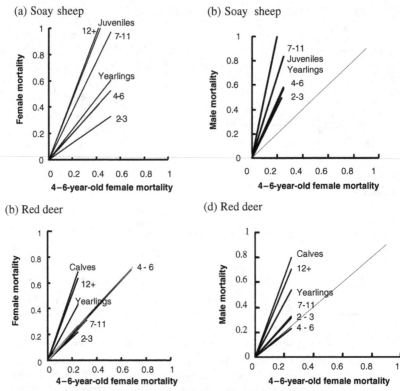

FIG. 10.2. Relative increases in mortality in different categories of sheep and deer. (a) Mortality rates for different age categories of female sheep plotted on the mortality of females in their prime (four to six years). Probability of mortality was the proportion of individually identified animals entering that age category that died before leaving it. All lines are forced through the origin. (b) Mortality of male sheep of different age categories plotted on the mortality of four- to six-year-old females. (c) Mortality of female deer of different ages plotted on the mortality of females in their prime (four to six years) (d) Mortality of male deer of different ages plotted on the mortality of four- to six-year-old females. When comparing relative changes in mortality in different age and sex categories, it is important to remember that the slopes shown here reflect changes in relative mortality and not the absolute level of mortality. For example, while the mortality of males rises more rapidly than that of prime females in the sheep though not in the deer, absolute levels of mortality are generally higher in males than females in both species (see Chapter 3 and Clutton-Brock et al. 1997a). (From Clutton-Brock and Coulson 2002).

there can be ten times as many surviving females as males. The high parasite loads of males may contribute to this (see Chapter 5). In addition, the relatively late mating season of the sheep means that males have little chance to replace reserves lost in the rut before the onset of winter. By contrast, the annual rut in red deer occurs earlier in the autumn than in the sheep so that male deer may be able to replace some of the reserves lost in the rut before the onset of acute food shortage in late winter (Clutton-Brock *et al.* 1997c).

FECUNDITY, EARLY GROWTH AND THE TIMING OF
REPRODUCTIVE EXPENDITURE

Differences in the timing of reproductive expenditure also affect the impact of environmental factors on fecundity in different species. In the sheep, there is no consistent relationship between population density and the proportion of mature females that give birth each year, whether or not they have reared lambs (Fig. 10.3a). However, high population density is associated with a decline in the proportion of juveniles that produce lambs, as well as in the proportion of older females that produce twins (Fig. 3.5b,f). In the deer, increasing population density lowers the fecundity of prime-aged females that have reared calves the previous year (Fig. 10.3b) and also increases average age at first conception from twenty-four to thirty-six months (Clutton-Brock *et al.* 1997a; Clutton-Brock and Coulson 2002).

So why – in contrast to red deer and many other ungulates (Fowler 1987) – is the fecundity of adult females so unresponsive to increasing population density in Soay sheep? Two separate factors appear to be important. First, as in many other northern ecosystems, food availability in the summer months appears to be superabundant, even when sheep numbers are high (see Chapter 4). This provides the sheep with a window of opportunity when food availability is unlikely to limit growth or reproduction (Clutton-Brock *et al.* 1997a). Second, unlike many other ungulates (including bighorns and red deer), Soay sheep wean their lambs by midsummer, giving mothers several months of abundant food when they can regain condition before the November rut. As a result, females can regain condition during the period when food is superabundant and enter the rut at much the same weight,

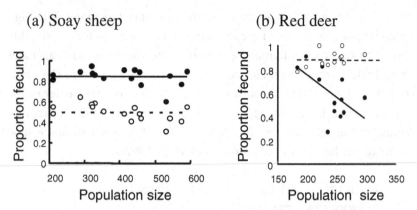

FIG. 10.3. Density-dependent changes in fecundity in females that had reared young in the previous breeding season (milks, solid circles) versus those that had not done so (yelds, open circles) in (a) sheep and (b) red deer. Data for both species confined to multiparous animals. (From Clutton-Brock and Coulson 2002.)

whether they have bred successfully or not (Fig. 10.4). As a result, most individuals conceive in all years (Clutton-Brock *et al.* 1997a). In red deer, which have a longer period of lactation than Soays, females that have raised calves are in relatively poor condition in October and fluctuations in food availability in late summer are more likely to affect their body weight and their probability of conceiving in the rut. The spatial structure of the Soay population – or, rather, its lack of structure – may also contribute to the absence of any close relationship between population density and fecundity. Regular association between mothers and their offspring ceases by the time lambs are two years old and, apart from divisions between hefts there are no obvious social subdivisions in the sheep population (see Chapter 2). In many other ungulates, individuals share home ranges with close relatives and their fecundity and breeding success decline as group size increases (Clutton-Brock *et al.* 1982a; Coulson *et al.* 1997). Group size is likely to interact with changes in population density to affect fecundity and survival (Clutton-Brock *et al.* 1987a, b), potentially strengthening the effects of food shortage associated with rising population density by focussing them on particular individuals (Clutton-Brock and Albon 1985).

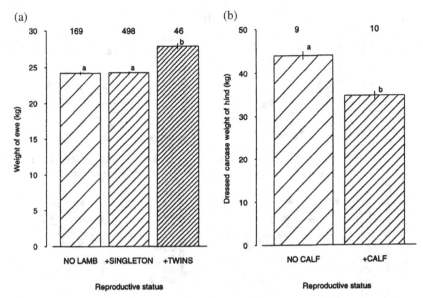

FIG. 10.4. Body weight in August in (a) Soay ewes two or more years old that have failed to breed, raised a single lamb, or raised twins; (b) multiparous red deer hinds that have failed to breed or have raised a single calf. Means that differ significantly at the 5% level are indicated by different superscripts. (From Clutton-Brock *et al.* 1997a.)

The timing of reproductive expenditure also has an important influence on early growth and recruitment. In the sheep, lambs born after crash years show low birth weights (Fig. 10.5a) and high neonatal mortality. Wet, windy winters also tend to reduce birth weights the following spring (Fig. 3.7) while spring temperature apparently has little effect (see Fig. 10.5b). In contrast, in red deer, birth weights are not consistently related to population density (Fig. 10.5c) or winter weather but are low in cold springs (Fig. 10.5d). Like differences in the effects of density on fecundity, these contrasts are probably caused by differences in the timing of reproductive expenditure (Clutton-Brock and Coulson 2002). The relatively short gestation period of the sheep (21.5 weeks) and their earlier birth season means that the last months of gestation (March, April) precede the main onset of spring growth (see Chapter 1), so that food availability and maternal condition may be determined largely by population density and climatic factors during the winter months. Because of their

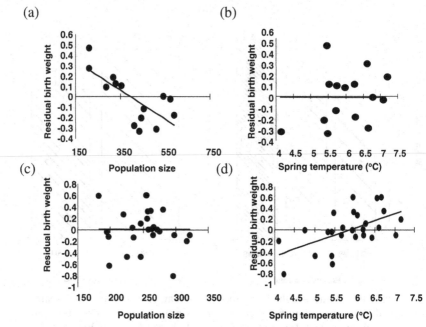

FIG. 10.5. Birth weight of Soay lambs and deer calves in relation to population density (a, c) and spring temperature (b, d). Spring temperature was measured as the mean average daily temperature in April and May. The y-axis (residual birth weight) shows the difference between the mean birth weight of each cohort and the population mean for the study period of both populations. Solid lines derived from linear regression models. (from Clutton-Brock and Coulson 2002.)

relatively long gestation period (32 weeks), red deer bear their young in late May or June, so that the last months of gestation occur during early spring, when grass growth is strongly influenced by differences in climatic conditions between years (Clutton-Brock et al. 1997a). As a result, spring temperature is the dominant factor affecting birth weight and subsequent growth.

The contrasting effects of density and density-independent factors on birth weight in the sheep and deer have important consequences for inter-cohort differences in growth, survival and breeding success in the two species. As a result of the relationship between population density and birth weight, cohorts of female sheep born at high and low density differ in their age-specific body weight and survival through their first year of life (Figs. 3.7, 3.8, 3.15), while inter-cohort differences in growth and juvenile survival in the deer are related to density-independent differences in spring climate (Albon *et al.* 1983a, b 1987; Albon and Clutton-Brock 1988). These differences can have important consequences for population dynamics. For example, in the deer, fluctuations in population size may be generated by the impact of density-independent changes several years earlier (Albon *et al.* 1987; Albon and Clutton-Brock 1988). The density-dependent cohort effects that occur in the sheep have a limited impact in population dynamics (see below). However, in other populations, similar effects could influence the quality and survival of animals at peak population density and might have a more important influence on population dynamics (see Clutton-Brock *et al.* 1992).

COMPARATIVE DEMOGRAPHY

Comparisons of the effects of density and of other environmental variables on different vital rates provide an opportunity to gener-alise about the ways in which animal numbers are regulated (Newton 1998). Unfortunately, comparisons are constrained by the availability of reliable data (McNaughton and Georgiadis 1985; Gaillard *et al.* 2000). In most studies, some demographic parameters are hard to measure unless large samples of recognisable individuals can be mon-itored throughout their lifespan. For example, unless individuals can be recognised, it is usually difficult to distinguish between changes in fecundity and neonatal mortality or to measure rates of emigration and immigration. Comparisons based on vital rates measured with contrasting levels of accuracy may be misleading and the reliability of contrasts within and between species (e.g. Gaillard *et al.* 1998, 2000; Eberhardt 2000) is often hard to assess. For example, when a study re-ports that first-winter mortality is density-dependent and neonatal mortality is not, might this be because neonatal mortality cannot be measured accurately? Similarly, where one study reports that fecun-dity is density-dependent while another reports that it is not, does this reflect a real biological difference – or a contrast in the accu-racy with which fecundity could be estimated? Studies where large numbers of individuals can be monitored throughout their lifespans consequently provide an opportunity to test some of the generalisations about population dynamics that are emerging from recent studies of herbivores (Gaillard *et al.* 1988; Sinclair 1989; Saether 1997; Eberhardt 2000).

In most populations that have been studied, survival rates vary less in adult females than in immatures, possibly because adults minimise risks of mortality in order to maximise their lifetime reproductive success (Festa-Bianchet and Jorgensen 1998). In addition, survival of old females is typically more variable and more strongly affected by environmental variation than survival of prime-aged animals (Gaillard *et al.* 2000). For example, survival is more strongly affected in older females by rainfall in kudu (Owen Smith 1990), by food availability in reindeer (Skogland 1985a) and by population density and winter weather in red deer (Clutton-Brock and Coulson 2002). These differences probably represent the consequences of ageing and are related to tooth wear in some cases (Skogland 1990; Gaillard *et al.* 1993).

Though the survival of juveniles and immatures is more variable than that of adults, adults are usually more numerous and are more effective breeders. As a result, a given percentage increase or decrease in adult numbers will have a proportionally larger effect on population growth than a similar change in juvenile numbers. Despite this, variation in immature numbers is responsible for a larger proportion of variation in population growth than variation in the number of prime-aged adults in some ungulate populations (Gaillard *et al.* 2000). Gaillard *et al.* have recently concluded that 'most studies suggest an overwhelming importance of the juvenile stage in accounting for between-year variation in population growth rate' (Gaillard *et al.* 2000, p. 383). In contrast, this study, as well as studies of several other ungulates (Gasaway *et al.* 1992, 1996; Crête *et al.* 1996; Albon *et al.* 2000) found that variation in adult survival contributed more to changes in population growth than variation in juvenile survival – though both were important as was the co-variance between them (Fig. 10.6).

These apparently contrasting conclusions may be caused partly by differences in the definitions of age categories. For example, Gaillard *et al.* (2000) include the effects of variation in fecundity within their

(a) (b)

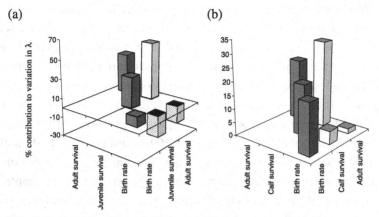

FIG. 10.6. Contributions of different age classes to variation in population size in (a) Soay sheep, (b) red deer. Dark columns show the contribution of a particular age class; lighter columns the contribution of positive co-variances between them; chequered columns the contribution of negative co-variances. (From Clutton-Brock and Coulson 2002.)

estimates of the impact of immature numbers and exclude the effects of variation in the number of older animals from their category of prime-age adults. Another possibility is that variation in the relative importance of changes in juvenile and adult survival is related to the stage of population growth. When the red deer in the North Block of Rum were released from culling, recruitment of juvenile females (which was strongly influenced both by changes in fecundity and changes in winter calf mortality) was the dominant component of changes in population size (Brown *et al.* 1993). However, after the population reached ecological carrying capacity and numbers had stabilised, mortality of adult females, reinforced by co-variation between adult and juvenile mortality, contributed most to changes in population size (Albon *et al.* 2000). If the stage of population growth is important, relative contributions of juvenile and adult mortality might be expected to vary between food-limited populations and populations maintained below ecological carrying capacity by predators (Gaillard *et al.* 2000).

It is sometimes suggested that vertebrate populations are usually limited by density-dependent mortality in the non-breeding season, while fecundity varies in a density-independent fashion in relation to fluctuations in environmental conditions in the breeding season (Lack 1966; Gaillard et al. 1998). Studies of a number of ungulates including Soay sheep confirm that population density is limited by density-dependent mortality in the non-breeding season, as Lack suggested, though juvenile mortality in summer can also play an important role (Hoefs and Bayer 1983; Skogland 1985a; Fryxell 1987; Sinclair 1989; Ballard et al. 1991; Choquenot 1991; Gasaway et al. 1996; Gaillard et al. 1998). In contrast, Lack's suggestion that fecundity rates usually vary in a density-independent fashion appears to be over-simple. The sheep and deer studies clearly show that fecundity, as well as winter mortality, is affected by a combination of density-independent and density-dependent variables and that the relative effects of density and weather differ between age and sex categories (see also Gaillard et al. 2000). For example, research on several ungulates shows that conditions in autumn, when food availability is relatively low and is likely to be density-dependent, can influence the condition, fecundity and conception date of females with downstream effects on birth dates and neonatal survival (see Chapters 2 and 3). Moreover, in animals with long gestation periods, foetal growth may be delayed by low food availability in late winter, generating density-dependent variation in birth weight, early growth and juvenile survival in the summer despite superabundant food supplies. Comparison of the sheep and deer reveals that these effects vary between species in relation to the timing of reproductive events (Fig. 10.5), indicating that the relative impact of density and climate on the same reproductive parameters probably differs widely between species.

Other suggested generalisations concern the timing of different density-dependent effects. It is sometimes argued that most density-dependent changes occur at relatively high population densities (Fowler 1981; Gaillard et al. 1998) but this, too, also appears to be an over-simplification. While changes in mortality and fecundity in the sheep were largely confined to years when population density was high (see Chapter 3), in the deer, density-dependent changes in age at first breeding, fecundity in milk hinds, juvenile and yearling survival and antler growth started to appear soon after numbers began to increase and showed smaller changes in the later stages of population growth (Clutton-Brock *et al.* 1982a, 1985b; Clutton-Brock and Albon 1989).

Similarly, on energetic grounds, Eberhardt (1977, 2000) predicted that, as density rises, juvenile survival should change first, followed by age at first reproduction, adult fecundity and, finally, adult survival (Gaillard *et al.* 1998). Several studies show that age at first breeding is relatively sensitive to changes in population density (Clutton-Brock and Albon 1989; Gaillard *et al.* 2002) and comparisons of density-dependent changes in a range of mammals provides support for this paradigm (Eberhardt 2000). However, the relative timing of changes in different vital rates appears to vary between species. In the sheep, changes in different vital rates occurred in approximate synchrony, though there were consistent differences between age and sex categories in the timing of death that reflected their relative condition at the onset of winter (Clutton-Brock *et al.* 1997a). In the deer, age at first breeding and adult fecundity were among the first parameters to change (Clutton-Brock *et al.* 1985b, 1997a), followed by juvenile survival and, eventually, adult mortality. Here, too, differences reflect variation in the timing of changes in energy requirements and condition throughout the year (Clutton-Brock and Albon 1989; Clutton-Brock *et al.* 1997a). Finally, even where average values follow the suggested pattern, they may not do so in all categories of animals. For example, in red deer, the fecundity of milk hinds and the antler growth of yearling males were among the first parameters to change as density increased (Clutton-Brock *et al.* 1982a).

COMPARATIVE STABILITY

Our analysis of the effects of population density, winter weather and population structure explains why the sheep population suffers intermittent crashes, but does not explain why sheep numbers are less stable than those of most other northern ungulates. At least three factors may be involved. First, the constraints on dispersal imposed by the size of the island probably contribute to the magnitude of winter mortality. Unusually high densities of rodents occur when local dispersal is prevented and severe population crashes are not uncommon

(Finnerty 1980; Cockburn 1988). Like Soay sheep, several other island populations of ungulates show persistent fluctuations, including feral goats on Rum (Boyd 1981), mouflon on the Kerguelen Islands (Boussès *et al.* 1991) and reindeer on some Arctic and Antarctic islands (Leader-Williams 1988), though this is not universal.

Second, the unusually high fecundity of the Soay sheep is probably important, for comparative analysis of data from different ungulate populations suggests that the maximum rate of population decline increases with the maximum rate of population increase (T. Coulson, unpublished). Conception in the first year of life is unusual in ungulates (Western 1979; Western and Ssemakula 1982) and most species (including bighorn sheep and red deer) show substantially lower fecundity than Soay sheep.

Third, in contrast to many other ungulates (Fowler 1987), fecundity is not strongly affected by rising population density (Chapter 3) and most adult sheep breed each year regardless of density (Fig. 10.3). Although the proportion of juveniles that breed and the proportion of adults that twin are somewhat reduced when sheep numbers are high (Fig. 3.5), these changes are most pronounced *after* years when mortality has been high, so that they do little to slow population growth, except in the breeding season immediately after a crash (Chapter 3). In the intervening years, neither fecundity nor neonatal mortality is closely related to population density and emigration to other parts of the island is rare, so that numbers rise rapidly.

Our work with the sheep suggests that large fluctuations in resource-limited ungulate populations may be most likely to occur in species living at high latitudes which show high potential rates of increase combined with weak relationships between density and recruitment. The saiga antelope is one of the few ungulates with a potential reproductive rate higher than that of Soay sheep: females commonly conceive in their first year of life, most adult females produce twins and fecundity appears to be unrelated to population density (Milner-Gulland 1994; Coulson *et al.* 2000). Numbers show large and persistent fluctuations, with up to 60% of individuals dying from starvation in some years though variation in hunting pressure may also be involved. All animal populations may suffer occasional periods of high mortality as a result of density-independent changes in climate, disease, predation or human interference (Caughley and Sinclair 1994). However, we would expect persistent fluctuations caused by over-compensatory density-dependence to be rare in animals with lower reproductive rates and longer lactation periods as well as in species that depend on food supplies showing little seasonal variation.

This view of instability in ungulate populations may help to explain some of the differences between the dynamics of large and small mammals (Caughley and Krebs 1983; Gaillard *et al.* 1998). Most rodents differ from most ungulates in conceiving in the same year that they are born, in producing litters and in breeding several times per year (Finnerty 1980; Krebs *et al.* 1986; Smith *et al.* 1988). Although, in some cases, rodent populations may increase during the summer months until further recruitment is constrained by density-dependent processes operating during the summer, midsummer populations commonly exceed the level that can be supported by winter food supplies, leading to overcompensatory winter mortality. Unlike small birds and rodents, Soay sheep populations that have suffered heavy mortality in the previous winter cannot increase rapidly enough to reach ecological carrying capacity in summer. However, as, in some rodents, high growth rates allow populations that are close to ecological carrying capacity in summer to increase to a level substantially above the number than can be supported in winter, leading to overcompensatory mortality in years when summer numbers are relatively high. The cycles shown by some populations of rodents probably have different causes from the fluctuations of the sheep and several studies suggest that interactions between food availability and predation play an important role (Finnerty 1980; Krebs *et al.* 1986; Smith *et al.* 1988; Hanski *et al.* 1991; Lambin *et al.* 2000).

10.3 Hard conditions and soft selection

THE OPPORTUNITY FOR SELECTION IN FEMALES AND MALES

Fluctuations in numbers can affect the opportunity for selection, the relative intensity of selection on particular traits and the evolution of life-histories (Endler 1986; Stearns 1992; Van Tienderen 2000). Cohorts of female sheep born at high density are likely to show relatively low average fitness (since many are likely to die without breeding) but relatively high variance in lifetime reproductive success (since those that survive their first year are likely to breed successfully in several years). Fluctuations in female numbers also have important consequences for the opportunity for selection on males (Coltman *et al.* 1999c). Males born immediately after population crashes show low birth weights and reduced neonatal survival but high mating success in their first rut (Chapter 6). The relatively high success of rams born at low density extends into their second year of life, and males born in years of low density have greater lifetime breeding success than those born at high density (Coltman *et al.* 1999c).

Fluctuations in the adult sex ratio also have consequences for reproductive behaviour. When adult sex ratios are relatively equal, multiple males compete to mate with oestrous females, and consorting males are less effective at defending receptive females. As a result, dominant males pursue a more mobile strategy, searching for oestrous females and waiting for opportunities to mate with them, instead of defending them for protracted periods. Females mate frequently with multiple males and sperm competition is intense, as indicated by the high proportion of twins sired by different males (74%: Chapters 6 and 9). In contrast, when population density is low and the sex ratio is strongly biased towards females, dominant males can guard oestrous females more effectively and are able to maintain exclusive access to individual females for protracted periods (see Chapter 6). A different situation is found in male red deer, where animals born at low density show high survival and encounter relatively equal adult sex ratios as adults while those born at high density encounter female biased sex ratios and reduced breeding competition as adults (Clutton-Brock *et al.* 1997a; Marshall *et al.* 1998; Rose *et al.* 1998). Negative relationships between population density, intensity of sexual selection and the development of secondary sexual characters were predicted by Geist (1971), who used them to explain geographical variation in the degree of sexual dimorphism.

SELECTION ON WEIGHT AND GROWTH

Our results provide extensive evidence of selection on a variety of phenotypic traits (including differences in body weight, leg length, horn type and coat colour) operating through several different components of fitness (see Chapter 7). In both sexes, there is strong selection for juvenile and adult weight across a wide range of age classes. The benefits of weight to females are larger in years when population density and winter mortality are high and smaller when they are low, indicating that selection on weight is density-dependent.

There is also strong selection for early growth. As in many other wild animals (Clutton-Brock and Albon 1989; Lindström 1999; Unsworth *et al.* 1999), early growth rates have an important influence on the survival of animals throughout their first year of life, when mortality is relatively high. Early-born lambs and those born below average weight show higher mortality than late- or heavy-born lambs, both during their first months of life and, as juveniles, at the end of their first winter. Females born light also show reduced weights at four months and sixteen months and are less likely to breed in their first year of life (Clutton-Brock *et al.* 1992). Factors associated with birth weight (including the lamb's birth date, the age of the mother and whether the lamb is a singleton or a twin and the mother's genotype) are consequently likely to influence the survival of lambs and juveniles. However, the affects of early development on survival wane with increasing age. After the sheep begin their reproductive careers, the effects of early development are less important than those of adult weight and reproductive status, as in bighorn sheep (Festa-Bianchet *et al.* 1997).

Like variation in many other quantitative traits (Mousseau and Roff 1987; Roff 1997), variation in growth and adult body weight is heritable (see Chapter 7). However, despite repeated episodes of selection for body size operating through survival and fecundity, there is no evidence of a change in average body weight or size during the period of our study. Several different mechanisms may contribute to this, including the relatively low additive genetic variance of body weight relative to the magnitude of environmental effects and the intermittent nature of selection pressures. In addition, it is possible that selection operates principally on the environmental component of variation, leading to situations where no evolutionary change occurs despite strong directional selection on a heritable trait (Stinchcombe *et al.* 2002). In red deer, for example, there is strong directional selection for antler size operating through variation in the lifetime breeding success of males. However, although antler size is heritable, there is no positive genetic correlation between antler size and fitness and antler size has declined over the last twenty-five years (Kruuk *et al.* 2002b). Whether there is a positive genetic correlation between body size and fitness in Soay sheep is not yet known but it is possible that similar effects could explain the lack of consistent change in body weight or size over the period of this study.

SELECTION ON HORN TYPE

The complexity of selection pressures is illustrated by our research on selection on horn type and coat colour, both of which show heritable variation (Appendix 2). Mating success is typically higher in males with normal horns than in those with scurred horns. However, when population density is high, the difference in mating success between horned and scurred males declines because horned males cannot defend females so effectively (Chapters 9). In addition, at the peak of the rut, the proportion of lambs that successful males with large horns sire declines as a consequence of sperm depletion (Preston *et al.* 2001). There is also selection on horn type operating through survival and female fecundity. When winter mortality is high, males and females with scurred horns both show higher winter survival than other animals (see Chapter 7). Among females, animals with scurred horns also show higher breeding success when population density is high as a result of higher rates of offspring survival, higher body weights and heavier offspring birth weights (Clutton-Brock *et al.* 1997b). The situation is further complicated by a difference in the effects of horn phenotype on juvenile and adult females: selection for scurredness is strongest in juveniles while non-scurred individuals are favoured among older adults (see Chapter 7).

Despite the relatively strong selection pressures affecting horn type in the St Kilda population, both horned and scurred phenotypes have been preserved in the population for several hundred years and probably much longer (see Chapter 1). Over the period of this study, there is no evidence of a change in the relative frequency of horned and scurred females though there has been a weak trend for the proportion of scurred males born to increase (see Chapter 7). While the mechanisms maintaining variation in horn type are not yet fully understood (see Appendix 2), our research suggests that three separate mechanisms may be involved. First, the relative benefits of horned versus scurred phenotypes to male mating success may decline as their frequency in the population changes (Chapter 9). Second, the costs of scurred horns to males may be balanced by survival benefits in both sexes. And, third, selection pressures favouring one phenotype over the others may never be consistent over long enough periods to eliminate the others from the population. There is no reason to suppose that these three mechanisms are alternatives and all three may contribute to maintaining variation in horn type. In addition, the proposed inheritance system for horns could play a role in limiting change, since some phenotypes result from different genotypes (see Appendix 2).

SELECTION ON COAT COLOUR

Selection on coat colour differs from selection on horn type (Chapter 7). There are no obvious differences in mating success between light- and dark-coloured males. Selection for coat colour operating through survival varies between years, with dark coats being favoured in some years, no differences between light and dark phenotypes in others and selection for light coats in females (especially juveniles) in some years (see Chapter 7). As in the case of horn variation, there have been no consistent changes in the frequency of dark and light phenotypes during our study. In contrast to horn type, the strength of selection is not concordant between the sexes across years and, in females, varies between age classes. An initial indication that selection on coat colour might favour light coats at low density and dark at high density (Moorcroft *et al.* 1996) has not been supported by more recent data. However, the strength of selection for dark coat colour appears to increase in years when winter mortality is high (see Chapter 7), possibly because dark individuals are consistently heavier than light ones. There is also some indication that light coats could be maintained by intermittent selection on younger animals.

PATTERNS OF SELECTION

Our analyses of selection have produced three main generalisations. First, most selection pressures are 'soft' – or even slushy. Almost all the relationships between variation in phenotype or genotype and survival or fecundity are affected by variation in population density and climate (see Chapters 7 and 8). In a number of cases, selection pressures only appear when population density is high or winter conditions are adverse (like differences in survival between horn phenotypes in adult females). In a few cases, even the direction of selection pressures changes between years (see Chapter 7). The intensity of selection also varies widely between age and sex categories. It tends to be strongest in the categories that show highest mortality so that selection is usually stronger in juveniles than in adults, and in males than in females – though, in years and classes of animals when mortality approaches 100%, the intensity of selection is inevitably reduced. Environmental factors sometimes interact with this variation, generating differences in the strength of selection pressures between categories of animals which vary in magnitude between years.

Second, our research suggests that, as in other natural populations, many traits are partly heritable, including body weight and some life-history characteristics (Milner *et al.* 1999a, b; Réale *et al.* 1999; Réale and Festa-Bianchet 2000; Kruuk *et al.* 2000; Coltman *et al.* 2001a). In addition, more outbred individuals show improved survival (see section 8.4). In addition, as in the red deer population on Rum (Pemberton *et al.* 1988, 1991, 1996), polymorphisms at the molecular level (including variation in the Major Histocompatibility Complex, MHC, transferrin (Tf) and adenosine deaminase (Ada)) are associated with differences in survival or fecundity (section 8.5). These relationships are varied: in some cases, heterozygotes at particular loci show the highest performance while, in others, homozygous individuals outperform heterozygotes. In addition, some genotypes appear to have different effects on fecundity and survival, suggesting that countervailing selection may be operating.

Third, our research shows that phenotypic and genetic differences between sheep are often associated with differences in parasite load (see Chapters 5 and 8). For example, relatively homozygous individuals carry twice as many parasites as less inbred sheep in years when population density is high (Coltman *et al.* 1999b) (section 8.4). Parasite load also varies between individuals with different genotypes at the MHC and gamma interferon loci (section 8.5). Since over-winter survival differs between genotypes and is affected by variation in parasite load (Chapter 5), the relationships between genotype and parasite load could be responsible for differences in survival between genotypes. However, an alternative possibility is that genotype has independent effects on parasite load and survival (section 8.4). This can be excluded in at least one case: treatment with anthelminthic boluses removes the difference in survival between sheep with high and low heterozygosity, indicating that inbreeding affects winter survival through its influence on parasite load.

There is still much that we do not understand, including the reasons why horn phenotype and coat colour are associated with particular components of fitness and the selection pressures that maintain variation in these two traits. Only by further investigation of the costs and benefits of different phenotypes under varying environmental conditions shall we be able to distinguish between the different mechanisms that may contribute to maintaining variation in these two traits.

THE MAINTENANCE OF GENETIC DIVERSITY

Despite the small size of the founder population and intermittently strong selection pressures, the sheep show appreciable levels of genetic variation and morphometric traits show modest heritabilities (Milner *et al.* 1999a; Coltman *et al.* 2001a). Rigorous selection over several thousand years might have been expected to produce a more uniform genotype (Fisher 1930). Why has it not done so?

Several different processes have probably contributed to slow or halt the erosion of genetic diversity. Many selection pressures are only apparent under particular circumstances or in particular categories of animals – or under particular circumstances within particular categories (see Chapter 7). As a result, short-term changes in environmental factors (like the fluctuations in population size), medium-term changes (such as decadal changes in winter weather) and longer-term fluctuations (such as the results of global warming) may all reduce the rate at which genetic diversity is eroded (Clarke 1972; Clarke and Beaumont 1992). The tendency for sperm depletion to restrict the mating success of the most successful males and the relatively high breeding success of juvenile males in years when the adult sex ratio is strongly biased may also have the same effect. Finally, contrasting effects of particular traits on different components of fitness on males and females may help to preserve genetic diversity (see Chapters 7 and 8).

Other selective mechanisms may delay the loss of diversity in the population (see Chapter 8). Relatively inbred sheep show higher parasite loads and lower winter survival than individuals with higher levels of heterozygosity, with the result that population crashes increase rather than reduce average levels of heterozygosity (Chapter 8). Fluctuations in parasite load may lead to negative frequency dependence in selection on particular alleles at specific loci (as, possibly, in the case of the MHC loci) while other polymorphisms may be maintained by heterozygote advantage, as appears to be the case at the Ada locus (see Section 8.5). Multiple mechanisms may often contribute to the maintenance of diversity in particular traits, their relative importance varying between traits.

Our analyses emphasise the complexity of interactions between environmental variation and selection pressures. It is probably safe to assume that, in reality, these interactions are more complex still, for we have focussed only on two environmental factors (population density and winter weather) and two ways of categorising animals (age and sex). The view of selection that our studies reveals bears little resemblance to the consistency envisaged by classical evolutionary texts and supports predictions that selection in natural populations will seldom be unaffected by environmental parameters (Wallace 1975; Kreitman *et al.* 1992). We are left with an impression of a process showing persistent variation in strength, commonly activated only under harsh environmental conditions and often affecting different categories of animals to different extents. It is not difficult to believe that, combined with short- and medium-term variation in the environment, selection would be slow to remove genetic variation from the population.

10.4 Adaptation in changing environments

COSTS OF REPRODUCTION

Fluctuations in population size also exert an important influence on the costs and benefits of different reproductive strategies. In females, the survival costs of breeding in the first year of life show a dramatic increase in years when population density and mortality are high, especially among individuals of relatively low body weight (Fig. 2.14). In contrast, when population density and mortality are low, there are no obvious costs of early breeding or twinning to survival. Adult females that conceive twins are also substantially more likely to die when population density is high, especially if they are relatively light. Here, too, when population density and mortality are low, there are no obvious costs of twinning and reproductive success tends to be positively correlated with subsequent survival or breeding success, indicating that variation in phenotypic quality is obscuring any costs of reproduction (Clutton-Brock and Harvey 1979).

Similar interactions between population density and the costs of reproduction have been described in other ungulates, though the parameters affected vary between studies. In bighorn sheep, the effects of successful reproduction on subsequent survival and fecundity are small and inconsistent, though females that raise lambs successfully are less likely to do so the following year, especially if they are light or population density is high (Festa-Bianchet *et al.* 1998). Positive correlations between reproductive success and subsequent survival or breeding success are not uncommon when conditions are favourable (Festa-Bianchet 1998; Bérubé et al. 1999). And, in red deer, successful breeding depresses autumn body weight and fecundity (especially when population density is high) but has no consistent effects on the birth weight or survival of calves (Clutton-Brock et al. 1983, 1989).

In males, the mortality of all age categories is high in years when population density is high and weather conditions are adverse (Fig. 10.2). Castrates (which do not rut) and scurred males (which seldom defend females) show higher survival than horned males (Chapter 9). Juvenile males that are prevented from rutting in their first autumn (by treatment with progestogens) also show substantially higher survival than control animals (see Chapter 9). Once again, these effects are most pronounced in years when population density and mortality are high and are enhanced in individuals of low body weight.

Our results show that the costs of reproduction are highest under adverse environmental conditions and in categories of animals that are already at some disadvantage. However, this is not invariably the case. By reducing survival differentials, very high levels of mortality can reduce the intensity of selection and the costs of reproduction. Under such conditions, selection may even favour increased expenditure on reproduction by categories of animals that have little to lose (Marrow et al. 1996).

BENEFITS OF RAPID REPRODUCTION

The heavy costs of breeding in the first year of life when population density is high have raised doubts as to whether early breeding represents an adaptive life-history strategy (Boyd and Jewell 1974). Might the relatively early age at breeding in Soay sheep and the relatively high incidence of twinning be a non-adaptive relict of human selection? Might this even explain the unusually high levels of mortality in crash years? As Chapter 9 shows, there is little support for this argument. While juveniles of both sexes show relatively low breeding success, the chance that they will die even if they do not breed has to be taken into account. Moreover, breeding in the first year of life has

substantial benefits when mortality is low, increasing lifetime breeding success and accelerating the entry of an individual's progeny into
the population (sections 9.2 and 9.3). Twinning has similar benefits:
calculations of optimal breeding strategies for animals of different
weights show that, across all years, benefits exceed net costs in most
categories of animals.

While early breeding and twinning are advantageous on average,
their costs outweigh their benefits in years when population density and mortality are high (Clutton-Brock *et al.* 1996; Marrow *et al.*
1996). In an environment that fluctuates within the lifespan of individuals, flexible reproductive strategies would offer substantial advantages. Like bighorns (Festa-Bianchet and Jorgensen 1998) and red
deer (Clutton-Brock *et al.* 1983), Soays show some capacity to adjust
reproduction to density-related changes in bodyweight: the probability that females conceive in their first year or that older females will
conceive twins declines with increasing density, while the frequency
of abortion shortly before birth increases (see section 3.3). However,
it would be to the advantage of all juveniles and many lighter adults
to avoid conception altogether in years when density is high and substantial mortality is to be expected (Marrow *et al.* 1996). The sheep
conspicuously fail to do this, conceiving only slightly fewer lambs in
years when summer density is high (Chapter 3). Although 'prospec-
tive' adjustment of fecundity would be beneficial, it seems likely that
the mechanisms controlling conception do not permit it. Specifically,
fecundity is probably controlled by body condition in autumn (Albon
et al. 1983b; Gunn *et al.* 1986) and this mechanism may not allow
fecundity to be adjusted to future variation in the costs of breeding
unless these are correlated with body weight in the previous summer.

Consideration of the optimal reproductive strategies of the sheep
emphasises the close connection between early reproduction, high
potential rates of population growth and over-compensatory density-
dependence in mortality (see Chapter 3). As well as increasing the
costs of breeding, over-compensatory density-dependence reduces the
life expectancy of animals whether they breed or not (see Chapter 2).
This is likely to favour the evolution of early breeding and increased
litter size (Harvey and Zammuto 1985; Stearns 1992), thereby raising

the growth rate of the population and further increasing the chance of demographic instability. Where seasonal fluctuations in resource availability free reproduction from density-dependent constraints in summer (see Chapter 3), over-compensation in winter mortality may favour early reproduction and increased litter size. Both relatively high population growth rates and demographic instability are not uncommon in vertebrate populations at high latitudes (Saether 1997; Post and Stenseth 1998, 1999) and the early reproduction and high twinning rates found in some palearctic ungulates, particularly in saiga antelope (Milner-Gulland 1994, 1997), in roe deer (Bramley 1970; Gaillard et al. 1998) and moose (Peterson 1955; Peek 1962), may be the result of evolutionary processes of this kind.

10.5 Extrapolations to other populations

So are the processes that we have documented in Soay sheep on Hirta likely to be important in other populations of ungulates? Or are they dominated to such an extent by the unusual history of the sheep and the constraints on their movements that they have little bearing on the ecology of wild ungulates? There are certainly reasons for thinking that several of the characteristics of the Soay sheep may be a consequence of their unusual evolutionary history. The small size, early weaning ages and low age at first breeding of Soay sheep are all typical of animals that have been subjected to artificial selection (J. Clutton-Brock 1981). In most wild sheep, mothers suckle lambs through the summer and females usually breed for the first time in their second or third year of life (Geist 1971; Schaller 1977; Festa-Bianchet 1988b). As we have argued, it is likely that the early weaning age of Soays is responsible for the lack of density-dependence in fecundity and, together with their capacity to become pregnant in their first year, is responsible for their unusually high rate of population growth (see Chapter 3). Consequently, artificial selection may have originally helped to create a phenotype that is likely to show persistent fluctuations under the conditions that prevail on St Kilda. These fluctuations may, in turn, generate selection pressures favouring rapid reproduction and so help to maintain the unusual life-history of the sheep. Their history may also have contributed to the phenotypic and genetic diversity of the population for polymorphism in coat colour and horn development are not common in wild ungulates, though both occur (Schaller 1977; Walther *et al.* 1983; Whitehead 1995). However, while some of the characteristics of the sheep are unusual, all aspects of their life-histories (including the timing of their reproductive cycle, their age at first breeding and the relative birth weights of lambs) fall within the range of variation recorded for wild ungulates (see Chapter 2).

The habitat of the sheep may also contribute to their unusual dynamics. By restricting dispersal, the boundaries of Hirta and the other islands may exaggerate variation in local population density, in the same way that fencing areas of suitable habitat increases local fluctuations in rodent numbers (Finnerty 1980; Cockburn 1988). In addition, many resident ungulate populations in more natural ecosystems are limited by predation (Sinclair 1977; Gasaway *et al.* 1992) and the absence of predators probably contributes to the relatively high survival of lambs (see Chapter 2) and thus to the rate of population growth.

While the history of the sheep is unusual in several respects, most of the processes that we have explored in this study probably occur in a greater or lesser extent in many other organisms. Winter mortality is the key factor limiting population size in many vertebrates (Sauer and Boyce 1983; Skogland 1985b; Caughley 1994; Caughley and Sinclair 1994; Kendall *et al.* 1998; Newton 1998) and is commonly affected both by population density and by density-independent factors (Novellie 1986; Bjornstad *et al.* 1995; Saether 1997; Portier *et al.* 1998; Post and Stenseth 1999). Many other vertebrate populations appear to be limited by starvation interacting with the effects of predation, parasite load and disease (Caughley 1994a,b; Newton 1998). Persistent fluctuations in population size are not uncommon in vertebrates, especially in populations at relatively high latitudes (Finnerty 1980; Krebs *et al.* 1986; Cockburn 1988; Bjornstad *et al.* 1995; Kendall *et al.* 1998) and probably have diverse causes, including interactions between survival and predation (Peterson *et al.* 1984), parasite density (Hudson 1986; Hudson *et al.* 1992a), climatic variation (Saether 1997; Post and Stenseth 1999), social factors (Cockburn 1988) and resource availability (Krebs *et al.* 1986; Bjornstad *et al.* 1995). Though several of the best-documented examples of instability in ungulates involve island populations (Boyd 1981; Leader-Williams 1988; Boussès *et al.* 1991), fluctuations in population size also occur in less circumscribed populations (Peterson *et al.* 1984; Skogland 1990; Fryxell *et al.* 1991; Messier 1991).

The factors affecting survival in the Soay sheep – birth date, birth weight, age, sex, condition and parasite load – have similar effects in other mammals, including bighorn sheep (Festa-Bianchet 1998; Festa-Bianchet *et al.* 1998; Portier *et al.* 1998; Loison *et al.* 1999a) and red deer (Clutton-Brock and Albon 1982, 1989; Clutton-Brock *et al.* 1987a). Higher mortality in juveniles relative to adults and in males relative to females is widespread in polygynous mammals (Clutton-Brock *et al.* 1997a; Loison *et al.* 1999b) and several studies indicate that these differences increase when density is high or weather conditions are adverse, generating female-biased adult sex ratios after periods of high mortality (Klein 1968; Sobanskii 1979; Clutton-Brock *et al.* 1982b). Differences in population density and in winter weather between years are known to affect pre- and postnatal growth rates in other vertebrates, generating variation in body size and breeding success between cohorts (Albon *et al.* 1991; Rose *et al.* 1998; Lindström 1999; Post and Stenseth 1999).

Interactions between environmental factors and the intensity of selection are also likely to be widespread in naturally regulated animal populations though relatively few studies have been in a position to assess them (Wallace 1975; Endler 1986; Grant 1986; Clutton-Brock *et al.* 1988b; Kreitman *et al.* 1992). Compared with 'strong' or continuous selection pressures, intermittent selection probably helps to delay the loss of diversity in many other animals while countervailing selection on particular traits has been demonstrated in several populations (Berry and Murphy 1970; Hall and Purser 1979; Grant 1986; Pemberton *et al.* 1991, 1996) and may also play an important role in maintaining genetic diversity. Persistent fluctuations in population size caused by intermittent periods of high mortality would be expected to favour the evolution of rapid breeding (Harvey and Zammuto 1985; Stearns 1992) and there is no evidence to suggest that either early breeding or frequent twinning is maladaptive on St Kilda. Our analysis of

Soay life-histories (see Chapter 9) shows that their existing reproductive strategies could only be improved if it were possible to adjust fecundity 'prospectively' – an adaptation that, to our knowledge, no non-human mammal has been shown to have mastered.

While we believe that most of the effects of intermittent starvation that we have described in the Soay sheep occur in many other food-limited populations, we are not surprised that they are more pronounced in Soays on St Kilda than in most other populations. The effects of starvation in the Soays are savage and are focussed during late winter. They are exacerbated by constraints on dispersal and high parasite loads (see section 10.2). As a result, differentials in survival or breeding success may be present in the Soay population of St Kilda that are absent in populations living in areas where food is more abundant. For example, it may be that the larger costs of reproduction in Soays compared to bighorn sheep (Festa-Bianchet *et al.* 1998) are a consequence of environmental differences between the two populations rather than of any intrinsic difference in reproductive strategies.

10.6 Summary

In summary, fluctuations in the numbers of Soay sheep on St Kilda are the result of over-compensatory density-dependent mortality in winter, caused by the lack of any substantial reduction in fecundity at high density resulting from superabundant summer resources and early weaning. Population crashes occur in years when density is high and winter weather conditions are unfavourable. Both density-dependent increases in parasite load and the high costs of reproduction contribute to their magnitude. Fluctuations in sheep numbers generate differences in development and breeding success between cohorts, but these appear to have few important consequences for population dynamics.

During periods of high mortality, selection pressures intensify. Individual differences in phenotype and genotype interact with the effects of sex, age, reproductive history and local resource availability to affect components of fitness. Temporal changes in the intensity of selection, combined with differences between sex and age categories, lead to complex interactions between environmental factors and variation in phenotype (or genotype) which may play an important role in delaying the loss of genetic diversity in the population.

The persistent fluctuations in sheep numbers lead to large fluctuations in the costs and benefits of different reproductive strategies. On average, conditions favour the evolution of early reproduction and high reproductive rates, maintaining a life-history which contributes to the instability of the population and reinforces selection for early reproduction. Since increasing density has little effect on autumn body weight or on fecundity, the capacity of the sheep to adjust fecundity to population density is limited. Though it would be to their advantage to adjust fecundity 'prospectively' to avoid breeding in years when mortality is likely to be high, the mechanisms controlling fecundity probably preclude adjustments of this kind.

Our research on St Kilda over the last twenty years and our experience of other long-term studies leave us with four firm convictions. First, attempts to understand and predict changes in population size need to be based on an understanding of the mechanisms underlying them, in much the same way that an understanding of many physiological changes requires knowledge of the underlying molecular processes. In particular, an understanding of variation in the magnitude and timing of energetic requirements may be necessary to predict how populations will respond to changing environmental conditions. Further estimates of the effects of changes in density and other environmental variables at the level of individuals and experiments to identify the causal processes involved both have an important role to play.

Second, it seems likely that most (if not all) selection pressures are strongly influenced by population density, as well as by density-independent factors. So, too, are the costs and benefits of different reproductive strategies and, hence, their net payoffs. Studies of selection based on short-term data from natural populations are likely to generate misleading conclusions, since they may fail to measure strong selection pressures or heavy costs that only occur under occasional adverse conditions. Less commonly, they may overestimate the strength of selection or the costs of particular strategies if data happen to have been collected under unusually adverse circumstances. For the same reasons, protracted studies under a realistic range of conditions are needed for reliable assessment of the costs and benefits of particular evolutionary strategies. Measures of the intensity of selection or the costs of reproduction in captive animals maintained with ad lib access to food are likely to generate misleading results that do not reflect the selection pressures operating in natural populations for similar reasons. This caveat applies at least as much to studies of sexual selection as it does to studies of survival strategies, for the costs of mating competition and of many secondary sexual characteristics commonly vary with environmental conditions.

Third, we have become progressively aware of both the benefits and the limitations of working on model systems in relatively simple environments. The accessibility of the sheep and the associated possibility of monitoring the growth and breeding success of large numbers of individuals throughout their lifespans make it feasible to answer a wider range of questions than if we had worked on a more wide-ranging animal living in a more complex environment. Since the consequences of intermittent starvation are probably similar in many resource-limited populations, the ecological mechanisms that we have documented probably affect many other animal populations where they cannot be investigated directly. Nevertheless, there are obvious limitations to extrapolating from island populations, like St Kilda, to more natural populations. In particular, the lack of predators and limitations on dispersal are likely to influence many aspects of population dynamics and selection (see Chapter 3). We believe that this points to the need for detailed research on simple model systems as well as on more natural populations.

Finally, our work underlines the benefits of collaborative studies involving scientists from different disciplines. This study has involved population demographers, reproductive physiologists, behavioural ecologists, geneticists, botanists and epidemiologists and its scope reflects a broader range of expertise than any one scientist could provide. The integration of research on population dynamics with studies of energetics, epidemiology and genetics has demonstrated relationships that we should never have anticipated had we conducted independent studies on separate populations. Moreover, the use of a single population where long-term records of individual life-histories were available has made it possible for new studies rapidly to reach a point where they were able to investigate novel questions. As the aims and objectives of empirical research become more closely focussed, the importance of long-term studies that concentrate expertise from several related fields on the biology of particular populations where individuals can be recognised and where the biology of individuals can be monitored is likely to increase. Populations that provide access to long-term records of the life-histories of individuals represent natural laboratories where it is feasible to answer ecological and evolutionary questions with a level of precision that is not possible where individuals cannot be consistently recognised. Their maintenance needs to be recognised as a priority by all ecologists.

10.7 Acknowledgements

I am very grateful to Tim Coulson, Jos Milner, Ian Stevenson, Steve Albon, Josephine Pemberton, Bryan Grenfell, Ken Wilson and Marco Festa-Bianchet for their comments on previous drafts of this chapter.

Appendix 1
The flora of St Kilda

M. J. Crawley   *Imperial College London*

The following systematic list of the vascular plants of St Kilda shows the grazing impacts of Soay sheep in different habitats. Nomenclature follows Stace (1997), and data are from M. J. Crawley (unpublished data) and Pankhurst and Mullin (1991).

The key to the habitats is as follows: I, sea cliffs and ungrazed places like inaccessible cleit-tops; II, freshwater wetlands and *Iris* mires; III, stream banks; IV, fertile grasslands within the Head Dyke; V, heaths and drier heathy grasslands; VI, mires and wet heathy grasslands; VII, *Plantago* sward and short seaside turf; VIII, walls, cleits and inland cliffs; IX, summit heaths and exposed places above 250 m, including the extensive *Nardus–Racomitrium* grasslands; X, The Village, the Army Base and in the vicinity of cleit doors.

The column headed 'Dafor' gives ranked abundance. The key is: d, dominant; a, abundant; f, frequent; o, occasional; r, rare; l, locally (prefix).

In the column headed 'Seen', plants are recorded on Hirta since 1993 by MJC unless noted by No (these are from published records in Pankhurst and Mullin (1991)). Those ruderals that might persist as dormant seeds in the soil seed bank are noted as ?SB.

In the column headed 'Grazing response' the key is: I, increasers (unpalatable species that increase in abundance under moderate to heavy grazing by sheep); D, decreasers (highly palatable species that decline in abundance under grazing; many species exhibit reduced rates of flowering and seed production under grazing, but do not decline markedly in abundance). Unmarked species are either avoided, show no pronounced responses in either direction, or are species about which we have too little information to record their grazing responses.

311

Soay Sheep: Dynamics and Selection in an Island Population, ed. T. H. Clutton-Brock and J. M. Pemberton.
Published by Cambridge University Press. © T. H. Clutton-Brock and J. M. Pemberton 2003.

Table A1.1. The vascular plants of St Kilda

Family	Genus and species	I	II	III	IV	V	VI	VII	VIII	IX	X	Dafor	Seen	Grazing response
Lycopodiaceae	Huperzia selago	.	.	.	.	r	.	.	.	o	.	r	.	.
Selaginellaceae	Selaginella selaginoides	.	.	o	.	o	o	.	.	o	.	o	.	.
Equisetaceae	Equisetum arvense	.	.	.	.	o	o	.	.	.	o	o	.	.
	Equisetum palustre	.	f	.	.	.	.	.	.	.	o	o	.	.
Ophioglossaceae	Ophioglossum vulgatum	.	.	.	r	.	.	r	.	.	.	r	.	.
	Ophioglossum azoricum	.	.	.	.	.	.	.	.	.	.	r	No	.
	Botrychium lunaria	.	.	.	r	.	.	r	.	.	.	r	.	.
Hymenophyllaceae	Hymenophyllum wilsonii	f	.	o	.	.	.	.	o	.	.	o	.	.
Polypodiaceae	Polypodium vulgare	f	.	o	.	.	.	.	f	.	.	f	.	.
Dennstaedtiaceae	Pteridium aquilinum	.	.	.	o	o	.	.	.	.	.	la	.	Not increasing
Aspleniaceae	Asplenium adiantum-nigrum	o	.	.	.	.	.	.	o	.	.	f	.	.
	Asplenium marinum	o	.	.	.	.	.	.	o	.	.	o	.	.
Woodsiaceae	Athyrium filix-femina	f	o	f	.	.	.	.	a	.	.	f	.	.
	Cystopteris fragilis	o	o	o	.	.	.	.	o	.	.	o	.	.
Dryopteridaceae	Dryopteris aemula	r	.	.	.	.	.	.	r	.	.	r	.	.
	Dryopteris dilatata	f	o	o	.	.	o	.	f	.	.	f	.	.
	Dryopteris expansa	r	.	.	.	.	.	.	r	.	.	o	.	.
Blechnaceae	Blechnum spicant	f	o	o	o	o	f	r	o	o	.	f	.	I
Ranunculaceae	Ranunculus acris	f	.	.	a	.	.	r	.	.	f	a	.	I
	Ranunculus repens	.	.	.	r	.	.	.	r	.	r	r	.	.

Family	Species	1	2	3	4	5	6	7	8	9		Luxuriant if ungrazed
	Ranunculus flammula	o	a	·	·	·	·	·	r	l	·	·
	Ranunculus ficaria	·	·	o	f	·	o	·	o	f	·	·
Urticaceae	*Urtica dioica*	o	·	·	·	·	·	·	lf	la	·	·
Chenopodiaceae	*Atriplex prostrata*	·	·	·	·	r	·	r	r	r	?SB	·
	Atriplex glabriuscula	o	·	·	·	r	·	·	·	o	·	·
	Salsola kali	·	·	·	·	r	·	·	r	r	?SB	·
Portulacaceae	*Montia fontana*	a	·	o	·	·	·	·	f	f	·	·
Caryophyllaceae	*Honkenya peploides*	·	·	·	·	r	·	·	r	r	No	·
	Stellaria media	·	r	·	·	·	o	·	a	a	·	·
	Stellaria uliginosa	f	o	·	·	·	·	·	·	f	·	·
	Cerastium fontanum	o	f	·	r	·	o	·	f	a	·	I
	Cerastium diffusum	·	·	·	·	·	·	·	r	r	No	·
	Sagina subulata	·	·	·	·	·	·	·	r	r	No	·
	Sagina procumbens	o	·	o	·	r	o	·	o	f	?SB	I
	Spergula arvensis	·	·	·	·	·	·	·	r	r	?SB	·
	Lychnis flos-cuculi	o	·	·	·	·	·	·	·	o	·	·
	Silene uniflora	f	·	·	·	If	·	o	o	o	·	D
	Silene acaulis	·	·	·	·	r	·	o	o	o	·	·
Polygonaceae	*Persicaria maculosa*	·	·	·	·	·	·	·	·	·	?SB	·
	Polygonum arenastrum	·	·	·	·	·	·	·	r	r	?SB	·
	Polygonum aviculare	·	·	·	·	·	·	r	r	r	?SB	·
	Rumex acetosella	·	o	·	o	·	·	r	r	r	·	I
	Rumex acetosa	a	o	·	o	·	·	f	f	a	·	I

Table A1.1. (contd.)

Family	Genus and species	I	II	III	IV	V	VI	VII	VIII	IX	X	Dafor	Seen	Grazing response
	Rumex crispus	.	.	.	.	.	.	.	.	r	.	r	.	.
	Rumex obtusifolius	.	.	.	.	.	.	.	o	.	f	o	.	.
	Oxyria digyna	o	.	o	r	.	.	.	r	.	.	lf	.	.
Plumbaginaceae	*Armeria maritima*	f	.	.	.	.	.	f	o	.	.	f	.	D
Clusiaceae	*Hypericum pulchrum*	o	.	o	r	o	.	.	.	.	.	o	.	.
Droseraceae	*Drosera rotundifolia*	.	.	.	.	.	o	.	.	.	.	o	.	.
Violaceae	*Viola riviniana*	.	.	o	f	o	r	.	.	.	.	f	.	.
	Viola palustris	.	.	.	.	.	r	.	.	.	.	lf	.	.
Salicaceae	*Salix repens*	.	.	.	.	o	r	.	.	o	.	o	.	.
	Salix herbacea	.	.	.	.	.	.	.	.	r	.	r	.	.
Brassicaceae	*Cardamine hirsuta*	.	.	.	.	.	.	.	.	.	r	r	?SB	.
	Cochlearia officinalis	o	.	.	.	.	.	.	.	.	.	o	.	.
	Cochlearia danica	o	.	.	.	.	.	.	.	.	.	o	.	.
	Capsella bursa-pastoris	.	.	.	.	.	.	.	.	.	r	r	?SB	.
	Brassica rapa	.	.	.	.	.	.	.	.	.	r	r	?SB	.
Empetraceae	*Empetrum nigrum*	.	.	.	.	r	o	.	.	f	.	o	.	.
Ericaceae	*Calluna vulgaris*	.	.	.	.	o	f	.	.	o	.	d	.	Tolerant of grazing
	Erica tetralix	.	.	o	.	r	f	.	.	r	.	o	.	.
	Erica cinerea	o	.	o	.	o	r	.	.	r	.	o	.	.
	Vaccinium myrtillus	o	.	o	.	o	o	.	.	r	.	o	.	.

Family	Species									Tolerant of grazing
Primulaceae	*Primula vulgaris*	.	.	f	.	.	o	.	.	.
	Anagallis tenella	.	o	o	.	.	f	.	.	.
Crassulaceae	*Sedum rosea*	f	o	o	lf	lf	o	.	D	
Saxifragaceae	*Saxifraga oppositifolia*	r	o	o	.	.	o	.	.	
Rosaceae	*Filipendula ulmaria*	.	o	.	.	.	o	.	.	
	Potentilla anserina	.	.	.	f	.	o	f	.	.
	Potentilla erecta	.	.	o	o	.	f	r	f	.
Fabaceae	*Vicia hirsuta*	.	.	.	r	.	r	r	r	?SB
	Vicia sepium	.	.	.	.	.	r	r	r	No
	Trifolium repens	.	.	a	f	.	a	f	f	I
	Trifolium pratense	.	.	r	.	.	r	.	.	.
Onagraceae	*Epilobium palustre*	.	o	.	o	.	r	.	.	.
Linaceae	*Linum catharticum*	.	.	r	.	.	r	.	.	.
Polygalaceae	*Polygala vulgaris*	.	.	.	o	.	r	.	.	No
	Polygala serpyllifolia	.	o	o	o	.	f	.	.	.
Apiaceae	*Hydrocotyle vulgaris*	o	f	.	.	.	la	.	.	.
	Ligusticum scoticum	.	.	.	.	.	o	.	D	.
	Angelica sylvestris	f	.	r	.	a	f	.	D	.
Gentianaceae	*Centaurium erythraea*	.	.	.	r	.	r	r	.	.
	Gentianella campestris	.	.	r	o	.	o	r	o	.
Boraginaceae	*Myosotis arvensis*	.	.	.	.	.	r	r	?SB	.
Lamiaceae	*Galeopsis tetrahit*	.	.	.	.	.	r	r	?SB	.
	Prunella vulgaris	.	o	o	.	.	f	f	.	.

Table A1.1. (contd.)

Family	Genus and species	I	II	III	IV	V	VI	VII	VIII	IX	X	Dafor	Seen	Grazing response
	Thymus polytrichus britannicus	o	.	o	f	f	o	r	.	o	.	f	.	.
Callitrichaceae	Callitriche stagnalis	.	a	o	.	.	.	.	.	.	.	la	.	.
Plantaginaceae	Plantago coronopus	f	.	.	o	.	a	a	.	.	.	a	.	I
	Plantago maritima	f	.	.	o	o	.	f	.	.	.	f	.	.
	Plantago major	.	.	.	.	.	.	.	.	.	o	o	.	Trampling tolerant
	Plantago lanceolata	f	.	.	f	o	.	f	.	.	.	f	.	.
Scrophulariaceae	Veronica officinalis	o	.	.	o	o	.	.	o	r	.	o	.	.
	Euphrasia arctica subsp. borealis	.	.	.	r	o	.	.	.	.	.	o	.	.
	Euphrasia tetraquetra	.	.	.	r	.	.	.	.	.	.	o	.	.
	Euphrasia nemorosa	.	.	.	o	o	.	.	.	.	.	o	.	.
	Euphrasia frigida	.	.	.	r	o	.	.	.	.	.	o	.	.
	Euphrasia foulaensis	.	.	.	r	.	.	.	.	.	.	r	.	.
	Euphrasia micrantha	.	.	.	r	o	.	.	.	.	.	r	.	.
	Euphrasia scottica	.	.	.	r	o	.	.	.	.	.	r	.	.
	Rhinanthus minor	.	.	.	r	.	.	.	.	.	.	r	.	.
	Pedicularis sylvatica	.	.	o	.	o	f	.	r	.	.	o	.	.
Lentibulariaceae	Pinguicula vulgaris	o	.	o	.	o	f	r	r	.	.	o	.	.
Rubiaceae	Galium saxatile	o	.	.	o	f	o	.	.	.	.	f	.	I

Family	Species										Notes
Caprifoliaceae	*Sambucus nigra*	.	.	.	o	.	.	.	r	r	.
	Lonicera periclymenum	.	.	.	.	o	.	.	r	o	.
Dipsacaceae	*Succisa pratensis*	o	.	o	o	.	f	o	.	f	f
Asteraceae	*Cirsium vulgare*	f	.	.	.	la	.	.	f	f	I
	Cirsium arvense	.	.	r	.	.	.	.	r	1	I
	Leontodon autumnalis	.	.	a	o	r	r	a	.		Tolerant of grazing
	Sonchus oleraceus	.	.	.	.	.	.	r	r	.	?SB
	Sonchus asper	.	.	.	.	.	.	r	r	.	
	Antennaria dioica	.	.	.	.	r	.	r	r	.	
	Bellis perennis	.	.	lf	.	lf	.	f	lf	.	Trampling tolerant
	Artemisia vulgaris	.	r	.	.	.	.	r	r	.	No
	Achillea ptarmica	r	r	.	.	.	.	.	.	.	
	Achillea millefolium	.	.	o	.	r	o	o	.	.	?SB
	Chrysanthemum segetum	.	.	.	.	.	r	r	.	.	?SB
	Tripleurospermum maritimum	f	.	.	.	la	.	la	.	.	D
	Senecio jacobaea	.	.	r	.	.	.	r	o	.	I but uncommon
	Senecio aquaticus	r	.	.	.	.	.	r	r	.	
	Senecio vulgaris	.	f	.	f	.	.	r	r	.	?SB
Potamogetonaceae	*Potamogeton polygonifolius*	.	.	.	.	.	.	la	.		

Table A1.1. (contd.)

Family	Genus and species	I	II	III	IV	V	VI	VII	VIII	IX	X	Dafor	Seen	Grazing response
Juncaceae	Juncus squarrosus	.	.	o	.	o	f	.	.	o	.	o	.	I
	Juncus bufonius	.	f	o	.	.	o	.	.	.	.	o	.	.
	Juncus articulatus	.	f	o	.	.	o	.	.	.	.	o	.	.
	Juncus bulbosus	.	f	o	.	.	.	.	.	.	.	o	.	.
	Juncus effusus	.	o	.	.	o	.	.	.	.	r	o	.	.
	Luzula sylvatica	f	.	o	.	o	.	.	.	ld	.	ld	.	Unpalatable
	Luzula campestris	.	.	.	o	o	.	r	.	.	.	o	.	I
	Luzula multiflora	.	.	o	r	o	f	.	o	.	.	o	.	.
Cyperaceae	Eriophorum angustifolium	.	.	.	.	.	f	.	.	ld	.	r	.	Unpalatable
	Trichophorum caespitosum	.	.	.	.	o	f	.	.	la	.	la	.	Unpalatable
	Eleocharis palustris	.	o	.	.	.	.	.	.	.	.	la	.	.
	Eleocharis uniglumis	.	o	.	.	.	.	.	.	.	.	o	.	.
	Schoenus nigricans	.	o	.	.	.	o	.	r	.	.	o	.	.
	Carex arenaria	.	.	.	.	.	.	.	.	.	.	r	.	.
	Carex echinata	.	.	o	.	o	f	.	.	o	.	f	.	.
	Carex demissa	.	.	f	.	o	o	.	.	.	.	f	.	.
	Carex flacca	.	.	o	o	f	o	o	.	o	.	f	.	.
	Carex panicea	.	.	o	.	o	f	.	.	o	.	f	.	.
	Carex binervis	.	.	o	o	f	o	.	.	o	.	f	.	.

		1	2	3	4	5	6	7	8		Notes
	Carex hostiana	.	.	.	.	o	o	.	r	.	.
	Carex pilulifera	.	.	r	.	f	o	o	f	.	.
	Carex nigra	.	o	.	.	o	o	o	f	.	.
	Carex bigelowii	.	.	.	.	.	la	la	la	.	.
	Carex pulicaris	o	.	.	o	f	f	f	f	.	.
Poaceae	*Nardus stricta*	a	.	d	.	f	a	a	ld	.	I
	Festuca rubra	a	.	d	.	.	.	a	d	.	Extreme D
	Festuca ovina	.	.	.	f	o	f	o	f	.	.
	Festuca vivipara	o	o	.	f	o	o	f	f	.	.
	Lolium perenne	.	.	lf	.	.	.	f	la	.	Trampling tolerant
	Poa annua	.	.	.	.	.	o	r	r	.	.
	Poa trivialis	.	.	r	.	.	f	f	lf	.	.
	Poa humilis	.	o	a	o	.	.	a	a	.	D
	Poa pratensis	o	.	r	.	.	o	r	o	.	.
	Arrhenatherum elatius	.	.	r	.	.	o	r	r	.	Extreme D
	Avena strigosa	.	.	.	.	.	.	.	.	No	.
	Koeleria macrantha	.	.	r	.	.	r	r	r	No	.
	Deschampsia flexuosa	.	o	.	.	f	r	o	r	.	.
	Holcus lanatus	a	.	a	f	o	o	r	a	.	Grazing tolerant
	Aira praecox	f	.	r	.	o	f	r	o	.	D
	Anthoxanthum odoratum	f	o	a	f	f	o	o	a	.	Grazing tolerant
	Agrostis capillaris	o	.	a	.	f	o	r	a	.	Unpalatable to Soay

Table A1.1. (contd.)

Family	Genus and species	I	II	III	IV	V	VI	VII	VIII	IX	X	Dafor	Seen	Grazing response
	Agrostis stolonifera	f	a	o	f	r	r	o	o	.	f	la	.	.
	Agrostis canina	o	.	.	o	o	o	r	r	o	.	f	.	.
	Alopecurus geniculatus	.	f	o	.	.	.	.	.	.	.	o	.	.
	Elytrigia repens	.	.	.	.	.	.	r	.	.	r	r	No	.
	Danthonia decumbens	.	.	o	o	f	f	r	r	o	.	f	.	.
	Molinia caerulea	.	.	.	.	o	a	.	.	o	.	la	.	Unpalatable I
Liliaceae	Narthecium ossifragum	.	.	.	.	r	f	.	f	f	.	f	.	.
Iridaceae	Iris pseudacorus	.	d	.	la	.	.	.	.	.	.	ld	.	Very unpalatable
Orchidaceae	Hammarbya paludosa	.	.	.	.	.	.	r	.	.	.	r	No	.
	Coeloglossum viride	.	.	.	.	.	r	.	.	.	.	r	No	.
	Dactylorhiza fuchsii	.	.	.	r	.	.	r	.	.	.	r	No	.
	Dactylorhiza maculata	.	.	.	r	o	o	.	.	o	.	o	.	.

Appendix 2

Inheritance of coat colour and horn type in Hirta Soay sheep

D. W. Coltman *University of Sheffield*

and

J. M. Pemberton *University of Edinburgh*

Coat colour

Four coat colour morphs are recognised in Hirta Soay sheep, dark wild, light wild, dark self and light self, where 'wild' (more correctly 'wild-type') refers to the mouflon pattern with pale belly and rump, and 'self' means having the same colour all over the body (see Chapter 2).

The morph frequencies are the same in males and females (data not shown) and the two components, colour and pattern, vary independently, i.e. the ratio of wild : self is the same within dark and light morphs and the ratio of dark : light is the same within wild and self morphs (Table A2.1) ($\chi^2 = 0.0137$, 1 df, ns).

An inheritance model for colour (dark/light) was proposed by Doney *et al.* (1974) and supported by preliminary data from test matings in the Soay flock brought from St Kilda to the Animal Breeding Research Organisation in Edinburgh in 1963 (Ryder *et al.* 1974). These authors proposed one autosomal locus, known as the Brown or B locus, with two alleles, at which the dark allele is dominant to the light allele, and a second locus, the Agouti or A locus, at which wild is dominant to self. Data from the Hirta Soays support this model, as shown by the analysis below (Helyar 2000), in which we have adopted the international nomenclature for sheep colour loci alleles (Sponenberg 1997).

Frequencies of the four morphs in Hirta Soays, along with their putative genotypes, are shown in Table A2.1. At the A locus the A^+

321

Soay Sheep: Dynamics and Selection in an Island Population, ed. T. H. Clutton-Brock and J. M. Pemberton.
Published by Cambridge University Press. © T. H. Clutton-Brock and J. M. Pemberton 2003.

Table A2.1. *Distribution of coat colour and pattern morphs, and their proposed genotypes, in 1193 Hirta Soay sheep with paternity inferred at 80% confidence from molecular data*

		Wild			Self	
	Genotype(s)	Observed number	Frequency	Genotype(s)	Observed number	Frequency
Dark	$A^+A^+B^+B^+$	805	0.675	$A^aA^aB^+B^+$	43	0.036
	$A^+A^+B^+B^b$			$A^aA^aB^+B^b$		
	$A^+A^aB^+B^+$					
	$A^+A^aB^+B^b$					
Light	$A^+A^+B^bB^b$	327	0.274	$A^aA^aB^bB^b$	18	0.015
	$A^+A^aB^bB^b$					

allele denotes the dominant wild pattern and the A^a allele the recessive self pattern. At the B locus the B^+ allele denotes the dominant dark allele (sometimes called black) and the B^b allele the recessive light (or brown) allele. Newborn lambs are difficult to phenotype for colour (although the colour of the skin around the eye is usually a good guide (J. Pilkington, pers. comm.)) so individuals reported here were classified when at least four months of age.

Taking the B locus first, and assuming Hardy–Weinberg equilibrium (which is indicated by molecular data; see section 8.2), the frequency of the B^b (light) allele can be estimated as $(0.274 + 0.015)^{0.5}$ or 0.538. The frequency of the B^+ (dark) allele must therefore be 0.462. Table A2.2a shows the expected (from the estimated allele frequencies) and observed outcome of matings on Hirta for 1192 lambs with known maternity, and paternity inferred at 80% confidence from molecular analysis (Pemberton et al. 1999).

Contingency table analysis suggests a deviation from the expected values ($\chi^2 = 12.91$, 4 df, $0.05 > p > 0.01$). Furthermore, this test ignores the light × light cross yielding dark offspring, where expected number is zero, though we have apparently found eleven cases. However, in this data set the father was inferred with only 80% confidence, so some 240 paternal identities may be incorrect. When the

Table A2.2. *Observed and expected outcome for all combinations of matings at the colour locus, (a) for 1132 offspring with paternity inferred at 80% confidence and (b) for 556 offspring with paternity inferred at 95% confidence*

Parent 1 phenotype	Parent 2 phenotype	Offspring phenotype	Observed number	Expected frequency from cross	Expected number from cross
(a) 1132 offspring with paternity inferred at 80% confidence					
Dark	Dark	Dark	559	0.878	582.3
Dark	Dark	Light	105	0.122	81.0
Dark	Light	Dark	278	0.651	296.2
Dark	Light	Light	117	0.349	158.8
Light	Light	Dark	11	0	0
Light	Light	Light	62	1.000	73.0
(b) 556 offspring with paternity inferred at 95% confidence					
Dark	Dark	Dark	285	0.878	285.0
Dark	Dark	Light	40	0.122	39.5
Dark	Light	Dark	120	0.651	130.9
Dark	Light	Light	81	0.349	70.1
Light	Light	Dark	3	0	0
Light	Light	Light	27	1.000	30.0

data is restricted to those cases in which paternity was inferred at 95% confidence (Table A2.2b), the inheritance model is a much better fit ($\chi^2 = 1.369$, 4 df, ns) and there are only three cases of the unexpected light $\times$ light cross yielding a dark offspring.

Now taking the A locus, the frequency of the A^a (self) allele can be estimated as $(0.036 + 0.015)^{0.5}$ or 0.226. The frequency of the A^+ (wild) allele must therefore be 0.774. Table A2.3 shows the expected (from the estimated allele frequencies) and observed outcome of matings on Hirta for 1192 lambs with known maternity and paternity inferred at 80% confidence from molecular analysis.

Contingency table analysis shows that the model fits well ($\chi^2 = 1.61$, 3 df, ns). These figures exclude the one case when two selfs mated to produce a self offspring, as expected by the model. Similarly, the re-

Table A2.3. *Observed and expected outcome for all combinations of matings at the pattern locus, (a) for 1132 offspring with paternity inferred at 80% confidence and (b) for 516 offspring with paternity inferred at 95% confidence*

Parent 1 phenotype	Parent 2 phenotype	Offspring phenotype	Observed number	Expected frequency from cross	Expected number from cross
(a) 1132 offspring with paternity inferred at 80% confidence					
Wild	Wild	Wild	1026	0.967	1031.8
Wild	Wild	Self	41	0.033	35.2
Wild	Self	Wild	105	0.819	101.5
Wild	Self	Self	19	0.181	22.5
Self	Self	Wild	0	0.000	0
Self	Self	Self	1	1.000	1.0
(b) 516 offspring with paternity inferred at 95% confidence					
Wild	Wild	Wild	448	0.967	443.9
Wild	Wild	Self	11	0.033	15.1
Wild	Self	Wild	47	0.819	45.8
Wild	Self	Self	9	0.181	10.2
Self	Self	Wild	0	0.000	0
Self	Self	Self	1	1.000	1.0

duced data set, consisting of only the 95% confident inferred paternal identities supports the model (Table A2.3b) ($\chi^2 = 1.32$, 3 df, ns).

Horn type

Horns are present in both sexes of wild species of sheep. However, it is thought that a mutation leading to the lack of horns, or polledness, occurred near the time of domestication (Clutton-Brock 1987) and is likely to have been maintained by the preference of early farmers (Kinsman 2001). Stunted or deformed horns, or scurs, also develop in many primitive breeds of British sheep (Kinsman 2001). All three horn phenotypes, normal, polled and scurred, are present in Soay sheep on Hirta, although the scurred phenotype appears very rarely in mainland flocks (Kinsman 2001).

Table A2.4. *Proposed genotype–phenotype relationships under a three-allele sex-specific expression model of horn inheritance, and frequencies of observed phenotypes in the Hirta population*

Phenotype	Rams			Ewes[a]		
	Genotype(s)	Observed	Frequency	Genotype(s)	Observed	Frequency
Polled				Ho^PHo^P Ho^PHo^L	171	0.284
Scurred	Ho^PHo^P	151	0.151	Ho^LHo^L Ho^+Ho^P	224	0.371
Normal	Ho^+- Ho^L-	848	0.849	Ho^+Ho^+ Ho^+Ho^L	208	0.345

[a]Note that because of the difficulty of discriminating scurred from polled ewes during early development, the ewe phenotype data shown are for ewes captured at one year or older

The inheritance of horn phenotypes in primitive sheep breeds has not yet been definitively worked out; however, a model suggested originally by Dolling (1961) for the Merino appears to have wide applicability to many other breeds including the Hebridean and other primitive British breeds (Kinsman 2001). This model proposes three alleles with differential expression in the sexes: Ho^L – sex-limited horns, Ho^+ – normal horns, and Ho^P – polledness. Following these authors, we propose the relationships between genotype and phenotype shown in Table A2.4.

According to this model, the Ho^P allele is recessive in rams, causing scurs, and dominant to Ho^L in ewes. The Ho^P allele is partially expressed in the presence of Ho^+ in ewes, causing an intermediate scurred phenotype. Ho^L is recessive in ewes to both the other alleles, and causes the scurred phenotype when homozygous.

The frequency of horn phenotypes shown in Table A2.4 can be used to predict allele frequencies. The frequency of the Ho^P allele can most easily be estimated from the male scurred class in rams as $0.151^{0.5} = 0.389$. Based on the frequency of polled ewes $Ho^L = 0.170$, leaving the frequency of $Ho^+ = 1 - 0.389 - 0.170 = 0.441$. Rechecking the

Table A2.5. *Observed and expected rates of inheritance of horn phenotypes in Soay sheep predicted under the three-allele sex-specific expression model*

Mother's phenotype	Father's phenotype	Offspring phenotype	Offspring sex	Number observed	Expected frequency from parental cross	Number expected
Scurred	Scurred	Scurred	Female	4	0.461	3.2
Scurred	Scurred	Polled	Female	2	0.539	3.8
Scurred	Scurred	Normal	Female	1	0	0
Scurred	Normal	Scurred	Female	45	0.384	38.4
Scurred	Normal	Polled	Female	27	0.243	24.3
Scurred	Normal	Normal	Female	28	0.372	37.2
Polled	Scurred	Scurred	Female	0	0	0
Polled	Scurred	Polled	Female	7	1.000	7.0
Polled	Scurred	Normal	Female	0	0	0
Polled	Normal	Scurred	Female	25	0.445	23.6
Polled	Normal	Polled	Female	25	0.434	23.0
Polled	Normal	Normal	Female	3	0.121	6.4
Normal	Scurred	Scurred	Female	0	0.782	3.1
Normal	Scurred	Polled	Female	2	0.218	0.9
Normal	Scurred	Normal	Female	2	0	0
Normal	Normal	Scurred	Female	22	0.263	19.2
Normal	Normal	Polled	Female	2	0.061	4.5
Normal	Normal	Normal	Female	49	0.676	49.4
Scurred	Scurred	Scurred	Male	3	0.461	7.4
Scurred	Scurred	Normal	Male	13	0.539	8.6
Scurred	Normal	Scurred	Male	27	0.129	22.5
Scurred	Normal	Normal	Male	147	0.871	151.5
Polled	Scurred	Scurred	Male	3	0.767	4.6
Polled	Scurred	Normal	Male	3	0.233	1.4
Polled	Normal	Scurred	Male	35	0.215	20.8
Polled	Normal	Normal	Male	62	0.785	76.2
Normal	Scurred	Scurred	Male	1	0	0
Normal	Scurred	Normal	Male	15	1.000	16.0
Normal	Normal	Scurred	Male	1	0	0
Normal	Normal	Normal	Male	112	1.000	113.0

predicted ewe phenotype frequencies leads to estimates of polled = 0.284, scurred = 0.372, and normal = 0.344. This model therefore matches the observed frequencies of phenotypes almost exactly.

From the population allele frequency estimates we can predict the expected phenotype of offspring of both sexes from all possible crosses in the population, and compare the observed phenotypes with the predictions. To test the predictions, we examined observed phenotypes of 666 lambs with known maternity and paternity inferred at 80% confidence from molecular data (Pemberton *et al.* 1999). Table A2.5 compares the observed and expected outcome for all types of cross and offspring sex.

A contingency analysis indicates that the observed and expected counts do not differ significantly ($\chi^2 = 24.51$, 17 df, $0.10 > p > 0.05$; expected cells less than 5 were pooled). In general the data fit the model acceptably well, except that there are more observed than expected cases when a polled female mated to a normal male produced a scurred male offspring. It is not possible to pursue this analysis among the 95% confident paternity data set at this stage, due to the reduction of sample size involved which leads to too many unacceptably low expected numbers.

We investigated other possible models of inheritance, including simpler two-allele models and more complicated models involving more than one locus with sex difference in expression, however none fits the data better than the three-allele model described above. This would therefore seem to be the most parsimonious model explaining the inheritance of horn phenotype given the available data.

Appendix 3
How average life tables can mislead

T. Coulson *University of Cambridge*

and

M. J. Crawley *Imperial College London*

Life tables provide information on the mean survival and reproductive rates of animals of different ages and sex (Keyfitz 1968; Pollard 1973; Caughley 1977). These rates can be presented in many ways, but the key information in any life table is the proportion of animals that survive to age x (l_x) and the number of recruits to the population that an animal of age x produces (m_x). Life tables can easily be used to calculate the mean number of progeny per individual per generation ($R_0 = \Sigma l_x m_x$), generation length ($T_c = \Sigma l_x m_x x / R_0$) and the population growth rate ($r = ln(R_0)/T_c$). When $R_0 = 1$ each individual replaces itself and the population growth rate (r) is zero. When $R_0 < 1$ individuals are failing to replace themselves, the population is declining and $r < 0$. Finally when $R_0 > 1$ individuals are more than replacing themselves and the population increases in size ($r > 0$). Life tables now exist for a wide range of species and are much used in population and evolutionary ecology to construct models of population dynamics, to estimate whether a population is increasing or decreasing in size and to estimate the strength of selection (Caswell 1989, 2001; Brault and Caswell 1993). If a population biologist wanted to find out details about a species' population biology he would almost certainly look for an available life table before doing anything else.

The detailed data we have collected on Soay sheep allow us to construct a particularly accurate life table for both males and females. Life tables based on all data collected between 1985 and 2002 give values of $R_0 = 1.02$ and $r = 0.01$ for females and $R_0 = 1.04$ and

328

Soay Sheep: Dynamics and Selection in an Island Population, ed. T. H. Clutton-Brock and J. M. Pemberton.
Published by Cambridge University Press. © T. H. Clutton-Brock and J. M. Pemberton 2003.

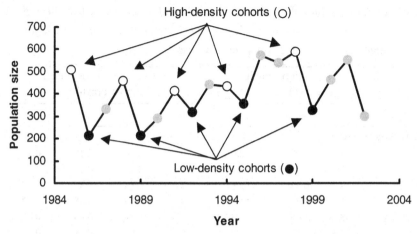

FIG. A3.1. Time series of the number of sheep living in Village Bay between 1985 and 2002. High-density (open circles) and low-density (solid black circles) cohorts used to construct life tables are indicated with arrows. At no point during the high-quality time series data (1985–present) do we observe two consecutive years when the population decreases in size. Data for 2001 and 2002 cohorts were not used because insufficient time had elapsed for accurate estimates of vital rates.

$r = 0.02$ for males; the male and female populations are very slightly increasing in size over the study period. We do not, however, display the average tables as they are of little use for populations like the Soay sheep population on Hirta, which fluctuate dramatically in size from year to year. These fluctuations mean that some cohorts have substantially higher values of R_0 than other cohorts. We demonstrate these differences by presenting male and female life tables, incorporating estimates of R_0 for cohorts of lambs born in high-density years immediately prior to a population decline (1985, 1988, 1991, 1994 and 1998) and for cohorts born in low-density years immediately following a population decline (1986, 1989, 1992, 1995 and 1999) (Fig. A3.1 and A3.2, Table A3.1). The substantial differences in R_0 between the life tables demonstrate how much variation can exist in survival and recruitment rates across animals of the same age born into different cohorts.

The most striking difference between cohorts is in the proportion of animals that survive their first year of life (comparisons within

Table A3.1. *Values of* l_x *and* m_x *for male and female sheep born at high and low densities. The columns* $l_x m_x$ *report the product of* m_x *and* l_x *for each age. The sum of the* $l_x m_x$ *column gives* R_0.

Cohorts born at high density (pre-crash)				Cohorts born at low density (post crash)			
Age	l_x	m_x	$l_x m_x$	Age	l_x	m_x	$l_x m_x$
Female sheep							
0	1.00	0.01	0.006	0	1.00	0.16	0.16
1	0.21	0.36	0.074	1	0.89	0.26	0.23
2	0.13	0.32	0.041	2	0.84	0.30	0.25
3	0.10	0.34	0.034	3	0.84	0.38	0.32
4	0.08	0.38	0.030	4	0.81	0.38	0.31
5	0.07	0.32	0.021	5	0.80	0.41	0.33
6	0.04	0.39	0.016	6	0.76	0.39	0.30
7	0.03	0.37	0.010	7	0.74	0.36	0.27
8	0.01	0.25	0.004	8	0.66	0.28	0.18
9	0.01	0.39	0.003	9	0.61	0.17	0.10
10	0.00	0.17	0.000	10	0.57	0.50	0.29
11	0.00	0.00	0.000	11	0.34	0.00	0.00
12	0.00	0.00	0.000	12	0.34	0.00	0.00
		R_0	0.238			R_0	2.741
		T_c	2.950			T_c	4.920
		r	− 0.486			r	0.205

Cohorts born at high density (pre-crash)				Cohorts born at low density (post crash)			
Age	l_x	m_x	$l_x m_x$	Age	l_x	m_x	$l_x m_x$
Male sheep							
0	1.00	0.11	0.110	0	1.00	0.35	0.347
1	0.32	0.39	0.122	1	0.80	0.32	0.253
2	0.08	0.12	0.009	2	0.71	0.21	0.151
3	0.03	0.40	0.010	3	0.70	0.44	0.308
4	0.01	0.30	0.003	4	0.70	0.61	0.429
5	0.01	0.50	0.005	5	0.64	0.36	0.229
6	0.01	0.88	0.009	6	0.64	0.96	0.615
7	0.01	0.00	0.000	7	0.21	0.50	0.107
8	0.01	0.00	0.000	8	0.21	0.50	0.107
9	0.01	0.00	0.000	9	0.00	0.50	0.001
		R_0	0.268			R_0	2.546
		T_c	0.970			T_c	3.790
		r	−1.356			r	0.247

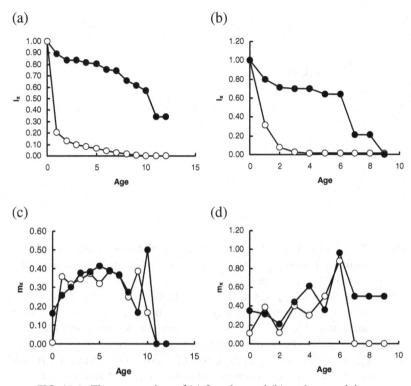

FIG. A3.2. The proportion of (a) females and (b) males surviving to different ages (l_x) for cohorts born at high (open circles) and low densities (closed circles). (c) m_x – mean female recruits per female and (d) male recruits per male for cohorts born at high-densities (open circles) and low densities (closed circles). l_x and m_x are calculated using population composition data from August each year and maternity and paternity information based on observation and genotyping. Estimates of m_x for males were made by multiplying the values of m_x calculated from paternity data by the ratio of the number of known mothers to the number of known fathers.

Figures A3.2a and A3.2b). For example, only 21% of females born at high density before a population decline survive to one year of age, while 89% survive from cohorts born at low density. This difference contributes substantially to differences in R_0 between these cohorts (Table A3.1).

References

Abbott, E. M. and Holmes, P. H. (1990) Influence of dietary protein on the immune responsiveness of sheep to *Haemonchus contortus*. *Research in Veterinary Science* **48**: 103–7.

Abbott, E. M., Parkins, J. J. and Holmes, P. H. (1985) Influence of dietary protein on the pathophysiology of ovine haemonchosis in Finn Dorset and Scottish Blackface lambs given a single moderate infection. *Research in Veterinary Science* **38**: 54–60.

Adam, C. L. and Findlay, P. A. (1997) Effect of nutrition on testicular growth and plasma concentrations of gonadotrophins, testosterone and insulin-like growth factor I (IGF-I) in pubertal male Soay sheep. *Journal of Reproduction and Fertility* **111**: 121–5.

Albers, G. A. A., Gray, G. D., Piper, L. R., Barker, J. S. F., Lejambre, L. F. and Barger, I. A. (1987) The genetics of resistance and resilience to *Haemonchus contortus* infection in young Merino sheep. *International Journal for Parasitology* **17**: 1355–63.

Albon, S. D. and Clutton-Brock, T. H. (1988) Climate and the population dynamics of red deer in Scotland. In *Ecological Change in the Uplands* (eds. M. B. Usher and D. B. A. Thompson), pp. 93–107. Oxford: Blackwell Scientific Publications.

Albon, S. D., Guinness, F. E. and Clutton-Brock, T. H. (1983a) The influence of climatic variation on the birth weights of red deer, *Cervus elaphus*. *Journal of Zoology* **200**: 295–8.

Albon, S. D., Mitchell, B. and Staines, B. W. (1983b) Fertility and body weight in female red deer: a density-dependent relationship. *Journal of Animal Ecology* **56**: 969–80.

Albon, S. D., Mitchell, B., Huby, B. J. and Brown, D. (1986) Fertility in female red deer (*Cervus elaphus*): the effects of body composition, age and reproductive status. *Journal of Zoology* **209**: 447–60.

Albon, S.D., Clutton-Brock, T.H. and Guinness, F.E. (1987) Early development and population dynamics in red deer. II. Density-independent effects and cohort variation. *Journal of Animal Ecology* **56**: 69–81.

Albon, S.D., Clutton-Brock, T.H. and Langvatn, R. (1991) Cohort variation in reproduction and survival: implications for population demography. In *The Biology of Deer* (ed. R.D. Brown), pp. 15–21. New York: Springer-Verlag.

Albon, S.D., Staines, H.J., Guinness, F.E. and Clutton-Brock, T.H. (1992) Density-dependent changes in the spacing behaviour of female kin in red deer. *Journal of Animal Ecology* **61**: 131–7.

Albon, S.D., Coulson, T.N., Brown, D., Guinness, F.E., Pemberton, J.M. and Clutton-Brock, T.H. (2000) Temporal changes in key factors and key age groups influencing the population dynamics of female red deer. *Journal of Animal Ecology* **69**: 1099–110.

Albon, S.D., Stien, A., Irvine, R.J., Langvatn, R., Ropstad, E. and Halvorsen, O. (2002) The role of parasites in the dynamics of a reindeer population. *Proceedings of the Royal Society B* **269**: 1625–32.

Allendorf, F.W. and Leary, R.F. (1986) Heterozygosity and fitness in natural populations of animals. In *Conservation Biology: The Science of Scarcity and Diversity* (ed. M.E. Soulé), pp. 57–76. Sunderland, MA: Sinauer Associates.

Altmann, D. (1970) Ethologische Studie en Mufflons *Ovis ammon musimon* (Pallas). *Der Zoologische Garten* **39**: 297–303.

Altmann, J., Hausfater, G. and Altmann, S.A. (1988) Determinants of reproductive success in savannah baboons. In *Reproductive Success* (ed. T.H. Clutton-Brock), pp. 403–18. Chicago, IL: University of Chicago Press.

Amos, W., Wilmer, J.W., Fullard, K., Burg, T.M., Croxall, J.P., Bloch, D. and Coulson, T. (2001) The influence of parental relatedness on reproductive success. *Proceedings of the Royal Society B* **268**: 2021–27.

Andersen, R. and Linnett, J.D.C. (1998) Ecological correlates of mortality of roe deer fawns in a predator-free environment. *Canadian Journal of Zoology* **76**: 1217–25.

Anderson, R.M. (1993) Epidemiology. In *Modern Parasitology: A Textbook of Parasitology* (ed. F.E.G. Cox), 2nd edn, pp. 75–116. Oxford: Blackwell Scientific Publications.

Anderson, R.M. and Gordon, D.M. (1982) Processes influencing the distribution of parasite numbers within host populations with special emphasis on parasite-induced host mortalities. *Parasitology* **85**: 373–98.

Anderson, R.M. and May, R.M. (1978) Regulation and stability of host–parasite population interactions. I. Regulatory processes. *Journal of Animal Ecology* **47**: 219–47.

(1979) Population biology of infectious diseases: Part I. *Nature* **280**: 361–7.

(eds.) (1982a) *Population Biology of Infectious Diseases*. Berlin: Springer-Verlag.

(1982b) Coevolution of hosts and parasites. *Parasitology* **85**: 411–26.

Armour, J., Jennings, F. W. and Urquhart, G. M. (1969) Inhibition of *Ostertagia ostertagi* in the early fourth larval stage. I. The seasonal incidence. *Research in Veterinary Science* **10**: 232–7.

Arnold, S. J. and Wade, M. J. (1984a) On the measurement of natural and sexual selection: theory. *Evolution* **38**: 709–19.

(1984b) On the measurement of natural and sexual selection: applications. *Evolution* **38**: 720–34.

Avise, J. C. (1994) *Molecular Markers, Natural History and Evolution*. New York: Chapman and Hall.

Bacon, C. W., Porter, J. K., Robbins, J. D. and Luttrell, E. S. (1977) *Epichloe typhina* from toxic tall fescue grasses. *Applied and Environmental Microbiology* **34**: 576–81.

Baker, R. L., Watson, T. G., Bisset, S. A. Vlassoff, A. and Douch, P. G. C. (1991) Breeding sheep in New Zealand for resistance to internal parasites: research results and commercial applications. In *Breeding for Disease Resistance in Sheep* (eds. G. D. Gray and R. R. Woolaston), pp. 19–32. Melbourne: Australian Wool Corporation.

Baker, R. L., Mwamachi, D. M., Audho, J. O. and Thorpe, W. (1994) Genetic resistance to gastrointestinal nematode parasites in red maasai sheep in Kenya. *Proceedings of the 5th World Congress on Genetics Applied to Livestock Production* **20**: 227–80.

Ball, D. M., Pedersen, J. F. and Lacefield, G. D. (1993) The tall fescue endophyte. *American Scientist* **81**: 370–9.

Ball, N. (1998) Genetic variation in St Kilda sheep. BSc thesis, University of Edinburgh.

Ballard, W. B., Miller, S. M. and Whitman, J. S. (1991) Population dynamics of moose in south-central Alaska. *Wildlife Monograph* **114**: 1–49.

Bancroft, D. R. (1993) Genetic variation and fitness in Soay sheep. PhD thesis, University of Cambridge.

(1995) A microsatellite polymorphism at the ovine pituitary andenylate cyclase activating polypeptide which can be co-amplified with two other loci. *Animal Genetics* **26**: 1–59.

Bancroft, D. R., Pemberton, J. M. and King, P. (1995a) Extensive protein and microsatellite variability in an isolated, cylic ungulate population. *Heredity* **74**: 326–36.

Bancroft, D. R., Pemberton, J. M., Albon, S. D., Robertson, A., MacColl, A. D. C., Smith, J. A., Stevenson, I. R. and Clutton-Brock, T. H. (1995b) Molecular genetic variation and individual survival during population crashes of an unmanaged ungulate population. *Philosophical Transactions of the Royal Society B* **347**: 263–73.

Barker, I. K. (1973) A study of the pathogenesis of *Trichostrongylus colubriformis* infection in lambs with observations on the contribution of gastrointestinal plasma loss. *International Journal of Parasitology* **3**: 743–57.

Barrington, R. M. (1866) Notes on the flora of St Kilda. *Journal of Botany* **24**: 213–16.

Bawden, R. J. (1969) The establishment and survival of *Oesophagostomum columbianum* in male and female sheep given high and low protein diets. *Australian Journal of Agricultural Research* **20**: 1151–9.

Bazely, D. R., Vicari, M., Emmerich, S., Filip, L., Lin, D. and Inman, A. (1997) Interactions between herbivores and endophyte-infected *Festuca rubra* from the Scottish islands of St Kilda, Benbecula and Rum. *Journal of Applied Ecology* **34**: 847–60.

Begon, M. (1992) Density and frequency dependence in ecology: messages for genetics. In *Genes in Ecology* (eds. R. J. Berry, T. J. Crawford and G. M. Hewitt), pp. 335–52. Oxford: Blackwell Scientific Publications.

Beh, K. J., Callaghan, M. J., Hulme, D. J., Windon, R. G., Leish, Z., Dilenno, K. D. and Lenane, I. (1998) Genes affecting resistance to *T. colubriformis* and *H. contortus* infestation in sheep. Poster at: *International Society for Animal Genetics*. Auckland, New Zealand.

Belgrano, A., Lindahl, O. and Hernroth, B. (1999) North Atlantic Oscillation primary productivity and toxic phytoplankton in Gullmar Fjord, Sweden (1985–1996). *Proceedings of the Royal Society B* **266**: 425–30.

Bell, A. W. (1984) Factors controlling placental and foetal growth and their effects on future reproduction. In *Reproduction in Sheep* (eds. D. R. Lindsay and D. T. Pearce), pp. 144–152. Cambridge: Cambridge University Press.

Bellman, R. (1957) *Dynamic Programming*. Princeton, NJ: Princeton University Press.

Bellows, T. S. (1981) The descriptive properties of some models for density-dependence. *Journal of Animal Ecology* **50**: 139–56.

Benzie, D. and Gill, J. C. (1974) Radiography of the skeletal and dental condition of the Soay sheep. In *Island Survivors: The Ecology of the Soay Sheep of St Kilda* (eds. P. A. Jewell, C. Milner and J. M. Boyd), pp. 326–37. London: Athlone Press.

Bercovitch, F. B. and Berard, J. D. (1993) Life history costs and consequences of rapid reproductive maturation in female rhesus macaques. *Behavioural Ecology and Sociobiology* **32**: 103–9.

Bernardo, J. (1993) Determinants of maturation in animals. *Trends in Ecology and Evolution* **8**: 166–73.

Berry, R. J. (1969) History in the evolution of *Apodemus sylvaticus* (Mammalia) at the edge of its range. *Journal of Zoology* **159**: 311–28.

(1970) Covert and overt variation, as exemplified by British mouse populations. In *Symposium of the Zoological Society*, pp. 3–26. London: Zoological Society.

Berry, R.J. and Murphy, H.M. (1970) The biochemical genetics of an island population of the house mouse. *Proceedings of the Royal Society B* **176**: 87–103.

Berry, R.J., Peters, J. and van Aarde, R.J. (1978) Sub-antarctic house-mice: colonisation, survival and selection. *Journal of Zoology* **184**: 127–41.

Bérubé, C. (1997) Les stratégies d'adaptation vitale chez les brebis du mouflon d'Amérique (*Ovis canadensis*): la reproduction en fonction de l'âge. PhD thesis, Université de Sherbrooke.

Bérubé, C., Festa-Bianchet, M. and Jorgenson, J.T. (1996) Reproductive costs of sons and daughters in Rocky Mountain bighorn sheep. *Behavioural Ecology* **7**: 60–8.

(1999) Individual differences, longevity and reproductive senescence in bighorn ewes. *Ecology* **80**: 2555–65.

Besier, R.B. and Dunsmore, J.D. (1993) The ecology of *Haemonchus contortus* in a winter rainfall climate in Australia: the survival of infective larvae on pasture. *Veterinary Parasitology* **45**: 293–306.

Birgersson B. and Ekvall, K. (1997) Early growth in male and female fallow deer fawns. *Behavioural Ecology* **8**: 493–9.

Birkhead, T.R. and Møller, A.P. (1992) *Sperm Competition in Birds: Evolutionary Causes and Consequences*. London: Academic Press.

Bishop, M.D., Kappes, S.M., Keele, J.W., Stone, R.T., Sunden, S.L.F., Hawkins, G.A., Toledo, S.S., Fries, R. and Gross, M.D. (1994) A genetic linkage map for cattle. *Genetics* **136**: 619–39.

Bishop, S.C., Bairden, K., McKellar, Q.A., Park, M. and Stear, M.J. (1996) Genetic parameters for faecal egg count following mixed, natural, predominantly *Ostertagia circumcincta* infection and relationships with live weight in young lambs. *Animal Science* **63**: 423–8.

Bisset, S.A., Vlassoff, A., Morris, C.A., Southey, B.R., Baker, R.L. and Parker, A.G.H. (1992) Heritability of and genetic correlations among faecal egg counts and productivity traits in Romney sheep. *New Zealand Journal of Agricultural Research* **35**: 51–8.

Bisset, S.A., Morris, C.A., Squire, D.R., Hickey, S.M. and Wheeler, M. (1994) Genetics of resilience to nematode parasites in Romney sheep. *New Zealand Journal of Agricultural Research* **37**: 521–34.

Bjørnstad, O.N., Falck, W. and Stenseth, N.C. (1995) A geographic gradient in small rodent density fluctuations: a statistical modelling approach. *Proceedings of the Royal Society B* **262**: 127–33.

Blitz, N.M. and Gibbs, H.C. (1972) Studies on the arrested development of *Haemonchus contortus* in sheep. II. Termination of arrested development and the spring rise phenomenon. *International Journal of Parasitology* **2**: 13–22.

Blood, D. A., Flook, D. R. and Wishart, W. D. (1970) Weights and growth of Rocky Mountain bighorn sheep in Western Alberta. *Journal of Wildlife Management* **34**: 451–5.

Bon, R., Dardaillon, M. and Estevez, I. (1993) Mating and lambing periods as related to age of female mouflon. *Journal of Mammalogy* **74**: 752–7.

Boussès, P., Barbanson, B. and Chapuis, J. L. (1991) The Corsican mouflon (*Ovis ammon musimon*) on Kerguelen archipelago: structure and dynamics of the population. In *Proceedings of the International Symposium, Ongulés / Ungulates 91* (eds. F. Spitz, G. Janeau, G. Gonzalez and S. Avlagnier), pp. 317–20. Toulouse, France.

Boyd, H. E. G. (1999) Early development of parasitism in Soay sheep. PhD thesis, University of Cambridge.

Boyd, I. L. (1981) Population changes and the distribution of a herd of feral goats (*Capra* sp.) on Rhum, Inner Hebrides, 1960–78. *Journal of Zoology* **193**: 287–304.

Boyd, J. M. (1953) The sheep population of Hirta, St Kilda, 1952. *Scottish Naturalist* **65**: 25–8.

Boyd, J. M. and Jewell, P. A. (1974) The Soay sheep and their environment: a synthesis. In *Island Survivors: The Ecology of the Soay Sheep of St Kilda* (eds. P. A. Jewell, C. Milner and J. M. Boyd), pp. 360–73. London: Athlone Press.

Boyd, J. M., Doney, J. M., Gunn, R. G. and Jewell, P. A. (1964) The Soay sheep of the island of Hirta, St Kilda: a study of a feral population. *Proceedings of the Zoological Society of London* **142**: 129–63.

Braisher, T. L. (1999). Genetic variation in trichostrongylid parasites of the Soay sheep of St Kilda. PhD thesis, University of Cambridge.

Bramley, P. (1970) Territoriality and reproductive behaviour of roe deer. *Journal of Reproduction and Fertility* **11**: 43–70.

Brault, S. and Caswell, H. (1993) Pod-specific demography of killer whales (*Orcinus orca*). *Ecology* **74**: 1444–54.

Briles, W. E., Stone, H. A. and Cole, R. K. (1977) Marek's disease: effects of B histocompatability alloalleles in resistant and susceptible chicken lines. *Science* **195**: 193–5.

Brown, D. (1988) Components of lifetime reproductive success. In *Reproductive Success* (ed. T. H. Clutton-Brock), pp. 439–53. Chicago, IL: University of Chicago Press.

Brown, D., Alexander, N. D. E., Marrs, R. W. and Albon, S. D. (1993) Structured accounting of the variance of demographic change. *Journal of Animal Ecology* **62**: 490–502.

Brown, J. H. and Heske, E. J. (1990) Control of a desert–grassland transition by a keystone rodent guild. *Science* **250**: 1705–7.

Brunsdon, R. V. (1962) The effect of nutrition on age resistance of sheep to infestation with *Nematodirus* spp. *New Zealand Veterinary Journal* **10**: 123–7.

(1964) The seasonal variations in the nematode egg counts of sheep: a comparison of the spring rise phenomenon in breeding and unmated ewes. *New Zealand Veterinary Journal* 12: 75–80.

(1970) The spring-rise phenomenon: seasonal changes in the worm burdens of breeding ewes and in the availability of pasture infection. *New Zealand Veterinary Journal* 18: 47–54.

Brunsdon, R.V. and Vlassoff, A. (1971) The post-parturient rise: a comparison of the pattern and relative generic composition of strongyle egg output from lactating and non-lactating ewes. *New Zealand Veterinary Journal* 19: 19–25.

Bryant, D.M. (1979) Reproductive costs in the house martin (*Delichon urbica*). *Journal of Animal Ecology* 48: 655–75.

Buchanan, F.C. and Crawford, A.M. (1993) Ovine microsatellites at the OarFCB11, OarFCB128, OarFCB193, OarFCB266, and OarFCB304 loci. *Animal Genetics* 24: 145.

Buchanan, F.C., Swarbrick, P.A. and Crawford, A.M. (1992) Ovine dinucleotide polymorphism at the MAF65 locus. *Animal Genetics* 23: 85.

Buchanan, M. (1995) *St Kilda: The Continuing Story of the Islands*. Edinburgh: HMSO.

Buechner, H.K., Morrison, J.A. and Leuthold, W. (1966) Reproduction in Uganda kob with special reference to behaviour. In *Comparative Biology of Reproduction in Mammals* (ed. I.W. Rowlands), pp. 69–88. London: Academic Press.

Bueno-de la Fuente, M.L., Moreno, V., Peréz, J.M., Ruiz-Martinez, I. and Soriguer, R.C. (1998) Oestrosis in red deer from Spain. *Journal of Wildlife Diseases* 34: 820–4.

Bunnell, F.L. (1980) Factors controlling lambing period of Dall's sheep. *Canadian Journal of Zoology* 58: 1027–31.

Burt, W.H. and Grossenheider, R.P. (1976) *A Field Guide to the Mammals*, 3rd edn. Boston, MA: Houghton Mifflin.

Byers, J.A. (1997) *American Pronghorn: Social Adaptations and the Ghost of Predators Past*. Chicago: University of Chicago Press.

Byers, J.A., Moodie, J.D. and Hall, N. (1994) Pronghorn females choose vigorous mates. *Animal Behaviour* 47: 33–43.

Byrne, K.A., Chikhi, L., Townsend, S.J., Cruickshank, R.H., Alderson, G.L.H. and Bruford, M.W. (in press) Extreme genetic diversity within and among European sheep types and its implications for breed conservation. *Molecular Ecology*.

Caballero, A. (1994) Developments in the prediction of effective population size. *Heredity* 73: 657–79.

Cabaret, J., Gasnier, N. and Jacquiet, P. (1998) Faecal egg counts are representative of digestive-tract strongyle worm burdens in sheep and goats. *Parasite – Journal de la Société Française de Parasitologie* 5: 137–42.

Cameron, C.D.T. and Gibbs, H.C. (1966) Effects of stocking rate and flock management on internal parasitism in lambs. *Canadian Journal of Animal Science* **46**: 121–4.

Campbell, R.N. (1974) St Kilda and its sheep. In: *Island Survivors: The Ecology of the Soay Sheep of St Kilda* (eds. P.A. Jewell, C. Milner and J.M. Boyd), pp. 8–35. London: Athlone Press.

Caro, T.M. (1995) *Cheetahs of the Serengeti Plains: Group Living in an Asocial Species.* Chicago, IL: University of Chicago Press.

Carson, H.L. (1959) Genetic conditions which promote or retard the formation of species. *Cold Spring Harbor Symposia in Quantitative Biology* **24**: 87–105.

(1982) Evolution of *Drosophila* on the newer Hawaiian volcanoes. *Heredity* **48**: 3–25.

Caswell, H. (1989) The analysis of lifetable response experiments. 1. Decomposition of effects on population growth. *Ecological Modelling* **43**: 33–44.

(2001) *Matrix Population Models: Construction Analysis and Interpretation,* 2nd edn. Sunderland, MA: Sinauer Associates.

Catchpole, E.A., Morgan, B.J.T., Coulson, T.N., Freeman, S. and Albon, S.D. (2000) Factors influencing Soay sheep survival. *Applied Statistics* **49**: 453–72.

Caughley, G. (1977) *Analysis of Vertebrate Populations.* New York: John Wiley.

(1994) Directions in conservation biology. *Journal of Animal Ecology* **63**: 215–44.

Caughley, G. and Krebs, C.J. (1983) Are big mammals simply little mammals writ large? *Oecologia* **59**: 7–17.

Caughley, G. and Sinclair, A.R.E. (1994) *Wildlife Ecology and Management.* Oxford: Blackwell Scientific Publications.

Chapman, A.P., Brook, B.W., Clutton-Brock, T.H., Grenfell, B. and Frankham, R. (2001) Population viability analyses on a cycling population: a cautionary tale. *Biological Conservation* **97**: 61–9.

Charlesworth, B. (1980) *Evolution in Age-Structured Populations.* Cambridge: Cambridge University Press.

Charlesworth, D. and Charlesworth, B. (1987) Inbreeding depression and its evolutionary consequences. *Annual Reviews of Ecology and Systematics* **18**: 237–68.

Cheyne, I.A., Foster, W.M. and Spence, J.B. (1974) The incidence of disease and parasites in the Soay sheep population of Hirta. In *Island Survivors: The Ecology of the Soay Sheep of St Kilda* (eds. P.A. Jewell, C. Milner and J.M. Boyd), pp. 338–59. London: Athlone Press.

Cheney, D.L., Seyfarth, R.M., Andelman, S.J. and Lee, P.C. (1988) Reproductive success in vervet monkeys. In *Reproductive Success* (ed. T.H. Clutton-Brock), pp. 384–402. Chicago, IL: University of Chicago Press.

Choquenot, D. (1991) Density-dependent growth, body condition and demography in feral donkeys: testing the food hypothesis. *Ecology* **72**: 805–13.

Clarke, B. (1972) Density-dependent selection. *American Naturalist* **106**: 1–13.

Clarke, B. and Beaumont, M. A. (1992) Density and frequency dependence: a genetical view. In *Genes in Ecology* (eds. R. J. Berry, T. J. Crawford and G. M. Hewitt), pp. 353–64. Oxford: Blackwell Scientific Publications.

Clarke, B. C. and Murray, J. J. (1969) Ecological genetics and speciation in land snails of the genus *Partula*. *Biological Journal of the Linnean Society* **1**: 31–42.

Clay, K. (1990) Fungal endophytes of grasses. *Annual Reviews of Ecology and Systematics* **21**: 275–297.

Clay, K. and Leuchtmann, A. (1989) Infection of woodland grasses by fungal endophytes. *Mycologia* **81**: 805–11.

Clayton, D. H. and Moore, J. (1997) *Host–Parasite Evolution: General Principles and Avian Models*. Oxford: Oxford University Press.

Clutton-Brock, J. (1981) *Domesticated Animals from Early Times*. London: William Heinemann and British Museum (Natural History).

Clutton-Brock, J. (1987) *A Natural History of Domesticated Mammals*. Cambridge: Cambridge University Press.

Clutton-Brock, J., Dennis-Bryan, K. and Armitage, P. L. (1990) Osteology of the Soay sheep. *Bulletin of the British Museum of Natural History (Zoology)* **56**: 1–56.

Clutton-Brock, T. H. (ed.) (1988a) *Reproductive Success: Studies of Individual Variation in Contrasting Breeding Systems*. Chicago, IL: University of Chicago Press.

Clutton-Brock, T. H. (1988b) Reproductive success. In *Reproductive Success* (ed. T. H. Clutton-Brock), pp. 472–85. Chicago. IL: University of Chicago Press.

(1991) *The Evolution of Parental Care*. Princeton, NJ: Princeton University Press.

Clutton-Brock, T. H. and Albon, S. D. (1982) Winter mortality in red deer *Cervus elaphus*. *Journal of Zoology* **198**: 515–19.

(1985) Competition and population regulation in social mammals. In *Behavioural Ecology* (eds. R. M. Sibly and R. H. Smith), pp. 557–76. Oxford: Blackwell Scientific Publications.

(1989). *Red Deer in the Highlands*. Oxford: Blackwell Scientific Publications.

Clutton-Brock, T. H. and Coulson, T. N. (2002) Comparative ungulate dynamics: the devil is in the detail. *Philosophical Transactions of the Royal Society B* **357**: 1285–98.

Clutton-Brock, T. H. and Harvey, P. H. (1979) Comparison and adaptation. *Proceedings of the Royal Society B* **205**: 547–65.

Clutton-Brock, T. H. and Iason, G. (1986) Sex ratio variation in mammals. *Quarterly Review of Biology* **61**: 339–74.

Clutton-Brock, T. H., Harvey, P. H. and Rudder, B. (1977) Sexual dimorphism, socionomic sex ratio and body weight in primates. *Nature* **269**: 797–800.

Clutton-Brock, T. H., Albon, S. D. and Guinness, F. E. (1981) Parental investment in male and female offspring in polygynous mammals. *Nature* **289**: 487–9.

Clutton-Brock, T. H., Guinness, F. E. and Albon, S. D. (1982a) *Red Deer: Behavior and Ecology of Two Sexes*. Chicago, IL: University of Chicago Press.

Clutton-Brock, T. H., Albon, S. D. and Guinness, F. E. (1982b) Competition between female relatives in a matrilocal mammal. *Nature* **300**: 178–80.

Clutton-Brock, T. H., Guinness, F. E. and Albon, S. D. (1983) The costs of reproduction to red deer hinds. *Journal of Animal Ecology* **52**: 367–83.

Clutton-Brock, T. H., Albon, S. D. and Guinness, F. E. (1984) Maternal dominance, breeding success and birth sex ratios in red deer. *Nature* **308**: 358–60.

(1985a) Parental investment and sex differences in juvenile mortality in birds and mammals. *Nature* **313**: 131–3.

Clutton-Brock, T. H., Major, M. and Guinness, F. E. (1985b) Population regulation in male and female red deer. *Journal of Animal Ecology* **54**: 831–46.

Clutton-Brock, T. H., Albon, S. D. and Guinness, F. E. (1986) Great expectations: dominance, breeding success and offspring sex ratios in red deer. *Animal Behaviour* **34**: 460–71.

(1987a) Interactions between population density and maternal characteristics affecting fecundity and survival in red deer. *Journal of Animal Ecology* **56**: 857–71.

Clutton-Brock, T. H., Major, M., Albon, S. D. and Guinness, F. E. (1987b) Early development and population dynamics in red deer. *Journal of Animal Ecology* **56**: 53–67.

Clutton-Brock, T. H., Albon, S. D. and Guinness, F. E. (1989) Fitness costs of gestation and lactation in wild mammals. *Nature* **337**: 260–2.

Clutton-Brock, T. H., Price, O. F., Albon, S. D. and Jewell, P. A. (1991) Persistent instability and population regulation in Soay sheep. *Journal of Animal Ecology* **60**: 593–608.

(1992) Early development and population fluctuations in Soay sheep. *Journal of Animal Ecology* **61**: 381–96.

Clutton-Brock, T. H., Stevenson, I. R., Marrow, P., MacColl, A. D. C., Houston, A. I. and McNamara, J. M. (1996) Population fluctuations, reproductive costs and life history tactics in female Soay sheep. *Journal of Animal Ecology* **65**: 675–89.

Clutton-Brock, T. H., Illius, A., Wilson, K., Grenfell, B. T., MacColl, A. D. C. and Albon, S. D. (1997a) Stability and instability in ungulate populations: an empirical analysis. *American Naturalist* **149**: 195–219.

Clutton-Brock, T. H., Wilson, K. and Stevenson, I. R. (1997b) Density-dependent selection on horn phenotype in Soay sheep. *Philosophical Transactions of the Royal Society B* **352**: 839–50.

Clutton-Brock, T. H., Rose, K. E. and Guinness, F. E. (1997c) Density-related changes in sexual selection in red deer. *Proceedings of the Royal Society B* **264**: 1509–16.

Clutton-Brock, T. H., Coulson, T. N., Milner-Gulland, E. J., Thomson, D. and Armstrong, H. (2002) Sex differences in emigration and mortality affect optimal management of deer populations. *Nature* **415**: 633–7.

Cockburn, A. (1988) *Social Behaviour in Fluctuating Populations*. London: Croom Helm.

(1998) Evolution of helping behaviour in cooperatively breeding birds. *Annual Reviews of Ecology and Systematics* **29**: 141–77.

Coltman, D. W., Bowen, W. D. and Wright, J. M. (1998) Birth weight and neonatal survival of harbour seal pups are positively correlated with genetic variation measured by microsatellites. *Proceedings of the Royal Society B* **265**: 803–9.

Coltman, D. W., Bancroft, D. R., Robertson, A., Smith, J. A. and Pemberton, J. M. (1999a). Male reproductive success in a promiscuous mammal: behavioural estimates compared with genetic paternity. *Molecular Ecology* **8**: 1199–209.

Coltman, D. W., Pilkington, J. G., Smith, J. A. and Pemberton, J. M. (1999b) Parasite-mediated selection against inbred Soay sheep in a free-living, island population. *Evolution* **53**: 1259–67.

Coltman, D. W., Smith, J. A., Bancroft, D. R., Pilkington, J., MacColl, A. D. C., Clutton-Brock, T. H. and Pemberton, J. M. (1999c) Density-dependent variation in lifetime breeding success and in natural and sexual selection in Soay rams. *American Naturalist* **154**: 730–46.

Coltman, D. W., Pilkington, J., Kruuk, L. E. B., Wilson, K. and Pemberton, J. M. (2001a) Positive genetic correlation between parasite resistance and body size in a free-living ungulate population. *Evolution* **55**: 2116–25.

Coltman, D. W., Wilson, K., Pilkington, J. G., Stear, M. J. and Pemberton, J. M. (2001b) A microsatellite polymorphism in the gamma interferon gene is association with resistance to gastrointestinal nematodes in a naturally parasitized population of Soay sheep. *Parasitology* **122**: 571–82.

Coltman, D. W., Festa-Bianchet, M., Jorgenson, J. T. and Strobek, C. (2002) Age-dependent sexual selection in bighorn rams. *Proceedings of the Royal Society B* **269**: 165–72.

Coltman, D. W., Pilkington, J. and Pemberton, J. M. (2003) Fine-scale genetic structure in a feral ungulate population. *Molecular Ecology.* **12**: 733–42.

Connan, R. M. (1968) The post-parturient rise in faecal nematode egg count of ewes: its aetiology and epidemiological significance. *World Review Animal Production* **4**: 53–8.

Coop, R. L., Sykes, A. R. and Angus, K. W. (1982) The effect of three levels of intake of *Ostertagia circumcincta* larvae on growth rate, food intake and body composition of growing lambs. *Journal of Agricultural Science* **98**: 247–55.

Coop, R. L., Graham, R. B., Jackson, F., Wright, S. E. and Angus, K. W. (1985) Effect of experimental *Ostertagia circumcincta* infection on the performance of grazing lambs. *Research in Veterinary Science* **38**: 282–7.

Corbett, G. B. and Harris, S. (1991) *The Handbook of British Mammals*, 3rd edn. Oxford: Blackwell Scientific Publications.

Cornett, M. W., Frelich, L. E., Puettmann, K. J. and Reich, P. B. (2000) Conservation implications of browsing by *Odocoileus virginianus* in remnant upland *Thuja occidentalis* forests. *Biological Conservation* **93**: 359–69.

Côté, S. D. and Festa-Bianchet, M. (2001) Birth date, mass and survival of mountain goat kids: effects of maternal characteristics and forage quality. *Oecologia* **127**: 230–8.

Coulson, J. C. and Wooller, R. D. (1976) Differential survival rates among breeding kittiwake gulls, *Rissa tridactyla* L. *Journal of Animal Ecology* **45**: 205–13.

Coulson, T., Albon, S., Guinness, F., Pemberton, J. and Clutton-Brock, T. H. (1997) Population substructure, local density and calf winter survival in red deer (*Cervus elaphus*). *Ecology* **78**: 852–63.

Coulson, T. N., Albon, S. D., Pemberton, J. M., Slate, J., Guinness, F. E. and Clutton-Brock, T. H. (1998a) Genotype by environment interactions in winter survival in red deer. *Journal of Animal Ecology* **67**: 434–45.

Coulson, T., Albon, S., Pilkington, J. and Clutton-Brock, T. (1999a) Small-scale spatial dynamics in a fluctuating ungulate population. *Journal of Animal Ecology* **68**: 658–71.

Coulson, T., Pemberton, J. M., Albon, S. D., Beaumont, M., Marshall, T. C., Slate, J., Guinness, F. E. and Clutton-Brock, T. H. (1998b) Microsatellites reveal heterosis in red deer. *Proceedings of the Royal Society B* **265**: 489–95.

Coulson, T., Albon, S., Slate, J. and Pemberton, J. (1999b) Microsatellite loci reveal sex-dependent responses to inbreeding and outbreeding in red deer calves. *Evolution* **53**: 1951–60.

Coulson, T., Milner-Gulland, E. J. and Clutton-Brock, T. (2000) The relative roles of density and climatic variation on population dynamics and fecundity rates in three contrasting ungulate species. *Proceedings of the Royal Society B* **267**: 1771–9.

Coulson, T., Catchpole, E. A., Albon, S. D., Morgan, B. J. T., Pemberton, J. M., Clutton-Brock, T. H., Crawley, M. J. and Grenfell, B. T. (2001) Age, sex, density, winter weather and population crashes in Soay sheep. *Science* **292**: 1528–31.

Cox, C. R. and Le Boeuf, B. J. (1977) Female incitation of male competition: a mechanism in sexual selection. *American Naturalist* **111**: 317–35.

Crawford, A. M. and McEwan, J. C. (1998) *Identification of Animals Resistant to Nematode Parasite Infection.* New Zealand Provisional Patent.

Crawford, A. M., Buchanan, F. C. and Swarbrick, P. A. (1990) Ovine dinucleotide repeat polymorphism at the MAF18 locus. *Animal Genetics* **21**: 433–4.

Crawford, A. M., Dodds, K. G., Ede, A. J., Pierson, C. A., Montgomery, G. W., Garmonsway, H. G., Beattie, A. E., Davies, K., Maddox, J. F., Kappes, S. W., Stone, R. T., Nguyen, T. C., Penty, J. M., Lord, E. A., Broom, J. E., Buitcamp, J., Schwaiger, W., Epplen, J. T., Matthew, P., Matthews, M. E., Hulme, D. J., Beh, K. J., McGraw, R. A. and Beattie, C. W. (1995) An autosomal genetic linkage map of the sheep genome. *Genetics* **140**: 703–24.

Crawley, M. J. (1983) *Herbivory: The Dynamics of Animal–Plant Interactions.* Oxford: Blackwell Scientific Publications.

(1993) *GLIM for Ecologists.* Oxford: Blackwell Scientific Publications.

(1997) Plant–herbivore dynamics. In *Plant Ecology* (ed. M. J. Crawley), pp. 401–74. Oxford: Blackwell Scientific Publications.

Creel, S. and Creel, N. M. (in press) *The African Wild Dog.* Princeton, NJ: Princeton University Press.

Creighton, P., Eggen, A., Fries, R., Jordan, S. A., Hetzel, J., Cunningham, E. P. and Humphries, P. (1992) Mapping of bovine markers CYP21, PRL and bola DRBP1 by genetic-linkage analysis in reference pedigrees. *Genomics* **14**: 526–8.

Crête, M., Coutourier, S., Hearn, B. J. and Chubbs, T. E. (1996) Relative contribution of decreased productivity and survival to recent changes in the demography of the Riviere George caribou herd. *Rangifer* **9**: 27–36.

Crockett C. M. and Rudran, R. (1987a) Red howler monkey birth data. I. Seasonal variation. *American Journal of Primatology* **13**: 347–68.

(1987b) Red howler monkey birth data. II. Interannual, habitat and sex comparisons. *American Journal of Primatology* **13**: 369–84.

Crofton, H. D. (1954) The vertical migration of infective larvae of strongyloid nematodes. *Journal of Helminthology* **28**: 35–52.

Crnokrak, P. and Roff, D. A. (1999) Inbreeding depression in the wild. *Heredity* **83**: 260–70.

Cumming, D. H. M. (1981) Elephant and woodlands in Chizarira National Park, Zimbabwe. In *Problems in Management of Locally Abundant Wild Mammals* (eds. P. A. Jewell, S. Hilt and D. Hart), pp. 347–9. New York: Academic Press.

Cummins, L. J., Thompson, R. I., Yong, W. K., Riffkin, G. G., Goddard, M. E., Callinan, A. P. L. and Saunders, M. J. (1991) Genetics of *Ostertagia* selection lines. In *Breeding for Disease Resistance in Sheep* (eds. G. D. Gray and R. R. Woolaston), pp. 11–18. Melbourne: Australian Wool Corporation.

Darwin, C. (1859) *The Origin of Species by Natural Selection, or the Preservation of Favoured Races in the Struggle for Life*. London: John Murray.

de Gortari, M., Freking, B. A., Cuthbertson, R. P., Kappes, S. M., Keele, J. W., Stone, R. T., Leymaster, K. A., Dodds, K. G., Crawford, A. M. and Beattie, C. W. (1998) A second generation linkage map of the sheep genome. *Mammalian Genome* **9**: 204–9.

Diamond, M. (1966) Progestagen inhibition of normal sexual behaviour in the male guinea pig. *Nature* **209**: 1322–4.

Dobson, A. P. and P. J. Hudson. (1992) Regulation and stability of a free-living host–parasite system: *Trichostrongylus tenuis* in red grouse. II. Population models. *Journal of Animal Ecology* **61**: 487–98.

Dobson, C. (1964) Host endocrine interactions with nematode infections. I. Effects of sex, gonadectomy, and thyroidectomy on experimental infections in lambs. *Experimental Parasitology* **15**: 200–12.

Dolling, C. H. S. (1960a) Hornedness and polledness in sheep. I. The inheritance of polledness in the Merino. *Australian Journal of Agricultural Research* **11**: 427–38.

(1960b) Hornedness and polledness in sheep. II. The inheritance of horns in the Merino ewe. *Australian Journal of Agricultural Research* **11**: 618–27.

(1961) Hornedness and polledness in sheep. IV. Triple alleles affecting horn growth in the Merino. *Australian Journal of Agricultural Research* **12**: 356–61.

Doney, J. M., Ryder, M. L., Gunn, R. G. and Grubb, P. (1974) Colour, conformation, affinities, fleece and patterns of inheritance of the Soay sheep. In *Island Survivors: The Ecology of the Soay Sheep of St Kilda* (eds. P. A. Jewell, C. Milner and J. M. Boyd), pp. 88–125. London: Athlone Press.

Douglas, C. L. and Leslie, D. M. (1986) Influence of weather and density on lamb survival of desert mountain sheep. *Journal of Wildlife Management* **50**: 153–6.

Downey, N. E. and Conway, A. (1968) Grazing management in relation to trichostrongylid infestation in lambs. I. Influence of stocking rate on the level of infestation. *Irish Journal of Agricultural Research* **7**: 343–62.

Downey, N. E., Connolly, J. F. and O'Shea, J. (1972) Experimental ostergiasis and the effect of diet on resistance in sheep. *Irish Journal of Agricultural Research* **11**: 11–29.

Drent, R. H. and Daan, S. (1980) The prudent parent: energetic adjustments in avian breeding. *Ardea* **68**: 225–52.

Dudash, M. R. (1990) Relative fitness of selfed and outcrossed progeny in a self-compatible, protandrous species, *Sabatia angularis* L. (Gentianaceae): a comparison in three environments. *Evolution* **44**: 1129–39.

Dunn, A. M. (1978) *Veterinary Helminthology*, 2nd edn. London: Heinemann Medical.

Eberhardt, L. L. (1977) Optimal policies for conservation of large mammals with special reference to marine ecosystems. *Environmental Conservation* 4: 205–12.

(2000) A paradigm for population analyses of long-lived vertebrates. *Ecology* 83: 2841–54.

Ebert, D. and Hamilton, W. D. (1996) Sex against virulence: the coevolution of parasitic diseases. *Trends in Ecology and Evolution* 11: 79–82.

Ede, A. J., Pierson, C. A. and Crawford, A. M. (1995) Ovine microsatellites at the OarCP9, OarCP16, OarCP20, OarCP21, OarCP23, and OarCP26 loci. *Animal Genetics* 26: 129–30.

Ellner, S. and Hairston, N. G. (1994). Role of overlapping generations in maintaining genetic variation in a fluctuating environment. *American Naturalist* 143: 403–17.

Else, K. J., Finkelman, F. D., Maliszewski, C. R. and Grencis, R. K. (1994) Cytokine-mediated regulation of chronic intestinal helminth infection. *Journal of Experimental Medicine* 179: 347–51.

Elton, C., Ford, E. B. and Baker, J. B. (1931) The health and parasites of a wild mouse population. *Proceedings of the Zoological Society of London* 1931: 657–721.

Endler, J. A. (1980) Natural selection on color patterns in *Poecilia reticulata*. *Evolution* 34: 76–91.

(1986) *Natural Selection in the Wild*. Princeton, NJ: Princeton University Press.

(1992) Genetic heterogeneity and ecology. In *Genes in Ecology* (eds. R. J. Berry, T. J. Crawford and G. M. Hewitt), pp. 281–312. Oxford: Blackwell Scientific Publications.

Ens, B. J., Kerstein, M., Brenninkmeizer, A. and Hutshcer, J. B. (1992) Territory quality, parental effort and reproductive success of oystercatchers, *Haematopus ostralegus*. *Journal of Animal Ecology* 61: 703–15.

Ericsson, R. J. and Dutt, R. H. (1965) Progesterone and 6 α-methyl-17 α-hydroxyprogesterone acetate as inhibitors of spermatogenesis and accessory gland function in the ram. *Endocrinology* 77: 203–12.

Estes, R. D. (1992) *The Behavior Guide to African Mammals*. Berkeley, CA: University of California Press.

Fairbanks, W. S. (1993) Birth date, birth weight, and survival in pronghorn fawns. *Journal of Mammalogy* 74: 129–35.

Falconer, D. S. and Mackay, T. F. C. (1996) *Introduction to Quantitative Genetics*, 4th edn. Harlow: Longman.

Festa-Bianchet, M. (1986) Seasonal dispersion of overlapping mountain sheep ewe groups. *Journal of Wildlife Management* 50: 325–30.

(1988a) Nursing behaviour of bighorn sheep: correlates of ewe age, parasitism, lamb age, birth date and sex. *Animal Behaviour* 36: 1445–54.

(1988b) Age-specific reproduction of bighorn ewes in Alberta, Canada. *Journal of Mammalogy* **69**: 157–60.

(1989a) Individual differences, parasites, and the costs of reproduction for bighorn ewes (*Ovis canadensis*). *Journal Animal Ecology* **58**: 755–95.

(1989b) Survival of male bighorn sheep in Southwestern Alberta. *Journal of Wildlife Management* **53**: 259–63.

(1991a) The social system of bighorn sheep: grouping patterns, kinship and female dominance rank. *Animal Behaviour* **42**: 71–82.

(1991b) Numbers of lungworm larvae in feces of bighorn sheep: yearly changes, influence of host sex, and effects on host survival. *Canadian Journal of Zoology – Revue Canadienne de Zoologie* **69**: 547–54.

(1996) Offspring sex ratio studies of mammals: does publication depend upon the quality of the research or the direction of the results? *Ecoscience* **3**: 42–4.

(1998) Condition-dependent reproductive success in bighorn ewes. *Ecology Letters* **1**: 91–4.

Festa-Bianchet, M. and Jorgenson, J.T. (1998) Selfish mothers: reproductive expenditure and resource availability in bighorn ewes. *Behavioural Ecology* **9**: 144–50.

Festa-Bianchet, M., Jorgenson, J.T., Lucherini, M. and Wishart, W.D. (1995) Life-history consequences of variation in age of primiparity in bighorn ewes. *Ecology* **76**: 871–81.

Festa-Bianchet, M., Jorgenson, J.T., Bérubé, C.H., Portier, C. and Wishart, W.D. (1997) Body mass and survival of bighorn sheep. *Canadian Journal of Zoology* **75**: 1372–9.

Festa-Bianchet, M., Gaillard, J.-M. and Jorgenson, J.T. (1998) Mass- and density-dependent reproductive success and reproductive costs in a capital breeder. *American Naturalist* **152**: 367–79.

Field, A.C., Brambell, M.R. and Campbell, J.A. (1960) Spring rise in faecal worm-egg counts of housed sheep, and its importance in nutritional experiments. *Parasitology* **50**: 387–99.

Finnerty, J.P. (1980) *The Population Ecology of Small Mammals*. New Haven, CT: Yale University Press.

Fisher, J. (1948) St Kilda: a natural experiment. *New Naturalist*: 91–108.

Fisher, R.A. (1930) *The Genetical Theory of Natural Selection*. Oxford: Clarendon Press.

(1941) The negative binomial distribution. *Annals of Eugenics* **11**: 182–7.

Fletcher, J.P., Hughes, J.P. and Harvey, I.F. (1994) Life expectancy and egg load affect oviposition decisions of a solitary parasitoid. *Proceedings of the Royal Society B* **258**: 163–7.

Forchhammer, M.C., Post, E. and Stenseth, N.C. (1998a) Breeding phenology and climate. *Nature* **391**: 29–30.

Forchhammer, M.C., Stenseth, N.C., Post, E. and Langvatn, R. (1998b) Population dynamics of Norwegian red deer: density-dependence and climatic variation. *Proceedings of the Royal Society B* **265**: 341–50.

Forchhammer, M.C., Clutton-Brock, T.H., Lindström, J. and Albon, S.D. (2001) Climate and population density induce long-term cohort variation in a northern ungulate. *Journal of Animal Ecology* **70**: 721–9.

Fowler, C.W. (1981) Density-dependence as related to life-history strategy. *Ecology* **62**: 602–10.

(1987) A review of density-dependence in populations of large mammals. In *Current Mammalogy* (ed. H.H. Genoways), pp. 401–41. New York: Plenum Press.

Frankham, R. (1995) Conservation genetics. *Annual Reviews of Ecology and Sytematics* **29**: 305–27.

Frayha, G.J., Lawlor, W.K. and Dajani, R.M. (1971) *Echinococcus granulosus* in albino mice: effect of host sex and sex hormones on the growth of hydatid cysts. *Experimental Parasitology* **29**: 255–62.

Fretwell, S.D. and Lucas, H.L. (1970) On territorial behaviour and other factors influencing habitat distribution in birds. *Acta Biotheoretica* **19**: 16–36.

Fries, R., Eggen, A. and Womack, J.E. (1993) The bovine genome map. *Mammalian Genome* **4**: 405–28.

Fryxell, J.M. (1987) Food limitation and demography of a migratory antelope: the white–eared kob. *Oecologia* **72**: 83–91.

Fryxell, J.M., Hussell, D.J.T., Lambert, A.B. and Smith, P.C. (1991) Time lags and population fluctuations in white-tailed deer. *Journal of Wildlife Management* **55**: 377–85.

Futuyma, D.J. (1986) *Evolutionary Biology.* Sunderland, MA: Sinauer Associates.

Gadgil, M. and Bossert, W.H. (1970) Life historical consequences of natural selection. *American Naturalist* **104**: 1–24.

Gaillard, J.-M., Delormé, D. and Jullien, J.M. (1993) Effects of cohort, sex and birth date on body development of roe deer (*Capreolus capreolus*) fawns. *Oecologia* **94**: 57–61.

Gaillard, J.-M., Festa-Bianchet, M. and Yoccoz, N.G. (1998) Population dynamics of large herbivores: variable recruitment with constant adult survival. *Trends in Ecology and Evolution* **13**: 58–63.

Gaillard, J.-M., Festa-Bianchet, M., Yoccoz, N.G., Loison, A. and Toigo, C. (2000) Temporal variation in fitness components and population dynamics of large herbivores. *Annual Reviews of Ecology and Systematics* **31**: 367–93.

Gallant, B., Réale, D. and Festa-Bianchet, M. (2001) Does mass change of primiparous bighorn ewes reflect reproductive effort? *Canadian Journal of Zoology* **79**: 312–18.

Gasaway, W. C., Boertje, R. D., Grangaard, D. V., Kellyhouse, D. G. and Stephenson, R. O. (1992) The role of predation in limiting moose at low densities in Alaska and Yukon and implications for conservation. *Wildlife Monographs* **120**: 1–59.

Gasaway, W. C., Gasaway, T. K. and Berry, H. H. (1996) Persistent low densities of plains ungulates in Etosha National Park, Namibia: testing the food-regulating hypothesis. *Canadian Journal of Zoology* **74**: 1556–72.

Geist, V. (1964) On the rutting behavior of the mountain goat. *Journal of Mammalogy* **45**: 551–68.

(1971) *Mountain Sheep: A Study in Behavior and Evolution.* Chicago, IL: University of Chicago Press.

Georges, M. and Massey, J. M. (1992) *Polymorphic DNA Markers in Bovidae.* Patent W092/13102.

Gibbs, H. C. (1986) Hypobiosis in parasitic nematodes: an update. *Advances in Parasitology* **25**: 129–74.

Gibbs, H. L. and Grant, P. R. (1987) Oscillating selection on Darwin's finches. *Nature* **327**: 511–13.

Gomendio, M., Clutton-Brock, T. H., Albon, S. D., Guinness, F. E. and Simpson, M. J. (1990) Mammalian sex ratios and variation in costs of rearing sons and daughters. *Nature* **343**: 261–3.

Gomendio, M., Harcourt, A. H. and Roldán, E. R. S. (1998) Sperm competition in mammals. In *Sperm Competition and Sexual Selection* (eds. T. R. Birkhead and A. P. Møller), pp. 667–756. London: Academic Press.

Gordon, H. M. (1964) Studies on resistance to *Trichostrongylus colubriformis* in sheep: influence of a quantative reduction in the ration. *Australian Veterinary Journal* **40**: 55–61.

Gordon, I. J., Illius, A. W. and Milne, J. D. (1996) Sources of variation in the foraging efficiency of grazing ruminants. *Functional Ecology* **10**: 219–26.

Gosling, L. M. (1986) Selective abortion of entire litters in the coypu: adaptive control of offspring production in relation to quality and sex. *American Naturalist* **127**: 772–95.

Grace, J. and Easterbee, N. (1979) The natural shelter for red deer (*Cervus elaphus*) in a Scottish glen. *Journal of Applied Ecology* **16**: 37–48.

Grant, B. R. and Grant, P. R. (1989) Natural selection in a population of Darwin's finches. *American Naturalist* **133**: 377–93.

Grant, P. R. (1965) The adaptive significance of some size trends in island birds. *Evolution* **19**: 355–67.

(1968) Bill size, body size and ecological adaptations of bird species to competitive situations on islands. *Systematic Zoology* **17**: 319–33.

(1986) *Ecology and Evolution of Darwin's Finches.* Princeton, NJ: Princeton University Press.

(ed.) (1998) *Evolution on Islands.* Oxford: Oxford University Press.

Grant, P. R. and Grant, B. R. (2002) Unpredictable evolution in a 30-year study of Darwin's finches. *Science* **296**: 707–11.

Green, W. C. H. and Rothstein, A. (1991) Sex bias or equal opportunity? Patterns of maternal investment in bison. *Behavioral Ecology and Sociobiology* **29**: 373–84.

Grencis, R. K. (1997) Th2-mediated host protective immunity to intestinal nematode infections. *Philosophical Transactions of the Royal Society* **352**: 1377–84.

Grenfell, B. T. (1988) Gastrointestinal nematode parasites and the stability and productivity of intensive ruminant grazing systems. *Philosophical Transactions of the Royal Society B* **321**: 541–63.

(1992) Parasitism and the dynamics of ungulate grazing systems. *American Naturalist* **139**: 907–29.

Grenfell, B. T. and Dobson, A. P. (eds.) (1995) *Ecology of Infectious Diseases in Natural Populations*. Cambridge: Cambridge University Press.

Grenfell, B. T. and Gulland, F. M. D. (1995) Introduction: ecological impact of parasitism on wildlife host populations. *Parasitology* **111**: S3–S14.

Grenfell, B. T., Price, O. F., Albon, S. D. and Clutton-Brock, T. H. (1992) Overcompensation and population cycles in an ungulate. *Nature* **355**: 823–6.

Grenfell, B. T., Wilson, K., Isham, V. S., Boyd, H. E. G. and Dietz, K. (1995) Modelling patterns of parasite aggregation in natural populations: trichostrongylid nematode–ruminant interactions as a case study. *Parasitology* **111**: S135–S151.

Grenfell, B. T., Wilson, K., Finkenstadt, B. F., Coulson, T. N., Murray, S., Albon, S. D., Pemberton, J. M., Clutton-Brock, T. H. and Crawley, M. J. (1998) Noise and determinism in synchronized sheep dynamics. *Nature* **394**: 674–7.

Groeneveld, E. (1995) *REML VCE: A Multivariate Multi Model Restricted Maximum Likelihood (Co)Variance Component Estimation Package, Ver. 3.2 User's Guide.* Mariensee, Germany: Institute of Animal Husbandry and Animal Behaviour, Federal Research Center of Agriculture.

Groeneveld, E. and Kovac, M. (1990) A generalized computing procedure for setting up and solving mixed linear models. *Journal of Dairy Science* **73**: 513–31.

Grossman, C. J. (1985) Interactions between the gonadal steroids and the immune system. *Science* **227**: 257–61.

Grubb, P. (1974a) Mating activity and the social significance of rams in a feral sheep community. In *The Behaviour of Ungulates and Its Relation to Management* (eds. V. Geist and F. Walther), pp. 457–76. Morges: IUCN.

(1974b) The rut and behaviour of Soay rams. In *Island Survivors: The Ecology of the Soay Sheep of St Kilda* (eds. P. A. Jewell, C. Milner and J. M. Boyd), pp. 195–223. London: Athlone Press.

(1974c) Population dynamics of the Soay sheep. In *Island Survivors: The Ecology of the Soay Sheep of St Kilda* (eds. P. A. Jewell, C. Milner and J. M. Boyd), pp. 242–72. London: Athlone Press.

(1974d) Social organisation of Soay sheep and the behaviour of ewes and lambs. In *Island Survivors: The Ecology of the Soay Sheep of St Kilda* (eds. P. A. Jewell, C. Milner and J. M. Boyd), pp. 131–59. London: Athlone Press.

Grubb, P. and Jewell, P. A. (1973) The rut and the occurrence of oestrus in the Soay sheep on St Kilda. *Journal of Reproduction and Fertility, Supplement* **19**: 491–502.

(1974) Movement, daily activity and home range of Soay sheep. In *Island Survivors: The Ecology of the Soay Sheep of St Kilda* (eds. P. A. Jewell, C. Milner and J. M. Boyd), pp. 160–194. London: Athlone Press.

Guinness, F. E., Gibson, R. M. and Clutton-Brock, T. H. (1978a) Calving times of red deer (*Cervus elaphus* L.) on Rhum. *Journal of Zoology* **185**: 105–14.

Guinness, F. E., Albon, S. D. and Clutton-Brock, T. H. (1978b) Factors affecting reproduction in red deer (*Cervus elaphus*) hinds on Rhum. *Journal of Reproduction and Fertility* **54**: 325–34.

Guinness, F. E., Clutton-Brock, T. H. and Albon, S. D. (1978c) Factors affecting calf mortality in red deer. *Journal of Animal Ecology* **47**: 812–32.

Gulland, F. M. D. (1991) The role of parasites in the population dynamics of Soay sheep on St Kilda. PhD thesis, University of Cambridge.

(1992) Role of nematode infections in mortality of Soay sheep during a population crash on St Kilda. *Parasitology* **105**: 493–503.

(1995) Impact of infectious diseases on wild animal populations: a review. In *Ecology of Infectious Diseases in Natural Populations* (eds. B. T. Grenfell and A. P. Dobson), pp. 20–52. Cambridge: Cambridge University Press.

Gulland, F. M. D. and Fox, M. (1992) Epidemiology of nematode infections of Soay sheep (*Ovis aries* L.) on St Kilda. *Parasitology* **105**: 481–92.

Gulland, F. M. D., Albon, S. D., Pemberton, J. M., Moorcroft, P. and Clutton-Brock, T. H. (1993) Parasite-associated polymorphism in a cyclic ungulate population. *Proceedings of the Royal Society B* **254**: 7–13.

Gunn, R. G., Doney, J. M. and Smith, W. F. (1979) Fertility in Cheviot ewes. III. The effect of level of nutrition before and after mating on ovulation rate and early embryo mortality in South Country Cheviot ewes in moderate condition at mating. *Animal Production* **29**: 25–31.

Gunn, R. G., Doney, J. M., Smith, W. F. and Sim, D. A. (1986) Effects of age and its relationship with body size on reproductive performance in Scottish blackface ewes. *Animal Production* **43**: 279–83.

Gustafsson, L. (1986) Lifetime reproductive success and heritability: empirical support for Fisher's fundamental theorem. *American Naturalist* **128**: 761–4.

(1988) Inter- and intra-specific competition for nest holes in a population of the collared flycatcher *Ficedula albicollis*. *Ibis* **130**: 11–15.

Gwynne, D., Milner, C. and Hornung, M. (1974) The vegetation and soils of Hirta. In *Island Survivors: The Ecology of the Soay Sheep of St Kilda* (eds. P. A. Jewell, C. Milner and J. M. Boyd), pp. 36–87. London: Athlone Press.

Gyllensten, U., Ryman, N., Reuterwall, C. and Dratch, P. (1983) Genetic differentiation in four European subspecies of red deer (*Cervus elaphus* L.). *Heredity* **51**: 561–80.

Haldane, J. B. S. (1956) The relation between density regulation and natural selection. *Proceedings of the Royal Society B* **145**: 306–8.

Hall, J. G. (1974) The blood groups of Soay sheep. In *Island Survivors: The Ecology of the Soay Sheep of St Kilda* (eds. P. A. Jewell, C. Milner and J. M. Boyd), pp. 126–30. London: Athlone Press.

Hall, J. G. and Purser, A. F. (1979) Lamb production and haemoglobin type. *Animal Breeding Research Organisation Annual Report 1979.*

Haltenorth, T. and Diller, H. (1977) *A Field Guide to the Mammals of Africa.* London: Collins.

Hamilton, W. D. (1980) Sex versus non-sex versus parasite. *Oikos* **35**: 282–90.

Hamilton, W. D. and Zuk, M. (1982) Heritable true fitness and bright birds: a role for parasites? *Science* **218**: 384–7.

Hamilton, W. D. and Poulin, R. (1997) The Hamilton and Zuk hypothesis revisited: a meta-analytical approach. *Behaviour* **134**: 299–320.

Hamilton, W. D., Axelrod, R. and Tanese, T. (1990) Sexual reproduction as an adaptation to resist parasites (a review). *Proceedings of the National Academy of Sciences of the USA* **87**: 3566–73.

Hanski, I. (1987) Populations of small mammals cycle – unless they don't. *Trends in Ecology and Evolution* **2**: 55–6.

Hanski, I., Hansson, L. and Haltovien, H. (1991) Specialist predators, generalist predators and the microtine rodent cycle. *Journal of Animal Ecology* **60**: 353–67.

Hansson, B., Bensch, S., Hasselquist, D. and Akesson, M. (2001) Microsatellite diversity predicts recruitment of sibling great reed warblers. *Proceedings of the Royal Society B* **268**: 1287–91.

Harcourt, A. H., Harvey, P. H., Larson, S. G. and Short, R. V. (1981) Testis weight, body weight and breeding system in primates. *Nature* **293**: 55–7.

Harman, M. (1995) The History of St Kilda. In: *St Kilda: The Continuing Story of the Islands* (ed. M. Buchanan), pp. 1–23. Edinburgh: HMSO.

Harman, M. (1997) *An Isle Called Hirte.* Skye: Maclean Press.

Harvey, P. H. and Zammuto, R. M. (1985) Patterns of mortality and age at first reproduction in natural populations of mammals. *Nature* **315**: 319–20.

Hass, C. C. (1989) Bighorn lamb mortality: predation, inbreeding and population effects. *Canadian Journal of Zoology* **67**: 695–705.

Hassell, M. P., Lawton, J. H. and May, R. M. (1976) Patterns of dynamical behaviour in single-species poulations. *Journal of Animal Ecology* **45**: 471–86.

Hedrick, P.W. (1986). Genetic polymorphism in heterogeneous environments: a decade later. *Annual Reviews of Ecology and Systematics* **17**: 535–66.

Hedrick, P.W. and Thomson, G. (1983) Evidence for balancing selection at HLA. *Genetics* **104**: 449–56.

Hedrick, P.W., Ginevan, M.E. and Ewing, E.P. (1976) Genetic polymorphism in heterogeneous environments. *Annual Reviews of Ecology and Systematics* **7**: 1–33.

Helyar, S. (2000) The inheritance of coat colour and homtype in Soay sheep. M.Sc. thesis, University of Edinburgh.

Hill, A.V.S., Allsopp, C.E.M., Kwiatkowski, D., Anstey, N.M., Twumasi, P., Rowe, P.A., Bennett, S., Brewster, D., McMichael, A.J. and Greenwood, B.M. (1991) Common West African HLA antigens are associated with protection from severe malaria. *Nature* **352**: 595–600.

Hill, W.G. (1972) Effective population size with overlapping generations. *Theoretical Population Biology* **3**: 278–89.

(1981) Estimation of effective population size from data on linkage disequilibrium. *Genetical Research* **38**: 209–16.

Hochachka, W.M. and Dhondt, A.A. (2000) Density-dependent decline of host abundance resulting from a new infectious disease. *Proceedings of the National Academy of Sciences of the USA* **97**: 5303–6.

Hoefs, M. and Bayer, M. (1983) Demographic characteristics of an unhunted Dall sheep (*Ovis dalli dalli*) population in south-west Yukon, Canada. *Canadian Journal of Zoology* **61**: 1346–57.

Hoelzel, A.R., Le Boeuf, B.J., Reiter, J. and Campagna, C. (1999) Alpha-male paternity in elephant seals. *Behavioral Ecology and Sociobiology* **46**: 298–306.

Hogg, J.T. (1984) Mating in bighorn sheep: multiple creative strategies. *Science* **225**: 526–9.

(1987) Intrasexual competition and mate choice in Rocky Mountain bighorn sheep. *Ethology* **75**: 119–44.

(1988) Copulatory tactics in relation to sperm competition in Rocky Mountain bighorn sheep. *Behavioral Ecology and Sociobiology* **22**: 49–59.

Hogg, J.T. and Forbes, S.H. (1997) Mating in bighorn sheep: frequent male reproduction via a high-risk 'unconventional' tactic. *Behavioral Ecology and Sociobiology* **41**: 33–48.

Hoglund, J., Piertney, S.B., Alatalo, R.V., Lindell, J., Lundberg, A. and Rintamaki, P.T. (2002) Inbreeding depression and male fitness in black grouse. *Proceedings of the Royal Society B* **269**: 711–15.

Holmes, P.H. (1985) Pathogenesis of trichostrongylosis. *Veterinary Parasitology* **18**: 89–101.

Hong, C., Michel, J.F. and Lancaster, M.B. (1987) Observations on the dynamics of worm burdens in lambs infected daily with *Ostertagia circumcincta*. *International Journal of Parasitology* **17**: 951–6.

Houston, D. B., Robbins, C. T. and Stevens, V. (1989) Growth in wild and captive mountain goats. *Journal of Mammalogy* **70**: 412–16.

Howard, R. S. and Lively, C. M. (1994) Parasitism, mutation accumulation and the maintenance of sex. *Nature* **367**: 554–7.

Hudson, P. J. (1986) The effect of a parasitic nematode on the breeding production of red grouse. *Journal of Animal Ecology* **55**: 85–92.

(ed.) (1997) *Wildlife Diseases: The Epidemiology of Infectious Diseases in Wild Animals and How They Relate to Mankind.* London: British Ecological Society.

Hudson, P. J. and Dobson, A. P. (1989). Population biology of *Trichostrongylus tenuis*, a parasite of economic importance for red grouse management. *Parasitology Today* **5**: 283–91.

(1997) Transmission dynamics and host–parasite interactions of *Trichostrongylus tenuis* in red grouse (*Lagopus lagopus scoticus*). *Journal of Parasitology* **83**: 194–202.

Hudson, P. J., Dobson, A. P. and Newborn, D. (1985) Cyclic and non-cyclic populations of red grouse: a role for parasitism? In *Ecology and Genetics of Host–Parasite Interactions* (eds. D. Rollinson and R. M. Anderson), pp. 79–89. London: Academic Press.

(1992a) Do parasites make prey vulnerable to predation? Red grouse and parasites. *Journal of Animal Ecology* **61**: 681–92.

Hudson, P. J., Newborn, D. and Dobson, A. P. (1992b) Regulation and stability of a free-living host-parasite system *Trichostrongylus tenuis* in red grouse. I. Monitoring and parasite reduction experiments. *Journal of Animal Ecology* **61**: 477–86.

Hudson, P. J., Dobson, A. P. and Newborn, D. (1998) Prevention of population cycles by parasite removal. *Science* **282**: 2256–8.

(1999) Population cycles and parasitism. *Science* **286**: 2426.

Hudson, P. J., Rizzoli, A., Grenfell, B. T., Heesterbeek, H. and Dobson, A. P. (eds.) (2002) *The Ecology of Wildlife Diseases.* Oxford: Oxford University Press.

Hughes, A. L. and Nei, M. (1988) Pattern of nucleotide subtitution at major histocompatability complex class-I loci reveals overdominant selection. *Nature* **335**: 167–70.

(1989) Nucelotide substitution at major histocompatability complex class-II loci: evidence for overdominant selection. *Proceedings of the National Academy of Sciences of the USA* **86**: 958–62.

Hughes, C. (1998) Integrating molecular techniques with field methods in studies of social behavior: a revolution results. *Ecology* **79**: 383–99.

Hunter, R. F. (1964) Home-range behaviour in hill sheep. In *Grazing in Terrestrial and Marine Environments* (ed. D. J. Crisp), pp. 155–71. Oxford: Blackwell Scientific Publications.

Hunter, R. L. and Markert, C. L. (1957) Histochemical demonstration of enzymes separated by zone electrophoresis of starch gels. *Science* **125**: 1294–5.

Illius, A. W., Albon, S. D., Pemberton, J. M., Gordon, I. J. and Clutton-Brock, T. H. (1995) Selection for foraging efficiency during a population crash in Soay sheep. *Journal of Animal Ecology* **64**: 481–92.

Jarman, M. V. (1979) *Impala Social Behaviour: Territory, Hierarchy, Mating and the Use of Space.* Berlin: Verlag Paul Parey.

Jeffreys, A. J., Wilson, V. and Thein, S. L. (1985) Individual-specific 'fingerprints' of human DNA. *Nature* **316**: 76–9.

Jewell, P. A. (1986) Survival in a feral population of primitive sheep on St Kilda, Outer Hebrides, Scotland. *National Geographic Research* **2**: 402–6.

(1989) Factors that affect fertility in a feral population of sheep. *Zoological Journal of the Linnean Society* **95**: 163–74.

(1997) Survival and behaviour of castrated Soay sheep (*Ovis aries*) in a feral island population on Hirta, St Kilda, Scotland. *Journal of Zoology* **243**: 623–36.

Jewell, P. A. and Grubb, P. (1974) The breeding cycle, the onset of oestrus and conception in Soay sheep. In *Island Survivors: The Ecology of the Soay Sheep of St Kilda* (eds. P. A. Jewell, C. Milner and J. M. Boyd), pp. 224–41. London: Athlone Press.

Jewell, P. A., Milner, C. and Boyd, J. M. (eds.) (1974) *Island Survivors: The Ecology of the Soay Sheep of St Kilda.* London: Athlone Press.

Jochle, W. and Schilling, E. (1965) Experience with progestagens in farm animals. *Journal of Reproduction and Fertility* **10**: 287–8.

Johnson, A. R., Green, R. E. and Hirons, G. J. M. (1999) Survival rates of greater flamingoes in the west Mediterranean region. In *Bird Population Studies* (eds. C. M. Perrins, J.-D. Le Breton and G. J. M. Hirons), pp. 249–71. Oxford: Oxford University Press.

Johnston, L. (2000) *Scotland's Nature in Trust.* London: Poyser.

Jorgenson, J. T., Festa-Bianchet, M., Lucherini, M. and Wishart, W. D. (1993) Effects of body size, population density, and maternal characteristics on age at first reproduction in bighorn ewes. *Canadian Journal of Zoology* **71**: 2509–17.

Karsch, F. J. (1984) The hypothalamus and anterior pituitary gland. In *Hormonal Control of Reproduction* (eds. C. Austin and R. V. Short), 2nd edn, pp. 1–20. Cambridge: Cambridge University Press.

Keech, M. A., Bowyer, R. T., Ver Hoef. J. M., Boertje, R. D., Dale, B. W. and Stephenson, T. R. (2000) Life-history consequences of maternal condition in Alaskan moose. *Journal of Wildlife Management* **64**: 450–62.

Keen, R. E. and Spain, J. D. (1992) *Computer Simulation in Biology: A BASIC Introduction.* New York: Wiley–Liss.

Kellas, L. M. (1954) Observations on the reproductive activities, measurements, and growth rate of the dikdik (*Rynchotragus kirkii thomasi* Neumann). *Proceedings of the Zoological Society of London* **124**: 751–84.

Keller, L. F. (1998) Inbreeding and its fitness effects in an insular population of song sparrows (*Melospiza melodia*). *Evolution* **52**: 240–50.

Keller, L. F. and Waller, D. M. (2002) Inbreeding effects in wild populations. *Trends in Ecology and Evolution* **17**: 230–41.

Keller, L. F., Grant, P. R., Grant, B. R. and Petren, K. (2002) Environmental conditions affect the magnitude of inbreeding depression in survival of Darwin's finches. *Evolution* **56**: 1229–39.

Kelly, R. W. (1984) Fertilisation failure and embryonic wastage. In *Reproduction in Sheep* (eds. D. R. Lindsay and D. T. Pearce), pp. 127–33. Cambridge: Cambridge University Press.

Kendall, B. E., Prendergast, J. and Bjørnstad, O. N. (1998) The macroecology of population dynamics: taxonomic and biogeographic patterns in population cycles. *Ecology Letters* **1**: 160–4.

Kettlewell, H. B. D. (1973) *The Evolution of Melanism*. Oxford: Oxford University Press.

Keyfitz, N. (1968) *Introduction to the Mathematics of Populations*. Reading, MA: Addison-Wesley.

Keymer, A. E., Tarlton, A. B., Hiorns, R. W., Lawrence, C. E. and Pritchard, D. I. (1990) Immunogenetic correlates of genetic susceptibility to infection with *Heligmosomoides polygyrus* in outbred mice. *Parasitology* **101**: 69–73.

Kimura, M. (1968) Evolutionary rate at the molecular level. *Nature* **217**: 624–6.

King, J. and Nicholson, I. A. (1964) Grasslands of the forest and sub-alpine zones. In *The Vegetation of Scotland* (ed. J. H. Burnett), pp. 168–206. Edinburgh: Oliver and Boyd.

King, J. L. and Jukes, T. H. (1969) Non-Darwinian evolution. *Science* **164**: 788–98.

Kingdon, J. (1979) *East African Mammals*, vol. IIIB. London: Academic Press.

Kinsman, D. J. J. (2001) *Black Sheep of Windermere: A History of the St Kilda or Hebridean Sheep*. Windermere: Windy Hall Publications.

Klein, D. R. (1968) The introduction, increase, and crash of reindeer on St Matthew Island. *Journal of Wildlife Management* **32**: 350–67.

Klein, J. (1986) *The Natural History of the Major Histocompatibility Complex*. New York: Wiley.

Klein, J., Satta, Y., Ohuigin, C. and Takahata, N. (1993) The molecular descent of the major histocompatability complex. *Annual Review of Immunology* **11**: 269–95.

Kluijver, H. N. (1951) The population ecology of the great tit *Parus m. major* (L). *Ardea* **39**: 1–135.

Knight, R. A., Vegors, H. H. and Lindahl, I. L. (1972) Comparative gastrointestinal nematode parasitism in naturally infected ewe and ram lambs. *Journal of Parasitology* **58**: 1216.

Knol, B. W. and Egberink-Alink, S. T. (1989) Androgens, progestagens and agonistic behaviour. *Veterinary Quarterly* **11**: 94–101.

Knott, S. A., Sibly, R. M., Smith, R. H. and Møller, H. (1995) Maximum likelihood estimation of genetic parameters in life-history studies using the 'animal model'. *Functional Ecology* **9**: 122–6.

Koehn, R. K. and Hillbish, T. J. (1987) The adaptive importance of genetic variation. *American Scientist* **75**: 134–41.

Kojola, I. (1997) Behaviour correlates of female social status and birth mass of male and female calves in reindeer. *Ethology* **103**: 809–14.

Kollars, T. M., Durden, L. A., Masters, E. J. and Oliver, J. H. (1997) Some factors affecting infestation of white-tailed deer by blacklegged ticks and winter ticks (Acari: Ixodidae) in south-eastern Missouri. *Journal of Medical Entomology* **34**: 372–5.

Komdeur, J. (1996) Influence of helping and breeding experience on reproductive performance in the Seychelles warbler: a translocation experiment. *Behavioral Ecology* **7**: 326–33.

Komers, P. E. (1997) Behavioural plasticity in variable environments. *Canadian Journal of Zoology* **75**: 161–9.

Komers, P. E., Messier, F. and Gates, C. C. (1994a) Plasticity of reproductive behaviour in wood bison bulls. I. When subadults are given a chance. *Ethology Ecology and Evolution* **6**: 313–30.

(1994b) Plasticity of reproductive behaviour in wood bison bulls. II. On risks and opportunities. *Ethology Ecology and Evolution* **6**: 481–95.

Komers, P. E., Pelabon, C. and Stenstrom, D. (1997) Age at first reproduction in male fallow deer: age-specific versus dominance specific behaviors. *Behavioral Ecology* **8**: 456–62.

Komers, P. E., Birgersson, B. and Ekvall, K. (1999) Timing of estrus in fallow deer is adjusted to the age of available mates. *American Naturalist* **153**: 431–6.

Kossarek, L. M., Su, X., Grosse, W. M., Finlay, O., Barendse, W., Hetzel, D. J. S. and McGraw, R. A. (1993) Bovine dinucleotide repeat polymorphism – RM106. *Journal of Animal Science* **71**: 3180.

Krebs, C. J., Gilbert, B. S., Boutin, S., Sinclair, A. R. E. and Smith, J. N. M. (1986) Population regulation of snowshoe hares. I. Demography of food-supplemented populations in the southern Yukon, 1976–84. *Journal of Animal Ecology* **55**: 963–82.

Krebs, J. R. and Davies, N. B. (1993) *An Introduction to Behavioural Ecology*. Oxford: Blackwell Scientific Publications.

Kreitman, M., Shorrocks, B. and Dytham, C. (1992) Genes and ecology: two alternative perspectives using *Drosophila*. In: *Genes and Ecology* (eds. R.J. Berry, T.J. Crawford and G.M. Hewitt), pp. 281–312. Oxford: Blackwell Scientific Publications.

Kruuk, L.E.B., Clutton-Brock, T.H., Albon, S.D., Pemberton, J.M. and Guinness, F.E. (1999a) Population density affects sex ratio variation in red deer. *Nature* 399: 459–61.

Kruuk, L.E.B., Clutton-Brock, T.H., Rose, K.E. and Guinness, F.E. (1999b) Early determinants of lifetime reproductive success differ between the sexes in red deer. *Proceedings of the Royal Society B* 266: 1655–61.

Kruuk, L.E.B., Clutton-Brock, T.H., Slate, J., Pemberton, J.M., Brotherstone, S. and Guinness, F.E. (2000) Heritability of fitness in a wild mammal population. *Proceedings of the National Academy of Sciences of the USA* 97: 698–703.

Kruuk, L.E.B., Merilä, J. and Sheldon, B.C. (2001) Phenotypic selection on a heritable size trait revisited. *American Naturalist* 158: 557–71.

Kruuk, L.E.B., Sheldon, B.C. and Merilä, J. (2002a) Severe inbreeding depression in collared flycatchers (*Ficedula albicollis*). *Proceedings of the Royal Society B* 269: 1581–89.

Kruuk, L.E.B., Slate, J., Pemberton, J.M., Brotherstone, S., Guinness, F.E. and Clutton-Brock, T.H. (2002b) Antler size in red deer: heritability and selection but no evolution. *Evolution* 56: 1683–95.

Lack, D. (1947) *Darwin's Finches*. Cambridge: Cambridge University Press.
(1966) *Population Studies of Birds*. Oxford: Oxford University Press.
(1968) *Ecological Adaptation for Breeding in Birds*. London: Methuen.
(1971) *Ecological Adaptations in Birds*. London: Methuen.

Lagesen, K. and Folstad, I. (1998) Antler asymmetry and immunity in reindeer. *Behavioral Ecology and Sociobiology* 44: 135–42.

Lambin, X., Krebs, C.J., Moss, R., Stenseth, N.C. and Yoccoz, N.G. (1999) Population cycles and parasitism. *Science* 286: 2425a.

Lambin, X. Petty, S.J. and Mackinnon, J.L. (2000) Cyclic dynamics in field vole populations and generalist predation. *Journal of Animal Ecology* 69: 106–18.

Lande, R. and Arnold, S.J. (1983) The measurement of selection on correlated characters. *Evolution* 37: 1210–26.

Langvatn, R. and Loison, A. (1999) Consequences of harvesting on age structure, sex ratio and population dynamics of red deer *Cervus elaphus* in central Norway. *Wildlife Biology* 5: 213–23.

Laws, R.M. (1966) Age criteria for the African elephant, *Loxodonta a. africana*. *East African Wildlife Journal* 4: 1–27.

Laws, R.M. and Clough, G. (1966) Observations on reproduction in the hippopotamus *Hippopotamus amphibius* Linn. In *Comparative Biology of*

Reproduction in Mammals (ed. I. W. Rowlands), pp. 117–40. London: Academic Press.

Le Boeuf, B. J. and Reiter, J. (1988) Lifetime reproductive success in Northern elephant seals. In *Reproductive Success* (ed. T. H. Clutton-Brock), pp. 344–62. Chicago, IL: University of Chicago Press.

Leader-Williams, N. (1988) *Reindeer on South Georgia.* Cambridge: Cambridge University Press.

Leader-Williams, N., Smith, R. I. L. and Rothery, P. (1987) Influence of introduced reindeer on the vegetation of South Georgia: results from a long-term exclusion experiment. *Journal of Applied Ecology* **24**: 801–22.

Leberg, P. L. and Smith, M. H. (1993) Influence of density on growth of white-tailed deer. *Journal of Mammalogy* **74**: 723–31.

Ledig, F. T. (1986) Heterozygosity, heterosis, and fitness in outbreeding plants. In: *Conservation Biology: The Science of Scarcity and Diversity* (ed. M. E. Soulé) pp. 77–104. Sunderland, MA: Sinauer Associates.

Lee, A. K. and McDonald, I. R. (1985) Stress and population regulation in small mammals. *Oxford Reviews of Reproductive Biology* **7**: 261–304.

Lessells, C. M. and Boag, P. T. (1986) Unrepeatable repeatabilities: a common mistake. *Auk* **104**: 116–21.

Leuchtmann, A., Schardl, C. L. and Siegel, M. R. (1994) Sexual compatability and taxonomy of a new species of *Epichloe* symbiotic with fine fescue grasses. *Mycologia* **86**: 802–12.

Lewontin, R. C., Ginzburg, L. R. and Tuljapurkar, S. D. (1978) Heterosis as an explanation of large amounts of genetic diversity. *Genetics* **88**: 149–70.

Lincoln, G. A. (1989) Seasonal cycles in testicular activity in mouflon, Soay sheep and domesticated breeds of sheep: breeding seasons modified by domestication. *Zoological Journal of the Linnean Society* **95**: 137–47.

Lincoln, G. A. (1998) Reproductive seasonality and maturation through the complete life-cycle in mouflon rams *Ovis musimon. Animal Reproduction Science* **53**: 87–105.

Lincoln, G. A. and Richardson, M. (1998) Photo-neuroendocrine control of seasonal cycles in body weight, pelage growth and reproduction: lessons from the HPD sheep model. *Comparative Biochemistry and Physiology* **119**: 283–94.

Lincoln, G. A. and Short, R. V. (1980) Seasonal breeding: Nature's contraceptive. *Recent Progress in Hormone Research* **36**: 1–52.

Lincoln, G. A., Lincoln, C. E. and McNeilly, A. S. (1990) Seasonal cycles in the blood plasma concentrations of FSH inhibin and testosterone, and testicular size in rams of wild, feral and domesticated breeds of sheep. *Journal of Reproduction and Fertility* **88**: 623–33.

Lindsay, D. R. (1966) Modification of behavioural oestrus in the ewe by social and hormonal factors. *Animal Behaviour* **14**: 73–83.

Lindsay, D. R. and Pierce, D. T. (eds.) (1984) *Reproduction in Sheep.* Cambridge: Cambridge University Press.

Lindström, J. (1999) Early development and fitness in birds and mammals. *Trends in Ecology and Evolution* **14**: 343–7.

Lindström, J., Coulson, T., Kruuk, L., Forchhammer, M. C., Coltman, D. W. and Clutton-Brock, T. H. (2002) Sex ratio variation in Soay sheep. *Behavioral Ecology and Sociobiology* **53**: 25–30.

Linnell, J. D. C. and Andersen, R. (1995) Who killed bambi? The role of predation in the neonatal mortality of temperate ungulates. *Wildlife Biology* **1**: 209–23.

Litt, M. and Luty, J. A. (1989) A hypervariable microsatellite revealed by *in vitro* amplification of a dinucleotide repeat within the cardiac muscle actin gene. *American Journal of Human Genetics* **44**: 397–401.

Loison, A. and Langvatn, R. (1998) Short- and long-term effects of winter and spring weather on growth and survival of red deer in Norway. *Oecologia* **116**: 489–500.

Loison, A., Langvatn, R. and Solberg, E. J. (1999a) Body mass and winter mortality in red deer calves: disentangling sex and climate effects. *Ecography* **22**: 20–30.

Loison, A., Festa-Bianchet, M., Gaillard, J.-M., Jorgenson, J. T. and Jullien, J.-M. (1999b) Age-specific survival in five populations of ungulates: evidence of senescence. *Ecology* **80**: 2539–54.

Luckinbill, L. S. and Clare, M. J. (1985) Selection for life span in *Drosophila melanogaster. Heredity* **55**: 9–18.

Luckinbill, L. S., Arking, R., Clare, M. J., Cirocco, W. C. and Buck S. A. (1984) Selection for delayed senescence in *Drosophila melanogaster. Evolution* **38**: 996–1003.

MacDonald, D. W. (ed.) (1984) *The Encyclopedia of Mammals.* Oxford: Andromeda.

Maddox, J. F., Davies, K. P., Crawford, A. M., Hulme, D. J., Vaiman, D., Cribiu, E. P., Freking, B. A., Beh, K. J., Cockett, N. E., Kang, N., Riffkin, C. D., Drinkwater, R., Moore, S. S., Dodds, K. G., Lumsden, J. M., van Stijn, T. C., Phua, S. H., Adelson, D. L., Burkin, H. R., Broom, J. E., Buitkamp, J., Cambridge, L., Cushwa, W. T., Gerard, E., Galloway, S. M., Harrison, B., Hawken, R. J., Hiendleder, S., Henry, H. M., Medrano, J. F., Paterson, K. A., Schibler, L., Stone, R. T. and van Hest, B. (2001) An enhanced linkage map of the sheep genome comprising more than 1000 loci. *Genome Research* **11**: 1275–89.

Malthus, T. R. (1798). *An Essay on the Principle of Population, as It Affects the Future Improvement of Society.* London: John Murray.

Markusson, E. and Folstad, I. (1997) Reindeer antlers: visual indicators of individual quality? *Oecologia* **110**: 501–7.

Marrow, P., McNamara, J. M., Houston, A. I., Stevenson, I. R. and Clutton-Brock, T. H. (1996) State-dependent life-history evolution in Soay sheep: dynamic modelling of reproductive scheduling. *Philosophical Transactions of the Royal Society B* **351**: 17–32.

Marshall, T. C., Slate, J., Kruuk, L. E. B. and Pemberton, J. M. (1998) Statistical confidence for likelihood-based paternity inference in natural populations. *Molecular Ecology* **7**: 639–55.

Marshall, T. C., Coltman, D. W., Pemberton, J. M., Slate, J., Spalton, J. A., Guinness, F. E., Smith, J. A., Pilkington, J. G. and Clutton-Brock., T. H. (2002) Estimating the prevalence of inbreeding from incomplete pedigrees. *Proceedings of the Royal Society B.* **269**: 1533–9.

Martin, D. J. (1962) The diet of Scottish black-face sheep. PhD thesis, University of Glasgow.

Martin, M. (1753) *A Voyage to St Kilda, The Remotest of All the Hebrides or Western Isles of Scotland*. London: Browne.

May, R. M. (1976) Simple mathematical models with very complicated dynamics. *Nature* **261**: 459–67.

May, R. M. and Anderson, R. M. (1978) Regulation and stability of host–parasite population interactions. II. Destabilizing processes. *Journal of Animal Ecology* **47**: 248–67.

(1979) Population biology of infectious diseases: Part II. *Nature* **280**: 455–61.

(1983) Epidemiology and genetics in the coevolution of parasites and hosts. *Proceedings of the Royal Society of London B* **219**: 281–313.

Maynard Smith, J. and Slatkin, M. (1973) The stability of predator–prey systems. *Ecology* **54**: 384–91.

Mayr, E. (1963) *Animal Species and Evolution*. Cambridge, MA: Belknap Press.

McCullough, D. R. (1979) *The George Reserve Deer Herd: Population Ecology of a K-Selected Species*. Ann Arbor, MI: Michigan University Press.

McCurdy, D. G., Shutler, D., Mullie, A. and Forbes, M. R. (1998) Sex-biased parasitism of avian hosts: relations to blood parasite taxon and mating system. *Oikos* **82**: 303–12.

McEwan, J. C., Mason, P., Baker, R. L., Clarke, J. N., Hickey, S. M. and Turner, K. (1992) Effect of selection for productive traits on internal parasite resistance in sheep. *Proceedings of the New Zealand Society for Animal Production* **52**: 53–6.

McLaren, B. E. and Peterson, R. O. (1994) Wolves, moose, and tree-rings on Isle Royale. *Science* **266**: 1555–8.

McLennan, D. A. and Brooks, D. R. (1991) Parasites and sexual selection: a macroevolutionary perspective. *Quarterly Review of Biology* **66**: 255–86.

McNamara, J. M. (1991) Optimal life histories: a generalisation of the Perron–Froebenius theorem. *Theoretical Population Biology* **40**: 230–45.

McNamara, J. M. and Houston, A. I. (1992) State-dependent life-history theory and its implications for optimal clutch size. *Evolutionary Ecology* **6**: 170–85.

(1996) State-dependent life histories. *Nature* **380**: 215–21.

McNaughton, S. J. (1985) Ecology of a grazing ecosystem: the Serengeti. *Ecological Monographs* **55**: 259–94.

(1993) Grasses and grazers, science and management. *Ecological Applications* **3**: 17–20.

McNaughton, S. J. and Georgiadis, N. J. (1985) Ecology of African grazing and browsing mammals. *Annual Reviews of Ecology and Systematics* **17**: 39–65.

McVean, D. N. (1961) Flora and vegetation of the islands of St Kilda and North Rona in 1958. *Journal of Ecology* **49**: 39–54.

Meagher, S., Penn, D. J. and Potts, W. K. (2000) Male–male competition magnifies inbreeding depression in wild house mice. *Proceedings of the National Academy of Sciences of the USA* **97**: 3324–9.

Meikle, D. B., Drickamer, L. C., Vessey, S. H., Arthur, R. D. and Rosenthal, T. L. (1996) Dominance rank and parental investment in swine (*Sus scrofo domesticus*). *Ethology* **102**: 969–78.

Merilä, J., Kruuk, L. E. B. and Sheldon, B. C. (2001a) Cryptic evolution in a wild bird population. *Nature* **412**: 76–9.

(2001b) Natural selection on the genetical component of variance in body condition in a wild bird population. *Journal of Evolutionary Biology* **14**: 918–29.

Merilä, J., Sheldon, B. C. and Kruuk, L. E. B. (2001c) Explaining stasis: microevolutionary studies in natural populations. *Genetica* **112**: 199–222.

Messier, F. (1991) Detection of density-dependent effects on caribou numbers for a series of census data. *Rangifer* **7**: 36–45.

(1994) Ungulate population models with predation: a case study with the North American moose. *Ethology* **75**: 478–88.

Michalakis, Y. and Hochberg, M. E. (1994) Parasitic effects on host life-history traits: a review of recent studies. *Parasite-Journal de la Société Française de Parasitologie* **1**: 291–4.

Michel, J. F. (1974) Arrested development of nematodes and some related phenomena. *Advances in Parasitology* **12**: 279–343.

(1991) Body size and the evolution of mammalian life histories. *Functional Ecology* **5**: 588–93.

Millar, J. S. and Zammuto, R. M. (1983) Life histories of mammals: an analysis of life tables. *Ecology* **64**: 631–5.

Milner, C. and Gwynne, D. (1974) The Soay sheep and their food supply. In *Island Survivors: The Ecology of the Soay Sheep of St Kilda* (eds. P. A. Jewell, C. Milner and J. M. Boyd), pp. 273–325. London: Athlone Press.

Milner, J. M., Albon, S. D., Illius, A. W., Pemberton, J. M. and Clutton-Brock, T. H. (1999a) Repeated selection of morphometric traits in the Soay sheep on St Kilda. *Journal of Animal Ecology* **68**: 472–88.

Milner, J. M., Elston, D. and Albon, S. D. (1999b) Estimating the contributions of population density and climatic fluctuations to interannual variation in survival of Soay sheep. *Journal of Animal Ecology* **68**: 1235–47.

Milner, J. M., Pemberton, J. M., Brotherstone, S. and Albon, S. D. (2000) Estimating variance components and heritabilities in the wild: a case study using the 'animal model' approach. *Journal of Evolutionary Biology* **13**: 804–13.

Milner-Gulland, E. J. (1994) A population model for the management of the Saiga antelope. *Journal of Applied Ecology* **31**: 25–39.

(1997) A stochastic dynamic programming model for the management of the Saiga antelope. *Ecological Applications* **7**: 130–42.

Minchella, D. J. and Loverde, P. T. (1981) A cost of increased early reproductive effort in the snail *Biomphalaria glabrata*. *American Naturalist* **118**: 876–81.

Ministry of Agriculture Fisheries and Food (1971) *Manual of Veterinary Laboratory Parasitological Techniques*, Technical Bulletin 18. London: HMSO.

Miquelle, D. G. (1990) Why don't bull moose eat during the rut? *Behavioral Ecology and Sociobiology* **27**: 145–51.

Mitchell, B. and Lincoln, G. A. (1973) Conception dates in relation to age and condition in two populations of red deer in Scotland. *Journal of Zoology* **171**: 141–52.

Mitchell, B., McCowan, D. and Nicholson, I. A. (1976) Annual cycles of body weight and condition in Scottish red deer, *Cervus elaphus*. *Journal of Zoology* **180**: 107–27.

Mitton, J. B. (1997) *Selection in Natural Populations*. Oxford: Oxford University Press.

Molan, A. L. and James, B. L. (1984) The effects of sex, age and diet of mice and gerbils on susceptibility to *Microphallus pygmaeus* (Digenea: Microphallidae). *International Journal of Parasitology* **14**: 521–6.

Monaghan, P., Bolton, M. and Houston, D. C. (1995) Egg production constraints and the evolution of avian clutch size. *Proceedings of the Royal Society B* **259**: 189–91.

Montgomery, G. W., Henry, H. M., Dodds, K. G., Beattie, A. E., Wuliji, T. and Crawford, A. M. (1996) Mapping the *Horns (Ho)* locus in sheep: a further locus controlling horn development in domestic animals. *Journal of Heredity* **87**: 358–63.

Moorcroft, P. R., Albon, S. D., Pemberton, J. M., Stevenson, I. R. and Clutton-Brock, T. H. (1996) Density-dependent selection in a cyclic ungulate population. *Proceedings of the Royal Society B* **263**: 31–8.

Moore, J. (1984) Altered behavioral responses in intermediate hosts: an acanthocephalan parasite strategy. *American Naturalist* **123**: 572–7.

Moore, J. and Gotelli, N. J. (1996) Evolutionary patterns of altered behavior and susceptibility in parasitized hosts. *Evolution* **50**: 807–19.

Moore, S. L. and Wilson, K. (2002) Parasites as a viability cost of sexual selection in natural populations of mammals. *Science* **297**: 2015–18.

Moran, P. A. P. (1953) The statistical analysis of the Canadian lynx cycle. II. Synchronisation and meteorology. *Australian Journal of Zoology* **1**: 291–8.

Morgan, D. O., Parnell, I. W. and Rayski, C. (1950) Further observations on the seasonal variation in worm egg output in Scottish hill sheep. *Journal of Helminthology* **24**: 101–22.

(1951) The seasonal variations in the worm burden of Scottish hill sheep. *Journal of Helminthology* **25**: 177–212.

Moritz, C., McCallum, H., Donnellan, S. and Roberts, J. D. (1991) Parasite loads in parthenogenetic and sexual lizards (*Heteronotia binoei*): support for the Red Queen hypothesis. *Proceedings of the Royal Society B* **244**: 145–9.

Morris, C. A., Bisset, S. A., Baker, R. L., Watson, T. G., Johnson, D. J. and Wheeler, M. (1993) An investigation of sire by location interactions for faecal nematode egg counts in lambs. *Proceeding of the New Zealand Society of Animal Production* **53**: 231–3.

Mousseau, T. A. and Roff, D. A. (1987) Natural selection and the heritability of fitness components. *Heredity* **59**: 181–97.

Mulvey, M. and Aho, J. M. (1993) Parasitism and mate competition: liver flukes in white-tailed deer. *Oikos* **66**: 187–92.

Murdoch, W. W. (1994) Population regulation in theory and practice. *Ecology* **75**: 271–87.

Murray, S. (2002) *Birds of St. Kilda. Scottish Birds* **23**, (Supplement).

Näsholm, A. and Danell, Ö. (1996) Genetic relationships of lamb weight, maternal ability and mature ewe weight in Swedish finewool sheep. *Journal of Animal Science* **74**: 329–39.

Nevo, E., Beiles, A. and Ben-Schlomo, R. (1984) The evolutionary significance of genetic diversity: ecological, demographic and life-history correlates. In *Evolutionary Dynamics of Genetic Diversity* (ed. G. S. Mani), pp. 13–213. Berlin: Springer-Verlag.

Newman, J. A., Parsons, A. J., Thornley, J. H. M., Penning, P. D. and Krebs, J. R. (1995) Optimal diet selection by a generalist grazing herbivore. *Functional Ecology* **9**: 255–68.

Newsome, A. E., Catling, P. C., Cooke, B. D. and Smyth, R. (2001) Two ecological universes separated by the Dingo Barrier fence in semi-arid Australia: interactions between landscapes, herbivory and carnivory, with and without dingoes. *Rangeland Journal* **23**: 71–98.

Newton, I. (1985) Lifetime reproductive output of female sparrowhawks. *Journal of Animal Ecology* **54**: 241–53.

(1989) *Lifetime Reproduction in Birds*. London: Academic Press.

(1998) *Population Limitation in Birds*. London: Academic Press.

Novellie, P. (1986) Relationships between rainfall, population density and the size of the bontebok lamb crop in the Bontebok National Park. *South African Journal of Wildlife Research* **16**: 39–46.

Nunney, L. (1991) The influence of age structure and fecundity on effective population size. *Proceedings of the Royal Society B* **246**: 71–6.

(1993) The influence of mating system and overlapping generations on effective population size. *Evolution* **47**: 1329–41.

Ockendon, N. (1999) Using genetic markers to assess the extent and effects of inbreeding in a population of Soay sheep. PhD thesis, University of Edinburgh.

Ollason, J. and Dunnet, C. M. (1988) Variation in breeding success in fulmars. In *Reproductive Success* (ed. T. H. Clutton-Brock), pp. 263–78. Chicago, IL: University of Chicago Press.

Ollerenshaw, C. B. and Smith, L. P. (1969) Meteorological factors and forecasts of helminthic disease. *Advances in Parasitology* **7**: 283–323.

Orwin, D. F. G. and Whitaker, A. H. (1984) Feral sheep (*Ovis aries* L.) of Arapawa Island, Marlborough Sound, and a comparison of their wool characteristics with those of four other feral stocks in New Zealand. *New Zealand Journal of Zoology* **11**: 201–24.

Osterhaus, A. and Vedder, E. J. (1988) Identification of virus causing recent seal deaths. *Nature* **335**: 20.

Ots, and I. Hõrak, P. (1996) Great tit *Parus major* trade health for reproduction. *Proceedings of the Royal Society B* **263**: 1443–7.

Owen, M. (1984) Dynamics and age structure of an increasing population: the Svalbard barnacle goose *Branta leucopsis*. *Norsk Polarinstitut Skrifter* **181**: 37–47.

Owen-Smith, N. (1990) Demography of a large herbivore, the greater kudu *Tragelaphus strepsiceros* in relation to rainfall. *Journal of Animal Ecology* **59**: 893–913.

(1993) Age, size, dominance and reproduction among male kudus: mating enhancement by attrition of rivals. *Behavioral Ecology and Sociobiology* **32**: 177–84.

Owen-Smith, R. N. (1974) *The Behavioral Ecology of the White Rhinoceros*. Ann Arbor, MI: Michigan: University Microfilms.

Owens, I. P. F. (2002). Sex differences in mortality rate. *Science* **247**:2008–9.

Pacala, S. W. and Crawley, M. J. (1992) Herbivores and plant diversity. *American Naturalist* **140**: 243–60.

Pacala, S. W. and Dobson, A. P. (1988) The relation between the number of parasites/host and host age: population dynamic causes and maximum likelihood estimation. *Parasitology* **96**: 197–210.

Packer, C., Herbst, L., Pusey, A. E., Bygott, J. D., Hanky, J. B., Cairns, S. J. and Borgerhoff-Mulder, M. (1988) Reproductive success of lions. In

Reproductive Success (ed. T. H. Clutton-Brock), pp. 363–83. Chicago, IL: University of Chicago Press.

Pankhurst, R. and Mullin, J. (1991) *Flora of the Outer Hebrides*. London: Natural History Museum Publications.

Parker, G. A. and Maynard Smith, J. (1990) Optimality theory in evolutionary biology. *Nature* **348**: 27–33.

Parker, G. A. and Simmons, L. W. (1994) Evolution of phenotypic optima and copula duration in dungflies. *Nature* **370**: 53–6.

Parnell, I. W., Dunn, A. M. and Mackintosh, G. M. (1954) Some observations on the 'spring rise' in worm-egg counts of halfbred sheep in south-east Scotland. *British Veterinary Journal* **110**: 185–93.

Parsons, A. J., Johnson, I. R. and Harvey, A. (1988) Use of a model to optimise the interaction between frequency and severity of intermittent defoliation and to provide a fundamental comparison of the continuous and intermittent defoliation of grass. *Grass and Forage Science* **43**: 49–59.

Paterson, S. (1998) Evidence for balancing selection at the major histocompatability complex in a free-living ruminant. *Journal of Heredity* **89**: 289–94.

Paterson, S. and Pemberton, J. M. (1997) No evidence for Major Histocompatability Complex-dependent mating patterns in a free-living ruminant population. *Proceedings of the Royal Society B* **264**: 1813–19.

Paterson, S., Wilson, K. and Pemberton, J. M. (1998) Major histocompatability complex (MHC) variation associated with juvenile survival and parasite resistance in a large unmanaged ungulate population (*Ovis aries* L.). *Proceedings of the National Academy of Sciences of the USA* **95**: 3714–19.

Paton, G., Thomas, R. J. and Waller, P. J. (1984) A prediction model for parasitic gastro-enteritis in lambs. *International Journal of Parasitology* **14**: 439–45.

Paver, H. (1955) Some factors influencing the seasonal variation in worm egg counts in Scottish hill sheep. *Journal of Comparative Pathology* **65**: 220–35.

Pearce, D. T. and Oldham, C. M. (1984) The rain effect, its management and application to the management of sheep. In *Reproduction in Sheep* (eds. D. R. Lindsay and D. T. Pearce), pp. 26–34. Cambridge: Cambridge University Press.

Peek, J. M. (1962) Studies of moose in the Gravelly and Snowcrest Mountains. *Journal of Wildlife Management* **26**: 360–5.

Pemberton, J. M., Albon, S. D., Guinness, F. E., Clutton-Brock, T. H. and Berry, R. J. (1988) Genetic variation and juvenile survival in red deer. *Evolution* **42**: 921–34.

Pemberton, J. M., Albon, S. D., Guinness, F. E., Clutton-Brock, T. H. and Dover, G. (1991) Countervailing selection in different fitness components in female red deer. *Evolution* **45**: 93–103.

Pemberton, J. M., Albon, S. D., Guinness, F. E., Clutton-Brock, T. H. and Dover, G. A. (1992) Behavioral estimates of male mating success tested by DNA fingerprinting in a polygynous mammal. *Behavioral Ecology and Sociobiology* **3**: 66–75.

Pemberton, J. M., Smith, J. A., Coulson, T. N., Marshall, T. C., Slate, J., Paterson, S., Albon, S. D. and Clutton-Brock., T. H. (1996) The maintenance of genetic polymorphism in small island populations: large mammals in the Hebrides. *Philosophical Transactions of the Royal Society B* **351**: 745–52.

Pemberton, J. M., Coltman, D. W., Smith, J. A. and Pilkington, J. G. (1999) Molecular analysis of a promiscuous, fluctuating mating system. *Biological Journal of the Linnean Society* **68**: 289–301.

Penn, D. J., Damjanovich, K. and Potts, W. K. (2002) MHC heterozygosity confers a selective advantage against multiple-strain infections. *Proceedings of the National Academy of Sciences of the USA* **99**: 11260–4.

Penning, P. D., Parsons, A. J., Orr, R. J., Harvey, A. and Champion, R. A. (1995) Intake and behavior responses by sheep, in different physiological states, when grazing monocultures of grass or white clover. *Applied Animal Behaviour Science* **45**: 63–78.

Pennycuick, L. (1971) Frequency distributions of parasites in a population of three-spined sticklebacks, *Gasterosteus aculeatus* L., with particular reference to the negative binomial distribution. *Parasitology* **63**: 389–406.

Perrins, C. M. (1979) *British Tits*. London: Collins.

Perry, J. S. (1953) The reproduction of the African elephant, *Loxodonta africana*. *Philosophical Transactions of the Royal Society B* **237**: 93–149.

Petch, C. P. (1933) The vegetation of St Kilda. *Journal of Ecology* **21**: 92–100.

Peterson, R. (1955) *North American Moose*. Toronto: University of Toronto Press.

Peterson, R. O., Page, R. E. and Dodge, K. H. (1984) Wolves, moose and the allometry of population cycles. *Science* **224**: 1350–2.

Pfeffer, P. (1967) Le mouflon de Corse (*Ovis ammon musimon* Schrober, 1782): position systématique, écologie, et éthologie comparées. *Mammalia* **31**: 1–262.

Pickering, S. P. C. (1983) Aspects of the behavioural ecology of feral goats (*Capra domestica*). PhD thesis, University of Durham.

Pierson, C. A., Hanrahan, V., Ede, A. J. and Crawford, A. M. (1993) Ovine microsatellites at the OarVH34, OarVH41, OarVH58, OarVH61 and OarVH72 loci. *Animal Genetics* **24**: 224.

Pollard, J. H. (1973) *Mathematical Models for the Growth of Human Populations*. Cambridge: Cambridge University Press.

Poore, M. E. D., and Robertson, V. C. (1949) The vegetation of St Kilda in 1948. *Journal of Ecology* **37**: 82–99.

Pope, T. R. (2000) The evolution of male philopatry in neotropical monkeys. In *Primate Males* (ed. P. M. Kappeler), pp. 218–35. Cambridge: Cambridge University Press.

Portier, C., Festa-Bianchet, M., Gaillard, J.-M., Jorgenson, J.T. and Yoccoz, N.G. (1998) Effects of density and weather on survival of bighorn sheep lambs (*Ovis canadensis*). *Journal of Zoology* **245**: 271–8.

Post, E. and Forchhammer, M.C. (2002) Synchronisation of animal population dynamics by large-scale climate. *Nature* **420**: 168–71.

Post, E. and Stenseth, N.C. (1998) Large-scale climatic fluctuation and population dynamics of moose and white-tailed deer. *Journal of Animal Ecology* **67**: 537–43.

(1999) Climate variability, plant phenology and northern ungulates. *Ecology* **80**: 1322–39.

Potts, G.R., Tapper, S.C. and Hudson, P.J. (1984) Population fluctuations in red grouse: analysis of bag records and a simulation model. *Journal of Animal Ecology* **53**: 21–36.

Potts, W.K. and Wakeland, E.K. (1990) Evolution of diversity at the major histocompatibility complex. *Trends in Ecology and Evolution* **5**: 181–7.

Potts, W.K., Manning, C.J. and Wakeland, E.K. (1991) Mating patterns in semi-natural poulations of mice influenced by MHC genotype. *Nature* **352**: 619–21.

Poulin, R. (1994) The evolution of parasite manipulation of host behavior: a theoretical analysis. *Parasitology* **109**: S109–S118.

(1996) Sexual inequalities in helminth infections: a cost of being a male? *American Naturalist* **147**: 287–95.

Poulin, R. and Vickery, W.L. (1993) Parasite distribution and virulence: implications for parasite-mediated sexual selection. *Behavioral Ecology and Sociobiology* **33**: 429–36.

Powell, R.G. and Petroski, R.J. (1992) Alkaloid toxins in endophyte infected grasses. *Natural Toxins* **1**: 163–70.

Powers, D.A., DiMichele, L. and Place, A.R. (1983) The use of enzyme kinetics to predict differences in cellular metabolism, developmental rate and swimming performance between LDH-B genotypes of the fish *Fundulus heteroclitus*. *Current Topics in Biological and Medical Research* **10**: 147–70.

Preston, B.T., Stevenson, I.R., Pemberton, J.M. and Wilson, K. (2001) Dominant rams lose out by sperm depletion. *Nature* **409**: 681–2.

Preston, B.T., Stevenson, I.R., Pemberton, J.M., Coltman, D.W. and Wilson, K. (2003) Overt and covert competition in a promiscuous mammal: the importance of weaponry and testes size to male reproductive success. *Proceedings of the Royal Society B* **270**: 633–40.

Price, T.D. and Boag, P.T. (1987) Selection in natural populations of birds. In *Avian Genetics: A Population and Ecological Approach* (eds. F. Cooke and P.A. Buckley), pp. 257–87. London: Academic Press.

Price, T.D., Grant, P.R., Gibbs, H.L. and Boag, P.T. (1984) Recurrent patterns of natural selection in a population of Darwin's finches. *Nature* **309**: 787–9.

Prins, H. and Weyhauser, F. J. (1987) Epidemics in populations of wild ruminants: anthrax and impala, rinderpest and buffalo in Lake Manyara National Park. *Oikos* 98: 28–38.

Procter, B. G. and Gibbs, H. C. (1968) Spring rise phenomenon in ovine helminthiasis. II. In stabled ewes experimentally infected with *Haemonchus contortus*. *Experimental Parasitology* 23: 323–30.

Promislow, D. E. L. (1992) Costs of sexual selection in natural population of mammals. *Proceedings of the Royal Society B* 247: 203–10.

Promislow, D. E. L. and Harvey, P. H. (1991) Mortality rates and the evolution of mammal life histories. *Acta Œcologica* 12: 119–37.

Putman, R. J., Edwards, P. J., Mann, J. C. E., How, R. C. and Hill, S. D. (1989) Vegetational and faunal changes in an area of heavily grazed woodland following relief of grazing. *Biological Conservation* 47: 13–32.

Queller, D. C. and Goodnight, K. F. (1989) Estimating relatedness using genetic markers. *Evolution* 43: 258–75.

Raynaud, F. (1972) Physiologie du transport des spermatozoides dans le col utérin de la brebis dans diverses conditions expérimentales. Thèse, 3ème cycle, University of Paris.

Read, A. F. (1987) Comparative evidence supports the Hamilton and Zuk hypothesis on parasites and sexual selection. *Nature* 328: 68–70.
  (1988) Sexual selection and the role of parasites. *Trends in Ecology and Evolution* 3: 97–102.

Read, A. F., Albon, S. D., Antonovics, J., Apanius, V., Dwyer, G., Holt, R. D., Judson, O., Lively, C. M., Martin-Lof, A., McLean, A. R., Metz, J. A. J., Schmid-Hempel, P., Thrall, P. H., Via, S. and Wilson, K. (1995) Genetics and evolution of infectious diseases in natural populations. In *Ecology of Infectious Diseases in Natural Populations* (eds. B. T. Grenfell and A. P. Dobson) pp. 450–77. Cambridge: Cambridge University Press.

Réale, D. and Festa-Bianchet, M. (2000) Quantitive genetics of life-history traits in a long-lived wild mammal. *Heredity* 85: 593–603.

Réale, D., Boussès, P. and Chapuis, J.-L. (1996) Female-biased mortality induced by male sexual harassment in a feral sheep population. *Canadian Journal of Zoology* 74: 1812–18.

Réale, D., Festa-Bianchet, M. and Jorgenson, J. T. (1999) Heritability of body mass varies with age and season in wild bighorn sheep. *Heredity* 83: 526–32.

Reid, J. F. S. and Armour, J. (1972) Seasonal fluctuations and inhibited development of gastrointestinal nematodes of sheep. *Research in Veterinary Science* 13: 225–39.

Reid, W. V. (1987) The cost of reproduction in the glaucous-winged gull. *Oecologia* 74: 458–67.

Reiter, J. and Le Boeuf, B. J. (1991) Life history consequences of variation in age at primiparity in northern elephant seals. *Behavioral Ecology and Sociobiology* **28**: 153–60.

Richner, H. (1998) Host – parasite interactions and life-history evolution. *Zoology – Analysis of Complex Systems* **101**: 333–44.

Richner, H., Oppliger, A. and Christe, P. (1993). Effect of an ectoparasite on reproduction in great tits. *Journal of Animal Ecology* **62**: 703–10.

Riney, T. (1982) *Study and Management of Large Mammals*. New York: Wiley.

Robertson, A., Hiraiwa-Hasegawa, M., Albon, S. D. and Clutton-Brock, T. H. (1992) Early growth and sucking behaviour of Soay sheep in a fluctuating population. *Journal of Zoology* **227**: 661–71.

Robinette, W. L. and Archer, A. L. (1971) Notes on ageing criteria and reproduction of Thomson's gazelle. *East African Wildlife Journal* **9**: 83–98.

Roelke-Parker, M. E., Munson, L., Packer, C., Kock, R., Cleaveland, S., Carpenter, M., O'Brien, S. J., Pospischil, A., Hofmann-Lehmann R., Lutz, H., Mwamengele, G. L. M., Mgasa, M. N., Machange, G. A., Summers, B. A. and Appel, M. J. G. (1996) A canine distemper virus epidemic in Serengeti lions (*Panthera leo*). *Nature* **379**: 441–5.

Roff, D. A. (1992) *The Evolution of Life-Histories: Theory and Analysis*. London: Chapman and Hall.

(1997) *Evolutionary Quantitative Genetics*. London: Chapman and Hall.

Rogers, J. C. (1984) The association between the North Atlantic Oscillation and the Southern Oscillation in the Northern Hemisphere. *Monthly Weather Review* **112**: 1999–2015.

Roitberg, B. D., Sircom, J., Roitberg, C. A., van Alphen, J. J. M. and Mangel, M. (1993) Life expectancy and reproduction. *Nature* **364**: 108.

Rose, K. E., Clutton-Brock, T. H. and Guinness, F. E. (1998) Cohort variation in male survival and mating success in red deer, *Cervus elaphus*. *Journal of Animal Ecology* **67**: 979–86.

Rose, M. R. (1982) Antagonistic pleiotropy, dominance and genetic variation. *Heredity* **48**: 63–78.

(1984) Laboratory evolution of postponed senescence in *Drosophila melanogaster*. *Evolution* **38**: 1004–10.

Rose, M. R. and Charlesworth, B. (1981) Genetics of life-history in *Drosophila melanogaster*. I. Sib analysis of adult females. *Genetics* **97**: 173–86.

Røskaft, E., Espmark, Y. and Järvi, T. (1983) Reproductive effort and breeding success in relation to age by the rook *Corvus frugilegus*. *Ornis Scandinavica* **14**: 169–74.

Rossiter, S. J., Jones, G., Ransome, R. D. and Barratt, E. M. (2001) Outbreeding increases offspring survival in wild greater horseshoe bats (*Rhinolophus ferrumequinum*). *Proceedings of the Royal Society B* **268**: 1055–61.

Royama, T. (1992) *Analytical Population Dynamics*. London: Chapman and Hall.

Runyoro, V. A., Hofer, H., Chausi, E. B. and Moehlman, P. D. (1995) Long-term trends in the herbivore populations of the Ngorongoro Crater, Tanzania. In *Serengeti II: Dynamics, Management and Conservation of an Ecosystem* (eds. A. R. E. Sinclair and P. Arcese), pp. 146–68. Chicago, IL: University of Chicago Press.

Ryder, M. L. (1966) Coat structure in Soay sheep. *Nature* **211**: 1092–3.

(1968) Fleece structure in some native and unimproved breeds of sheep. *Zeitschrift fur Tierzuchtung und Zuchtungsbiologie* **85**: 143–170.

Ryder, M. L. and Stephenson, S. K. (1968) *Wool Growth*. London: Academic Press.

Ryder, M. L., Land, R. B. and Ditchburn, R. (1974) Coat colour inheritance in Soay, Orkney and Shetland sheep. *Journal of Zoology* **173**: 477–85.

Sabat, A. M. (1994) Mating success in brood-guarding male rock bass, *Ambloplites rupestris*: the effect of body size. *Environmental Biology of Fishes* **39**: 411–15.

Saether, B.E. (1997) Environmental stochasticity and population dynamics of large herbivores: a search for mechanisms. *Trends in Ecology and Evolution* **12**: 143–9.

Saether, B. E. and Heim, M. (1993) Ecological correlates of individual variation in age at maturity in female moose (*Alces alces*): the effects of environmental variability. *Journal of Animal Ecology* **62**: 482–9.

Saether, B. E., Andersen, R., Hjeljord, O. and Heim, M. (1996) Ecological correlates of regional variation in the life-history of the moose *Alces alces*. *Ecology* **77**: 1493–1500.

Sand, H. and Cederlund, G. (1996) Individual and geographical variation in age at maturity in female moose (*Alces alces*). *Canadian Journal of Zoology* **74**: 954–64.

Sauer, J. R. and Boyce, M. S. (1983) Density-dependence and survival of elk in north-eastern Wyoming. *Journal of Wildlife Management.* **47**: 31–7.

Sayer, J. A. and van Lavieren, L. P. (1975) The ecology of the Kafue lechwe population of Zambia before the operation of hydro-electric dams on the Kafue River. *East African Wildlife Journal* **13**: 9–37.

Schalk, G. and Forbes, M.R. (1997) Male biases in parasitism of mammals: effects of study type, host age, and parasite taxon. *Oikos* **78**: 67–74.

Schaller, G. B. (1972) *The Serengeti Lion*. Chicago, IL: University of Chicago Press.

(1977) *Mountain Monarchs: Wild Sheep and Goats of the Himalayas*. Chicago, IL: University of Chicago Press.

Schenkel, R. and Schenkel-Hulliger, L. (1969) *Ecology and Behaviour of the Black Rhinoceros* (Diceros bicornis L.). Hamburg: Verlag Paul Parey.

Schoener, T. W. (1968) The *Anolis* lizards of Bimini: resource partitioning in a complex fauna. *Ecology* **49**: 704–26.

Schmidt, P., Ludt, C., Kuhn, C. and Buitkamp, J. (1996) A diallelic tetranucleotide repeat, (GT(3))(5 or 6), within intron 1 of the ovine interferon-gamma gene. *Animal Genetics* **27**: 437–8.

Schuurs, A. and Verheul, H. A. M. (1990) Effects of gender and sex steroids on the immune-response. *Journal of Steroid Biochemistry and Molecular Biology* **35**: 157–72.

Schwaiger, F. W., Weyers, E., Buitkamp, J., Ede, A. J., Crawford, A. M. and Epplen, J.T. (1994) Interdependent MHC-DRB exon-plus-intron evolution in artiodactyls. *Molecular Biology and Evolution* **11**: 239–49.

Schwaiger, F. W., Gostomski, D., Stear, M. J., Duncan, J. L., McKellar, Q. A., Epplen, J. T. and Buitkamp, J. (1995) An ovine major histocompatability complex DRB1 allele is associated with low fecal egg counts following natural, predominantly *Ostertagia circumcincta* infection. *International Journal for Parasitology* **25**: 815–22.

Seigel, R. A., Huggins, M. M. and Ford, N. B. (1987) Reduction in locomotor ability as a cost of reproduction in gravid snakes. *Oecologia* **73**: 481–5.

Shaw, D. J. and Dobson, A. P. (1995) Patterns of macroparasite abundance and aggregation in wildlife populations: a quantitative review. *Parasitology* **111**: S111–S133.

Shaw, D. J., Grenfell, B. T. and Dobson, A. P. (1998) Patterns of parasite aggregation in wildlife host populations. *Parasitology* **117**: 597–610.

Shelby, R. A. and Dalrymple, L. W. (1987) Incidence and distribution of the tall fescue endophyte in the United States. *Plant Disease* **71**: 783–6.

Sheldon, B. C. and Verhulst, S. (1996) Ecological immunology: costly parasite defences and trade-offs in evolutionary ecology. *Trends in Ecology and Evolution* **11**: 317–21.

Simmons, M. J., Preston, C. R. and Engels, W. R. (1980) Pleiotropic effects on fitness and factors affecting viability in *Drosophila melanogaster*. *Genetics* **94**: 467–75.

Sinclair, A. R. E. (1974). The natural regulation of buffalo populations in East Africa. II. Reproduction, recruitment and growth. *East African Wildlife Journal* **12**: 169–83.

(1977) *The African Buffalo: A Study of Resource Limitation of Populations.* Chicago, IL: University of Chicago Press.

(1989) Population regulation in animals. In *Ecological Concepts* (ed. J. M. Chewett), pp. 197–241. Oxford: Blackwell Scientific Publications.

Sinclair, A. R. E. and Arcese, P. (eds.) (1995) *Serengeti II: Dynamics, Management and Conservation of an Ecosystem.* Chicago, IL: University of Chicago Press.

Sinclair, A. R. E. and Norton-Griffiths, M. E. (1979) *Serengeti: Dynamics of an Ecosystem.* Chicago, IL: University of Chicago Press.

Skogland, T. (1985a) The effects of density-dependent resource limitations on the demography of wild reindeer. *Journal of Animal Ecology* **54**: 359–74.

(1985b) Tooth-wear by food limitation and its life-history consequences in wild reindeer. *Oikos* **51**: 238–42.

(1990). Density-dependence in a fluctuating wild reindeer herd: maternal versus offspring effects. *Oecologia* **84**: 442–50.

Slate, J. and Pemberton, J. M. (2002) Comparing molecular measures for detecting inbreeding depression. *Journal of Evolutionary Biology* **15**: 20–31.

Slate, J., Coltman, D. W., Goodman, S. J., MacLean, I., Pemberton, J. M. and Williams, J. L. (1998) Bovine microsatellite loci are highly conserved in red deer (*Cervus elaphus*), sika deer (*Cervus nippon*) and Soay sheep (*Ovis aries*). *Animal Genetics* **29**: 307–15.

Slate, J., Kruuk, L. E. B., Marshall, T. C., Pemberton, J. M. and Clutton-Brock, T. H. (2000a) Inbreeding depression influences lifetime breeding success in a wild population of red deer (*Cervus elaphus*). *Proceedings of the Royal Society B* **267**: 1657–62.

Slate, J., Marshall, T. C. and Pemberton, J. M. (2000b) A retrospective assessment of the accuracy of the paternity inference program CERVUS. *Molecular Ecology* **9**: 801–8.

Smith, J. A. (1996) Polymorphism, parasites and fitness. PhD thesis, University of Cambridge.

Smith, J. A., Wilson, K., Pilkington, J. G. and Pemberton, J. M. (1999) Heritable variation in resistance to gastro-intestinal nematodes in an unmanaged mammal population. *Proceedings of the Royal Society B* **266**: 1283–90.

Smith, J. M. (1988) Determinants of lifetime reproductive success in the songsparrow. In *Reproductive Success* (ed. T. H. Clutton-Brock), pp. 154–72. Chicago, IL: University of Chicago Press.

Smith, J. N. M., Krebs, C. J., Sinclair, A. R. E. and Boonstra, R. (1988) Population biology of snowshoe hares. II. Interactions with winter food plants. *Journal of Animal Ecology* **57**: 269–86.

Smith, T. P. L., Lopez-Corrale, N., Grosz, M. D., Beattie, C. W. and Kappes, S. M. (1997) Anchoring of bovine chromosomes 4, 6, 7, 10 and 14 linkage group telomeric ends via FISH analysis of lambda clones. *Mammalian Genome* **8**: 333–6.

Smithies, O. (1955) Zone electrophoresis in starch gels: group variations in the serum proteins of normal human adults. *Biochemical Journal* **61**: 629–41.

Sobanskii, G. G. (1979) Selective elimination in the Siberian stag population in the Altais as a result of the early winter of 1976/77. *Soviet Journal of Ecology* **10**: 78–80.

Soulsby, E. J. L. (1982) *Helminths, Arthropods and Protozoa of Domesticated Animals*, 7th edn. London: Baillière Tindall.

Sowls, L. K. (1966) Reproduction in the collared peccary (*Tayassu tajacu*). In *Comparative Biology of Reproduction in Mammals* (ed. I. W. Rowlands), pp. 155–72. London: Academic Press.

Spinage, C. A. (1982) *A Territorial Antelope: The Uganda Waterbuck*. London: Academic Press.

Sponenberg, D. (1997) Genetics of colour and hair texture. In *The Genetics of Sheep* (eds. L. Piper and A. Ruvinsky) pp. 51–86. Wallingford: CAB International.

Sreter, T., Kassai, T. and Takacs, E. (1994) The heritability and specificity of responsiveness to infection with *Haemonchus contortus* in sheep. *International Journal of Parasitology* 24: 871–6.

Stace, C. (1997) *New Flora of the British Isles*, 2nd edn. Cambridge: Cambridge University Press.

Staines, B. W. (1976) The use of natural shelter by red deer (*Cervus elaphus*) in relation to weather in north-east Scotland. *Journal of Zoology* 180: 1–8.

Stear, M. J., Park, M. and Bishop, S. C. (1996) The key components of resistance to *Ostertagia circumcincta* in lambs. *Parasitology Today* 12: 438–41.

Stear, M. J., Bairden, K., Duncan, J. L., Holmes, P. H., McKellar, Q. A., Park, M., Strain, S., Murray, M., Bishop, S. C. and Gettinby, G. (1997) How hosts control worms. *Nature* 389: 27.

Stearns, S. C. (1989) Trade-offs in life-history evolution. *Functional Ecology* 3: 259–68.

(1992) *The Evolution of Life-Histories*. Oxford: Oxford University Press.

Steel, T. (1988) *The Life and Death of St Kilda*. Glasgow: Collins.

Stenseth, N. C. (1995) Snowshoe hare populations: squeezed from below and above. *Science* 269: 1061–2.

Stevenson, I. R. (1994) Male-biased mortality in Soay sheep. PhD thesis, University of Cambridge.

Stevenson, I. R. and Bancroft, D. R. (1995) Fluctuating trade-offs favour precocial maturity in male Soay sheep. *Proceedings of the Royal Society B* 262: 267–75.

Stien, A., Irvine, R. J., Ropstad, E., Halvorsen, O., Langvatn, R. and Albon, S.D. (2002) The impact of gastrointestinal nematodes on wild reindeer: experimental and cross-sectional studies. *Journal of Animal Ecology* 71: 937–45.

Stinchcombe, J. R., Rutter, M. T., Burdick, D. S., Tittin, P., Rausher, M. D. and Mauricio, R. (2002) Testing for environmentally induced bias in phenotypic estimates of natural selection: theory and practice. *American Naturalist* 160: 511–23.

Stryer, L. (1988) *Biochemistry*, 3rd edn. New York: W. H. Freeman.

Suryahadiselim and Gruner, L. (1985) Effects of flooding a sheep pasture on vertical dispersion of *Teladorsagia circumcincta*. *Annales de Parasitologie Humaine et Comparée* 60: 709–14.

Swarbrick, P. A., Buchanan, F. C. and Crawford, A. M. (1991) Ovine dinucleotide repeat polymorphism at the MAF35 locus. *Animal Genetics* 22: 369–70.

Swarbrick, P. A., Schmack, A. E. and Crawford, A. M. (1992) MAF45, a highly polymorphic marker for the pseudoautosomal region of the sheep genome, is not linked to the FecX' (Inverdale) gene. *Genomics* **13**: 849–51.

Swinton, J., Woolhouse, M. E. J., Begon, M. E., Dobson, A. P., Ferroglio, E., Grenfell, B. T., Guberti, V., Hails, R. S., Heesterbeek, J. A. P., Lavazza, A., Roberts, M. G., White, P. J. and Wilson, K. (2002) Microparasite transmission and persistence. In *The Ecology of Wildlife Diseases* (eds. P. J. Hudson, A. Rizzoli, B. T. Grenfell, H. Heesterbeek and A. P. Dobson), pp. 83–101. Oxford: Oxford University Press.

Sykes, A. R. and Coop, R. L. (1976) Intake and utilisation of food by growing lambs with parasitic damage to the small intestine caused by daily dosing with *Trichostrongylus colubriformis* larvae. *Journal of Agricultural Science* **86**: 507–15.

Sykes, A. R., Coop, R. L. and Angus, K. W. (1977) The influence of chronic *Ostertagia circumcincta* infection on the skeleton of growing sheep. *Journal of Comparative Pathology* **87**: 521–9.

Symons, L. E. A. (1985) Anorexia: occurrence, pathophysiology and possible causes in parasitic infections. *Advances in Parasitology* **24**: 103–33.

Tate, M. L., Dodds, K. G., Thomas, K. J. and McEwan, K. M. (1992) Genetic polymorphism of plasminogen and vitmain D binding protein in red deer, *Cervus elaphus* L. *Animal Genetics* **23**: 303–13.

Tautz, D. (1989) Hypervariability of simple sequences as a general source for polymorphic DNA markers. *Nucleic Acids Research* **17**: 6463–71.

Taylor, E. L. (1934) The epidemiology of winter outbreaks of parasitic gastritis in sheep, with special reference to outbreaks which occured during the winter of 1933-34. *Journal of Comparative Pathology and Therapeutics* **47**: 235–54.

Thamsborg, S. M., Jorgensen, R. J., Waller, P. J. and Nansen, P. (1996) The influence of stocking rate on gastrointestinal nematode infections of sheep over a 2-year grazing period. *Veterinary Parasitology* **67**: 207–24.

Thibault, C. (1973) Sperm transport and storage in vertebrates. *Journal of Reproduction and Ferility, Supplement* **18**: 39–53.

Thomas, A. S. (1960) Changes in vegetation since the advent of myxomatosis. *Journal of Ecology* **48**: 287-306.

Thomas, C. S. and Coulson, J. C. (1988) Reproductive success of kittiwake gulls, *Rissa tridactyla*. In *Reproductive Success* (ed. T. H. Clutton-Brock), pp. 251–62. Chicago, IL: University of Chicago Press.

Thouless, C. (1986) Feeding competition in red deer hinds. PhD thesis, University of Cambridge.

Tompkins, D. M. and Begon, M. (1999) Parasites can regulate wildlife populations. *Parasitology Today* **15**: 311–13.

Torgerson, P. R., Gulland, F. M. D. and Gemmell, M. A. (1992) Observations on the epidemiology of *Taenia hydatigena* in Soay sheep on St Kilda. *Veterinary Record* **131**: 218–19.

Torgerson, P. R., Pilkington, J., Gulland, F. M. and Gemmell, M. A. (1995) Further evidence for the long-distance dispersal of taeniid eggs. *International Journal of Parasitology* **25**: 265–7.

Turchin, P. (1995) Population regulation: old arguments and a new synthesis. In *Population Dynamics* (eds. N. Cappuccino and P. N. Price), pp. 19–40. London: Academic Press.

Turpie, S. (1999) The effects of inbreeding on the lifetime breeding success and offspring birth weight of Soay ewes (*Ovis aries* L.). BSc thesis, University of Edinburgh.

Unsworth, J. M., Pac, D. F., White, G. C. and Bartmann, R. M. (1999) Mule deer survival in Colorado, Idaho and Montana. *Journal of Wildlife Management* **63**: 315–26.

Vaiman, D., Osta, R., Mercier, D., Grohs, C. and Leveziel, H. (1992) Characterization of five new bovine dinucleotide repeats. *Animal Genetics* **23**: 537–41.

Van Tienderen, P. H. (2000) Elasticities and the link between demographic and evolutionary dynamics. *Ecology* **81**: 666–79.

Van Valen, L. (1973) A new evolutionary law. *Evolutionary Theory* **1**: 1–30.

Van Vuren, D. and Coblentz, B. E. (1989) Population characteristics of feral sheep on Santa Cruz Island. *Journal of Wildlife Management* **52**: 306–13.

Verme, L. J. (1989) Maternal investment in white-tailed deer. *Journal of Mammalogy* **70**: 438–42.

Vicari, M. and Bazely, D.R. (1993) Do grasses fight back? The case for anti-herbivore defences. *Trends in Ecology and Evolution* **8**:137–41.

Virtanen, R., Henttonen, H. and Laine, K. (1997) Lemming grazing and structure of a snowbed plant community: a long-term experiment at Kilpisjarvi, Finnish Lapland. *Oikos* **79**: 155–66.

Virtanen, R., Edwards, G. R. and Crawley, M. J. (2002) Red deer management and vegetation on the Isle of Rum. *Journal of Applied Ecology* **39**: 572–83.

Vogel, S. W. and Heyne, H. (1996) Rinderpest in South Africa: 100 years ago. *Journal of the South African Veterinary Association – Tydskrif van Die Suid-Afrikaanse Veterinere Vereniging* **67**: 164–70.

Vucetich, J. A., Waite, T. A. and Nunney, L. (1997) Fluctuating population size and the ratio of effective to census population size. *Evolution* **51**: 2017–21.

Wakelin, D. (1996) *Immunity to Parasites: How Parasitic Infections Are Controlled.* Cambridge: Cambridge University Press.

Wallace, B. (1975) Hard and soft selection revisited. *Evolution* **29**: 465–73.

Walther, F. R., Mungall, E. C. and Grau, G. A. (1983) *Gazelles and their Relatives: A Study in Territorial Behavior.* Park Ridge, NJ: Noyes Publications.

Waser, P. M., Elliott, L. F., Creel, N. M. and Creel, S. R. (1995) Habitat variation and mongoose demography. In *Serengeti II: Dynamics, Management and Conservation of an Ecosystem* (eds. A. R. E. Sinclair and P. Arcese), pp. 421–48. Chicago, IL: University of Chicago Press.

Watson, A., Moss, R., Parr, R., Mountford, M. D. and Rothery, P. (1994) Kin landownership, differential aggression betwen kin and non-kin and population fluctuations in red grouse. *Journal of Animal Ecology* **63**: 39–50.

Watson, T. G., Baker, R. L. and Harvey, T. (1986) Genetic variation in resistance or tolerance to internal nematode parasite strains of sheep at Rotomahona. *Proceedings of the New Zealand Society of Animal Production* **46**: 23–6.

Watt, W. B., Cassin, R. C. and Swan, M. S. (1983) Adaptation at specific loci. III. Field behavior and survivorship differences among Colias PGI genotypes are predictable from *in vitro* biochemistry. *Genetics* **103**: 725–39.

Watterson, G. A. (1978) The homozygosity test of neutrality. *Genetics* **88**: 405–17.

Weatherhead, P. J., Barry, F. E., Brown, G. P. and Forbes, M. R. L. (1995) Sex ratios, mating behavior and sexual size dimorphism of the northern water snake, *Nerodia sipedon*. *Behavioral Ecology and Sociobiology* **36**: 301–11.

Weber, J. L. and May, P. E. (1989) Abundant class of human DNA polymorphisms which can be typed using the polymerase chain reaction. *American Journal of Human Genetics* **44**: 388–96.

Wedekind, C. (1994) Mate choice and maternal selection for specific parasite resistance before, during and after fertilization. *Philosophical Transactions of the Royal Society B* **346**: 303–11.

Welch, D. and Scott, D. (1995) Studies in the grazing of heather moorland in north-east Scotland. VI. 20-year trends in botanical composition. *Journal of Applied Ecology* **32**: 596–611.

Welles, R. E. and Welles, F. B. (1961) *The Bighorn of Death Valley*. Washington, DC: US Department of the Interior.

West, S. A. and Godfray, H. C. J. (1997) Sex ratio strategies after perturbation of the stable age distribution. *Journal of Theoretical Biology* **186**: 213–21.

Westendorp, R. G. J. and Kirkwood, T. B. L. (1998) Human longevity at the cost of reproductive success. *Nature* **396**: 743–6.

Western, D. (1979) Size, life-history and ecology in mammals. *African Journal of Ecology* **17**: 185–204.

Western, D. and Ssemakula, J. (1982) Life-history parameters in birds and mammals and their evolutionary interpretation. *Oecologia* **54**: 281-90.

Westneat, D. F. (1987) Extra-pair fertilizations in a predominantly monogamous bird: genetic evidence. *Animal Behaviour* **35**: 877–86.

(1990) Genetic parentage in the indigo bunting: a study usng DNA fingerprinting. *Behavioral Ecology and Sociobiology* **27**: 67–76.

White, J. F. (1987) Widespread distribution of endophytes in the Poaceae. *Plant Disease* **71**: 340–2.

Whitehead, G. K. (1995) *Deer of the World*. London: International Book Distributors.

Wiggins, E. L. and Terrill, C. E. (1953) Variation in penis development in ram lambs. *Journal of Animal Science* **12**: 524–35.

Williamson, K. and Boyd, J. M. (1960) *St Kilda Summer*. London: Hutchinson.

Wilson, E. O. (1974) *Sociobiology: The New Synthesis*. Cambridge, MA: Belknap Press.

Wilson, K. and Grenfell, B. T. (1997) Generalized linear modelling for parasitologists. *Parasitology Today* **13**: 33–8.

Wilson, K., Grenfell, B. T. and Shaw, D. J. (1996) Analysis of aggregated parasite distributions: a comparison of methods. *Functional Ecology* **10**: 592-601.

Wilson, K., Bjørnstad, O. N., Dobson, A. P., Merler, S., Poglayen, G., Randolph, S. E., Read, A. F. and Skorping, A. (2002) Heterogeneities in macroparasite infections: patterns and processes. In *The Ecology of Wildlife Diseases* (eds. P. J. Hudson, A. Rizzoli, B. T. Grenfell, H. Heesterbeek and A. P. Dobson), pp. 6–44. Oxford: Oxford University Press.

Wilson, K., Owens, I. P. F. and Moore, S. L. (2003) Response to comment on 'Parasites as a viability cost in sexual selection in natural populations of mammals. *Science* **300**: 5616.

Wilson, W. and Orwin, D. F. G. (1964) The sheep population of Campbell Island. *New Zealand Journal of Science* **7**: 460–90.

Windon, R. G., Dineen, J. K. and Kelly, J. D. (1980) The segregation of lambs into 'responders' and 'non-responders': response to vaccination with irradiated *Trichostrongylus colubriformis* larvae before weaning. *International Journal of Parasitology* **10**: 65–73.

Wolff, J. O. (1992) Parents suppress reproduction and stimulate dispersal in opposite sex juvenile white-footed mice. *Nature* **359**: 409–10.

Woodgerd, W. (1964) Population dynamics of bighorn sheep on Wildhorse Island. *Journal of Wildlife Management* **28**: 381–91.

Woolaston, R. R. and Piper, L. R. (1996) Selection of Merino sheep for resistance to *Haemonchus contortus*: genetic variation. *Animal Science* **62**: 451–60.

Woolaston, R. R., Windon, R. G. and Gray, G. D. (1991) Genetic variation in resistance to internal parasites in Armidale experimental flocks. In *Breeding for Disease Resistance in Sheep* (eds. G. D. Gray and R. R. Woolaston), pp. 1–9. Melbourne: Australian Wool Corporation.

Woolley, P. (1966) Reproduction in *Antechinus* spp. and other dasyurid marsupials. In *Comparative Biology of Reproduction in Mammals* (ed. I. W. Rowlands), pp. 281–94. London: Academic Press.

Zar, J. H. (1996) *Biostatistical Analysis*. London: Prentice-Hall International.

Zumpe, D. and Michael, R. P. (1988) Effects of medroxyprogesterone acetate on plasma testosterone and sexual behavior in male cynomolgus monkeys (*Macaca fascicularis*). *Physiology and Behavior* **42**: 343–9.

Zumpe, D., Bonsall, R. W., Kutner, M. H. and Michael, R. P. (1991) Medroxyprogesterone acetate, aggression, and sexual behavior in male cynomolgus monkeys (*Macaca fascicularis*). *Hormones and Behavior* **25**: 394–409.

Index

Printed in the United States
by Bookmasters

Printed in the United States
By Bookmasters